60 0415666 3

AF750686

University of Bath Library

28 DAY LOAN

This book should be returned not later than the last date s
The loan period may be extended on request if no oth
to use the book.
Renewals 01225 3862 or http://www.bath.ac.uk/lib
If this book has a hold ed by another der it
must be returned by th n the r

Innovation Dynamism and Economic Growth

Masaaki Hirooka
Executive Director, Institute of Technoeconomics, Japan
Dr of Engineering and Dr of Economics from Kyoto University
Member, the Engineering Academy of Japan
An advisory editor for *Research Policy*, Elsevier
Former Vice-President and Professor of Economics,
Ryutsu Kagaku University and Professor of Economics,
Kobe University, Japan

Innovation Dynamism and Economic Growth

A Nonlinear Perspective

Masaaki Hirooka

Institute of Technoeconomics, Kyoto, Japan

Edward Elgar

Cheltenham, UK • Northampton, MA, USA

UNIVERSITY OF BATH
LIBRARY
8 -4 JUL 2006
338.45

© Masaaki Hirooka, 2006

Japanese edition published in 2003 by Nihon Keizai Shimbun Inc., Tokyo, Japan under the title *Technological Innovation and Economic Development: Elucidation of Nonlinear Dynamism*, authored by Masaaki Hirooka.
© Masaaki Hirooka, 2003, for Japanese edition

All rights reserved. No part of this publication may be reproduced, stored in a retrieval system or transmitted in any form or by any means, electronic, mechanical or photocopying, recording, or otherwise without the prior permission of the publisher.

Published by
Edward Elgar Publishing Limited
Glensanda House
Montpellier Parade
Cheltenham
Glos GL50 1UA
UK

Edward Elgar Publishing, Inc.
136 West Street
Suite 202
Northampton
Massachusetts 01060
USA

A catalogue record for this book
is available from the British Library

Library of Congress Cataloguing in Publication Data

Hirooka, Masaaki.
Innovation dynamism and economic growth: a nonlinear perspective/
Masaaki Hirooka.
p. cm.
Includes bibliographical references and index.
1. Technological innovations—Economic aspects. I. Title.
HC79.T4H57 2006
338'.064—dc22 2005044217

ISBN-13: 978 1 84376 578 3
ISBN-10: 1 84376 578 0

Printed and bound in Great Britain by MPG Books Ltd, Bodmin, Cornwall

Contents

Preface

> We can be but partially acquainted even with events which actually influence our course through life, and our final destiny. There are innumerable other events, if such they may be called, which come close upon us, yet pass away without actual results, or even betraying their near approach, by the reflection of any light or shadow across our minds. Could we know all the vicissitudes of our fortunes, life would be too full of hope and fear, exultation or disappointment, to afford us a single hour of true serenity. (Nathaniel Hawthorne (2001), *Twice-Told Tales*, p. 139)

Linear thought is usually in front of us even if we fail to recognize it. We move and act and, indeed, believe, supposing that the event that happened yesterday will do so again tomorrow. An event that has continued for several years, however, will not always continue indefinitely. Although events in society are often governed by nonlinear factors, we cannot escape from linear thinking and thus often we make mistakes and suffer from the unanticipated consequences of our actions. We can, however, grasp the nonlinear aspects through careful examination. This book is a first step towards elucidating such nonlinear aspects.

This book analyzes the nature of innovation in detail, clarifies its dynamism, and describes its concrete features as a trajectory. Through exploration of the dynamism of innovation, the causal relationship between innovation and economic development is elucidated. This is a comprehensive study designed to interpret the developmental scenario of a modern industrial society.

Schumpeter expressed with deep insight the dynamic economics of technological change by using a metaphor he called the 'gale of creative destruction'. This indicates a dynamic system of economy in which innovations break down the old customs and equilibrium of an economy to jump up to a more advanced economic system with higher value-added. This is a challenge to conventional economics, which presupposes a balanced economy that is continually in equilibrium. Since then, neo-Schumpeterians have developed an economics of technological change and have built up the present economics on the basis of empirical analyses.

Now, many economists have moved into evolutionary economics. This is a metaphor for the evolutionary theory of life applied to economics. It seems to inherit Schumpeterian thought in terms of the notion of a dynamic system with uncertainty. Evolutionary economics skillfully describes economic change in the age of innovation and is certainly a challenge to the analysis of innovative competition focusing on firm behaviors. Evolutionary economics, however, is apart from Schumpeterian analysis in that it is governed by the Darwinian approach based on gradualism, while Schumpeter's evolution takes place by an invasion of innovations with sigmoid functionality, which rather belongs to Gouldian 'punctuated evolutionary theory'. Further, evolutionary economics is based on deductive thinking while Schumpeter's is inductive.

This book analyzes the entire scheme of innovation and its correlation with economic development. It also clarifies the relation between the original Schumpeterian approach and evolutionary economics. Returning to the original approach by Schumpeter, the present study is carried out through inductive analysis of innovation and economic development on the basis of fact-finding. It is important to step into the field of technology development, which has been treated as the black box in economics, using a scientific approach to obtain a comprehensive picture of innovation.

This book is a substantial first step in clarifying the entity of innovation itself, beyond the domain of economics. It is an attempt to describe the abstract phenomenon of the development of innovation by using a concrete trajectory. It proposes a method for analyzing the process of knowledge transfer in the human community. In that sense, it is a kind of 'theory of field'. Furthermore, knowledge transfer from person to person is a discrete phenomenon in a complex system, and innovation is interpreted as a nonlinear complex system with fractals. The innovation is a phenomenon with maturation within a definite time span. Based on this nonlinear analysis, various interesting features of innovation are clarified and economic development is described from a new angle.

The author hopes that this approach will contribute to the development of a new branch of science and economics.

Masaaki Hirooka

Acknowledgments

Following a long career as a research scientist in the chemical industry, I have devoted myself for the past 15 years to studying the economics of technological change as a professor of economics, initially at Kobe University and then at Ryutsu Kagaku University. I have learned much about economics from many notable scholars, especially J. Stanley Metcalfe, Christopher Freeman and Nathan Rosenberg. When I retired from Kobe University in 1994, the three of them kindly contributed papers for the special issue of the *Journal of Economics and Business Administration* that Kobe University published in honor of my retirement. They have greatly encouraged me, and my work owes much to their help. I am sincerely grateful to them. I also appreciate the advice and comments offered me by Carlota Perez, Peter Allen, Rod Coombs, Ian Miles, Cesare Marchetti, Nebojša Nakićenović, Arnulf Grübler, Sir Geoffrey Allen, Franz Lehner, Rudolf Pariser, Geoffrey Eastmond, James C. Abegglen, Kei Takeuchi, Fumio Kodama, Yasunori Baba, Satoshi Sechiyama, Kazuo Yoshida, Kiichiro Yagi, Hiroyuki Uni, Shuji Hayashi, Keizo Nagatani, Hideyuki Takahashi, Koji Shinjo, Chihiro Watanabe, Kumiko Miyazaki, Hisashi Hosokawa and Yutaka Kuwahara. Taiji Hagiwara and Mitsushige Nitta are co-authors of several of my papers, and I appreciate their tremendous help and the rewarding discussions I had with them.

Most of my work was done at Kobe University and Ryutsu Kagaku University, and several of my papers on the economics of innovation have been published in the *Journal of Economics and Business Administration*, Kobe University, the *Kobe University Economic Review* and *Ryutsu Kagaku Daigaku Ronshu* (*Journal of the University of Marketing and Distribution Sciences*). Some papers have also appeared in the *Journal of Chemical Economics*, Japan, and the *Plastics Age Encyclopedia*, Japan. A comprehensive summary of my work was published in Japanese as a book entitled *Technological Innovation and Economic Development: Elucidation of Nonlinear Dynamism* by Nihon Keizai Shimbun Inc. in 2003. I owe a great debt of thanks to Tsuneo Taguchi for editing and brushing up the writing in my original book in Japanese. This book is the English version of that one, with permission. It has been revised and enhanced with additional materials and more discussion.

I am very grateful to Arnold Fisher of Kurdyla & Associates for his patient correction of my English manuscript. Finally, I would like to express my appreciation to Luke Adams, Jo Betteridge and Caroline Cornish of Edward Elgar Publishing Ltd for their kindness in publishing this book.

1. Nonlinear dynamism of innovation: executive summary

This book elucidates innovation itself, beyond the domain of economics, and explores how the economic scenario of technological change can be clarified from the viewpoints of science and technology. It empirically analyzes innovation dynamism based on fact-finding.

The main objectives are to clarify the nature of technological innovation and to propose a new methodology for analyzing it. The economics of technological change are reconsidered on the basis of the revealed innovation dynamics. The key to this undertaking is to use the nonlinear nature of innovations and to describe the economics of innovation using a new approach.

The developmental histories of individual sciences and technologies have been precisely described, and huge amounts of information and knowledge have been accumulated. The general behavior of science and technology, however, has not been well analyzed. This means that there is not yet a general methodology for describing the developmental trends of science and technology. Even less attention has been devoted to the interrelation of economic development and technological innovation from the science and technology points of view.

Consider how economists have treated technological innovation. If the origin of economics goes back to Adam Smith, then it started with the Industrial Revolution. Economists at the time recognized science and technology as important elements of economic activity. Smith stressed the importance of the division of labor by which the individual becomes more expert in a particular task. This division of labor considerably increased the quantity of science. Marx described how an invention becomes a branch of business and how the application of science to immediate production can be used to identify the inventions. He claimed that the bourgeoisie cannot exist without constantly revolutionizing the means of production – they have long recognized the central importance of technological innovation to economic progress. Since then, however, economists have mainly focused on the relationships between money, labor and goods and have ignored technological innovation. Schumpeter inherited Marx's thought and the notions of dialectical materialism, although he severely criticized Marx's theoretical system of economics and established his own dynamic theory of innovation.

Both men agreed that economic activities are part of a dynamic system in which the activities interact with each other. Schumpeter pointed out that innovation is the engine of economic development. When an innovation newly enters the economy, 'a gale of creative destruction' takes place, breaking the equilibrium of the old economic system. The result is a challenging dynamism of the economy. Following this thinking, neo-Schumpeterians have accumulated many studies, but not enough to establish a comprehensive theory. In particular, the stage of technological development is still a black box.

This book aims to clarify the dynamism of innovation starting from the origin of technological development and the causal relationship between innovation and economic development. Through this approach, a new analytical methodology can be developed that will enable the application of a systematic and dynamic method for evaluating the development processes of science and technology per se.

1.1 BACKGROUND AND GOALS OF THE BOOK

Studies on the economics of technological change have been well summarized by Rosenberg (1976), Freeman (1982), Freeman and Soete (1997) and Coombs et al. (1987), among others. These studies are based on the neo-Schumpeterian approach and are typical descriptions of the economics of technological change. They are well organized and describe the complete picture of innovation and economic development on the basis of empirical analysis. The evolutionary economics of Nelson and Winter (1982) explores a new Schumpeterian approach. However, although some of these studies, for example, Rosenberg (1976, 1982), Freeman (1982) and Nelson (2000, 2004), have addressed the relationships between science, invention and innovation, few have focused on technological development per se. The goal of this book is to unlock the black box of the economics of technological change.

Although the above studies did not address the development of technologies, they recognized the tendency of clusters of innovations to form trajectories. Relevant notions include 'technological regimes' and 'natural trajectory' by Nelson and Winter (1977), 'technological paradigm' and 'technological trajectory' by Dosi (1982), 'technology T-cluster' by Rosenberg and Frischtak (1984) and 'new technology system' by Freeman (1984b). These notions suggest that technologies in an innovation paradigm form clusters that make trajectories. There has, however, been no concrete description of how these clusters are formed and by what equation these trajectories can be expressed.

Concerning the expression of the diffusion of innovations, Griliches (1957) first proposed that the development of innovations can be expressed by a logistic equation, using the case of hybrid corn as an example. Bibliometric analysis has been used to clarify the trends of innovation paths, for example, by Pavitt (1980) and Andersen (2001). Nakićenović and Grübler (1991) and Grübler (1999) expressed the diffusion of various energy and transportation systems by logistic equations. Marchetti (1979, 1980) described various technology developments using a logistic equation. Interestingly, Marchetti (1985) indicated that the achievements in life related to various human activities, such as musical composition by famous composers and production of scientific papers by great scholars, can be expressed by a logistic S-curve. This implies that human activities, including knowledge development and transfer, commonly have a logistic nature. The present study extends this idea. The above ideas have led to the concept of trajectories composing innovation paradigms. That is, even the phase of technological development can be expressed by a logistic equation, meaning that an innovation paradigm as a whole can be expressed by logistic equations.

The notion that an innovation paradigm has a logistic nature indicates that the paradigm exists within a definite time span with nonlinear causality. These characteristics reveal various features of innovations. The nonlinear nature of innovation development leads to various new findings and provides a comprehensive picture of innovation and economic development.

This book describes a broad picture of innovation and economic development, including the latent technological development period, which has been a black box. Such a bird's-eye view of innovation provides useful insight into the economics of technological change. The book also re-examines Schumpeter's approach by focusing on a discussion of technological development per se and in relation to science, invention and innovation. Many of the studies on innovation and technological development used in this book are Japanese in origin due to their accessibility and suitability.

1.2 NONLINEAR DYNAMISM OF INNOVATION – STRUCTURE OF THE BOOK

This book consists of four parts. Part I describes the diffusion of innovations. This is the point of contact with the conventional economics of technological change, and the logistic nature of innovation diffusion is extensively discussed. Part II presents the core part of this study – the technological development period and a new concept of trajectory. Part III discusses the theoretical treatment of sigmoid equations with a nonlinear nature and describes discrete complex systems with fractals. The discussion expands to cover scientific disciplines and global issues of human society and estimates the mega-structure of modern industrialized society. In Part IV, it concludes with a comprehensive discussion of a new picture of innovation paradigms and re-examines the economics of technological change from specific angles derived from the present study. Furthermore, a wider concept of knowledge dynamics is also proposed.

Part I examines the diffusion of innovations and clarifies their logistic nature. Kondratiev business cycles are analyzed for various logistic curves of innovation diffusion, and the causal relationship between the driving force of cycle formation and innovation dynamism is clarified. On the upswing of Kondratiev business cycles, a bubble economy can form due to irrational exuberance brought about by innovation diffusion; it is followed by a crash and depression. The causal relationship between the bubble and depression is analyzed. A new concept, 'trunk innovation', is proposed for describing the pervasive innovations that bring about large value-added to the economy. 'Trunk innovation' is a broader concept than 'technological innovation' – it includes 'resources innovation' and 'organizational innovation', defined as 'neue Kombination' by Schumpeter. The various unique features of trunk innovations are introduced, and a longer trajectory having a 70 to 100 year time span, an 'infra-trajectory', across two Kondratiev cycles is identified.

Part II is the heart of this book. It deals with the technological development period before product commercialization. This period has not heretofore been well analyzed. Long before the commercialization of an innovation product, a radical invention or creative discovery initiates the innovation. Starting from this invention or discovery, a series of core technologies are developed, and they form a trajectory. This 'technology trajectory' is a cluster of core technologies and spans, for example, 30 years. After this period, innovation products using core technologies are developed, resulting in another trajectory, the 'development trajectory'. The subsequent diffusion of these products leads to a 'diffusion trajectory'. Thus, the innovation paradigm consists of three trajectories: technology, development and diffusion.

The technology trajectory has not heretofore been sufficiently analyzed because most technologies are usually developed in academia, exogenously to the economy. The results of basic research are transferred to industry, and new products are developed along the development trajectory. Thus, the development trajectory is a field of technology transfer from universities to firms and also of starting venture businesses. The development trajectory reflects the timing of the development of venture businesses and commercialization. It also reflects the timing of government support for new industries. Existing industries as well absorb new innovation technologies through this development trajectory. The nonlinear nature of innovation trajectories makes it possible to judge the appropriate timing for participating in the innovation. The behavior of firms in the course of innovation, international competitiveness of industries, technology fusion by existing industries and evolution of innovation paradigms are discussed in terms of innovation trajectories.

Part III summarizes the discussions on innovation paradigms and trajectories. The nature of a trajectory is discussed in terms of mathematical functions, and a logistic equation is examined to determine whether it is the most appropriate one for expressing an innovation. The development process of science, including natural sciences and economics, can also be expressed by logistic equations. Innovation is a discrete phenomenon involving knowledge transfer from one person to another, and the field of innovation knowledge is treated as a complex system. In fact, an innovation paradigm is composed of a series of small fractals. The self-organization phenomenon of innovations is also discussed. The discussion of innovation dynamism is expanded to include energy resources and global problems, and a mega-structure of the modern industrialized society is proposed. A discrete knowledge dynamics is proposed that expands innovation dynamics to cover wider knowledge transfer behavior in society.

As the final Part IV, a comprehensive discussion summarizes new concepts derived from the analysis of the nature of innovation on the basis of nonlinear dynamics and provides new insights into the economics of technological change.

1.3 PART I: DIFFUSION OF INNOVATION AND ECONOMIC GROWTH

The book begins with an examination of the diffusion of innovations and clarifies the logistic nature of diffusion. Part I analyzes the logistic nature of diffusion, the Kondratiev business cycles, the transitional sequence of bubble formation, crash, and depression, and the construction of infrastructures and institutional change induced by trunk innovations. These issues are discussed in terms of innovation dynamics with a nonlinear logistic nature.

Logistic Nature of Innovation Diffusion

The diffusion of innovations has been discussed extensively since Griliches (1957) identified its logistic nature, and many economists have confirmed this characteristic for many cases. The diffusion of innovations, however, often deviates from the logistic trends, and the equation has been amended or alternative equations have been proposed. This book reconfirms the logistic nature of diffusion for many cases and clarifies the intrinsic nature of innovation diffusion. In particular, it has been found that diffusion of innovations penetrates economic turbulence and retains its logistic nature with an intrinsic diffusion coefficient. This finding strongly indicates that the diffusion of innovations is a physical phenomenon with the following characteristics.

1. Under sound economic conditions, the diffusion of innovations obeys a logistic equation.
2. When the economy stagnates due to turbulence, diffusion is suppressed by the dropping demand and deviates from the original logistic locus.
3. When the economy recovers, the diffusion returns to the original locus and regains the same diffusion rate. This indicates that diffusion has an inherent diffusion coefficient and suggests a kind of physical phenomenon.

Kondratiev Business Cycles and Innovation Dynamism

Business fluctuates with the ups and downs of the economy. Kondratiev business cycles have been discussed with keen interest because of their large impact on the economy. Various theories have been put forward to explain the cause of business fluctuation, but no conclusive explanation has yet been provided. Schumpeter postulated that the long cycles are raised by innovations and examined this using various empirical data. While many neo-Schumpeterians have pursued this hypothesis, a persuasive argument for its validity has not yet been made.

The present study first identifies the most probable positioning of Kondratiev cycles reflecting various determinations made so far. Then, a trial is carried out to examine the causality of business cycles by innovation diffusion. First, loci of various innovations are expressed by S-curves and normalized so that the mature market level is unity, that is, dimensionless. These S-curves are projected on a map of Kondratiev long business cycles with the vertical scale being unity. The result is noteworthy in that all innovation diffusion curves gather along the upswings of Kondratiev waves. That is, innovations selectively develop on the upswing of prosperity, forming a cluster. This indicates that innovations provide higher value-added to the economy. This is important evidence that innovations contribute to the economy as engines of economic development.

Economic development is compared in three countries, the United Kingdom, the United States and Japan, on the same Kondratiev cycles. The United Kingdom developed economically during the first two cycles, and the United States followed with a slight delay. During the third Kondratiev cycle, the United States took the initiative and enjoyed rapid growth. Japan began industrializing 100 years after England, at the start of the Meiji Restoration in 1868. Interestingly, Japan industrialized during the upswing of a delayed Kondratiev cycle and the industrialization of all three countries was synchronized with upswings of the Kondratiev cycles. The time lag between the development of Japan and that of Western countries shortened year by year until it disappeared in the early 1970s.

Sequence of Bubble Economy, Crash and Depression

The study summarized above on Kondratiev business cycles produced interesting results and indicated that Kondratiev waves are induced by the clustering of innovations. On the other hand, the Great Depression has been discussed for a long time, but no conclusive explanation for it has yet been proposed. Interestingly, a crash occurred at every peak of a Kondratiev cycle after a bubble economy, and it was followed by a serious depression. The same trends observed in four Kondratiev cycles suggest that there must be a common cause. Why does a bubble economy form during the latter half of the upswing of a Kondratiev cycle? The answer must be connected to the clustering of innovations, which accelerates the speculation that can lead to economic euphoria. This is the cause of bubble economics, and the escalated fever of the economy is brought down at the peak of the Kondratiev cycle because the innovations have reached a mature stage, and there is no more room to grow. An economy that grows due to the clustering of innovations changes its institutional structures, and, after a crash, the twisted economy has difficulty recovering because a huge institutional change has occurred. This is the scenario of the sequence of bubble, crash and depression during a Kondratiev cycle. This scenario is ascribed to the nonlinear nature of innovation diffusion, and this postulation is supported by quite a lot of evidence from all four Kondratiev cycles.

Trunk Innovations and Institutional Change

Schumpeter (1912) defined five categories of innovations: new products, new processes, new markets, new resources and new organizations. While the economics of technological change has mainly treated the first two as technological innovations, actual innovations should be included in all five categories. The present study proposes that those innovations having a pervasive impact on the economy be termed 'trunk innovations'. A trunk innovation is a much wider concept than general purpose technology or generic technology because a trunk innovation is not always a technological innovation – it can also be a resource or organization innovation. Thus, trunk innovations include those of energy, resources, motive forces, transportation and information and communications. Trunk innovations form infrastructures in the economy and induce institutional change. Trunk innovations also have a pervasive impact – they induce subsequent innovations and deeply penetrate the economy.

In this sense, Kondratiev business cycles are again examined based on their relationship with trunk innovations, and it is shown that all trunk innovations have diffused along the upswings of Kondratiev cycles, during which time they formed clusters. During the first Kondratiev cycle, the construction of canals played an important role in stimulating the Industrial Revolution. During the second Kondratiev cycle, coal played an important role in developing railways and iron production. During the third Kondratiev cycle, electricity and oil were added as energy resources, motor cars as transportation, and wireless telegraphy and telephony as means of communication. During the fourth wave, the comprehensive structure of modern civilized society was established with the addition of jet engines, nuclear energy and highways. The twentieth century is especially noted for the rise of oil, which led to the creation of an advanced economy.

The induction effect of trunk innovations was extensively studied for the case of the evolution of retail business. Department stores were established in the latter half of the nineteenth century, starting with Bon Marché in Paris in 1858. This followed the development of railways and steamships because the transportation of goods from great distances became much easier, and transportation costs dropped to one-tenth of what they had been earlier. It is clear that department stores were made possible by this revolution in transportation. Chain stores and supermarkets emerged during the early twentieth century, closely following the development of motor vehicles and the construction of highways. After the Second World War, discount stores were brought about by innovations in mass production. Along with the development of computer and information technologies, POS (point-of-sales) systems revolutionized retail businesses. Convenience stores with small sales floors have become profitable enough with the advent of the POS system and the logistic evolution brought about by computer systems. These trends clearly indicate that trunk innovations lead to the evolution of retail businesses.

Another important aspect of this study is the notion of 'infra-trajectories'. The study shows that trunk innovations stimulate the economy, and that diffusion of these innovations creates upswings of the Kondratiev cycles. The trunk innovations reach a state of maturity at the peaks of the cycles. The demand for innovation products, however, increases again during the subsequent Kondratiev cycle. This rebound phenomenon can be interpreted to mean that the maturation during the first stage corresponds to the economic capacity at that time; the economy then evolves due to the development of innovations, and institutional adjustments work to enhance the demand corresponding to the improved economic fundamentals. This is the result of institutional change of the economy. This long innovation trajectory is termed an 'infra-trajectory', which has a time span of 70 to 100 years across two Kondratiev cycles. These figures coincide with those described by Perez (2002). The existence of a longer trajectory and its relationship with the Kondratiev cycles was also discussed by Grübler (1999).

Another interesting feature is that subsequent innovations induced by a trunk innovation form a cluster, which is distributed from the root of an upswing of a Kondratiev cycle to the peak of the subsequent cycle, making a bundle of infra-trajectories. This phenomenon could be ascribed to the self-organization mechanism of innovations.

1.4 PART II: DYNAMICS OF TECHNOLOGICAL DEVELOPMENT AND EMERGENCE OF NEW INDUSTRIES

There is a long latent period of technological development before the commercialization of innovation products. The diffusion period of innovation lasts for around 30 years, until the market enters a state of maturity. Before this diffusion period, the period of technological development, starting from the first basic invention, also takes around 30 years. Though the diffusion of innovation products and its relationship with economic development have been discussed intensively, the period of technological development has been a black box in economics, and no explicit analysis of this period has been carried out. The present study looks into this latent period of technological development and clarifies the dynamic behavior of innovations. This makes it possible to describe comprehensively the innovation paradigm, grasp the essence of innovation, and gain new insights into the economics of technological change.

How to Describe Technological Development

While the period of innovation diffusion can be described by trends in product sales, the period of technological development is not so easy to describe. While the business output of manufacturing industries is product sales, the output of technological development is patents and development of new products. For the purpose of describing technological development, bibliometric analysis has often been used. For instance, Marchetti (1979, 1980), Pavitt (1980) and Andersen (2001) carried out their analyses using this methodology. Marchetti and Andersen discovered that a series of technological developments plotted against the trends in the physical properties of the products or against the distribution of the number of patents can be described by a logistic equation. This is an important finding and explicitly indicates that technological development has a logistic nature.

We also confirmed a logistic relationship in the annual trends of the number of new products of synthetic dyestuffs and in the annual distribution of the number of patents for plastics and pesticides, as described in Chapter 6. An especially important finding was that the 'annual distribution' of innovation elements, as represented by the invention of synthetic dyestuff and pesticide products, fully met the logistic locus determined by the distribution of the 'number' of new products or the 'number' of patents. This indicates that if we line up the history of the development of innovation elements from the earliest to the latest, we can determine the logistic trajectory from the time span. This relationship depends on the nonlinear nature of the logistic curve converging towards a mature level, and the whole distribution is substantially included within a definite time span. Because we can determine the time span of the distribution of a cluster of inventions belonging to the same innovation by finding the first and last inventions, the innovation trajectory is defined a priori. While this procedure appears rough, our analysis of more than 40 examples showed that it produced well-defined results. This process has been used as the standard procedure for determining trajectories.

Technology Trajectory and Development Trajectory

An extensive study of innovation trajectories indicated that there are two kinds of trajectories observed during the technological development period: 'technology trajectory' and 'development trajectory'. A technology trajectory is a swarm of 'core technologies' belonging to an innovation, starting from a basic invention. A development trajectory is a swarm of 'new products' developed using these core technologies. These two kinds of trajectories are clearly distinguished, and the time span is definitely determined by the line-up of relevant elements arranged in a time series. An extensive examination of more than 40 innovations since the Industrial Revolution indicates that the trajectory span has been shortened year by year, from about 60–90 years during the Industrial Revolution to 25–35 years today. Following these trajectories, a diffusion trajectory appears that leads to a new market. Thus, the innovation paradigm consists of three trajectories, technology, development and diffusion, in this order. It is interesting to see that the diffusion trajectory starts immediately after the technology trajectory is completed, making a cascade. This phenomenon becomes the basis for estimating the future diffusion trajectory in technology forecasting.

Comprehensive Structures of Innovation Paradigm

We examined eight fields of the chemical industry since the Industrial Revolution in relation to the structure of paradigms, and the existence of three trajectories in the above order was first confirmed, as described in Chapter 6. Furthermore, the structure of innovation paradigms was fully examined for synthetic dyestuffs, biotechnology and electronics. The technology trajectory was composed of a series of basic research studies carried out mostly in academia. For synthetic dyestuffs, the technology trajectory was based on basic research on organic chemistry carried out at universities, mostly in Germany. After organic chemistry was established in academia, the industry could analyze the composition of organic substances and invented synthetic dyestuffs when the technology trajectory was almost complete. The technology trajectory of biotechnology was also an academic science trajectory in which all the basic research was carried out at universities, and no one in the early stage of technology trajectory knew of the feasibility of commercializing products produced using recombinant DNA technology. Only for electronics was the research for the technology trajectory carried out in industry, but the research staff in charge were scientists with an extensive background in basic research. The nature of a technology trajectory strongly inclined to basic science is one reason for the black box in economics. It is not easy to find the existence of a technology trajectory in advance because facts in a technology trajectory are latent in the economy. At least, it is fairly difficult to recognize the value of an innovation to the market during the course of the technology trajectory. Recognition usually occurs during the last part of the technology trajectory, when actual development of innovation products begins.

Role of Development Trajectory and Rise of Venture Business

As we have seen in the examples in the previous section, basic research along the technology trajectory is carried out mostly in academia, so universities play the main role. The knowledge obtained is transferred to industry during the development trajectory. In this sense, the development trajectory is the field of knowledge transfer from academia to industry and plays the most important role in the innovation paradigm. At the beginning of the development trajectory, new products begin to be developed, and venture firms emerge to launch new businesses related to the innovation. Interestingly, the window of opportunity for these new businesses is rather narrow – about 15 years during the first half of the development trajectory. This was found by examining three areas: synthetic dyestuffs, biotechnology and electronics. Three German chemical companies, BASF, Hoechst and Bayer, were set up as small venture firms during the first half of the development trajectory of synthetic dyestuffs in the 1860s. The founding of typical biotechnology companies, such as Cetus, Genentech, Biogen and Amgen, was concentrated between 1971 and 1983. Major electronics companies, like Intel, Microsoft, Apple Computer and Sun Microsystems, were founded between 1969 and 1982.

Utterback and Abernathy (1975) described the behavior of firms during innovation and proposed a concept of 'dominant design'. They argued that quite a few firms rush into a new innovation business and develop various kinds of new products. The new products fall within a certain pattern, a 'dominant design', and only one or a few firms become successful by monopolizing it. This is also the phenomenon of 'lock-in', proposed by Arthur (1985) and David (1985). The present study traces this trend 'on the trajectories of innovation paradigms' for typewriters, automobiles and synthetic dyestuffs. The number of firms rapidly increases along the development trajectory, but after a dominant design takes hold, it suddenly begins to decrease. In all three cases, the number of firms sharply increased for the first ten years during the development trajectory, reaching 50 to 60 per cent of the eventual maximum number of firms; this period corresponds to the first 10 per cent of the diffusion trajectory. This phenomenon clearly supports the so-called 'bandwagon effect' described by Schumpeter. The dominant design appears at about 80 to 90 per cent of the development trajectory, which is still in the middle of the diffusion trajectory. The number of firms reaches a peak at around 60 – 70 per cent of the diffusion trajectory, rather early in the maturation period, and then begins to decrease sharply. These insights indicate that the behavior of firms during the course of innovation can be more informatively described by trajectory analysis, and this analysis provides an important tool for studying innovation.

National Systems of Innovation and the Role of Government

The concept of a 'national system of innovation' has been discussed in relation to the economics of technological change by Lundvall (1985), Freeman (1987) and Nelson (1993) among others. Lundvall depicted the system as one in which an innovation develops through a comprehensive system of social institution, as described by Veblen. Freeman described it as more focused on new technology systems, reflecting Perez's notion of a 'techno-economic paradigm' in which the government promotes new industries. Freeman (1987) summarized the behavior of the Japanese Ministry of International Trade and Industry (MITI) in fostering high-tech industries based on the notion of a national system of innovation.

The innovation paradigm expressed by the three trajectories is effective for clarifying the role of government and relevant firms in the development of high-tech industries because the trajectories illustrate the timing of the movement and behavior of the government and the firms. Their use also facilitates analysis of the international competitiveness of industries.

National system of innovation in Japan

We analyzed the behavior of firms and government during the development of the electronics business in Japan using a logistic procedure, as shown in Chapter 8. It is interesting to learn that the Japanese electronics industry introduced American electronics technologies and that the government, through MITI, supported their development, which was carried out early in the course of the technology trajectory. The technology trajectory covers a period of 25 years, from 1948 to 1973. Sony Corporation introduced Western Electric's transistor technology and commercialized it in 1954. NEC introduced Fairchild's IC patent in 1962. Several companies developed electric calculators, first with transistors and then with IC devices, and Japan held more than 80 per cent of the world market in the early 1970s.

The activities of Japanese electronics firms were supported by the Japanese government. In 1957, Japan's MITI adopted the Provisional Act for the Promotion of the Electronics Industry, and in 1958, the Japan Association for the Promotion of the Electronics Industry was set up. In 1961, the Act for Collaborative Research Consortium by MITI was passed, and the Research Association for Large Computer Development was set up. These government actions stimulated the Japanese electronics industry. The most important action was the setting up of the VLSI Research Project between 1976 and 1979. This project was aimed at strengthening the basic technologies for very large scale integration (VLSI) circuits through collaboration between five leading electronics companies based on subsidies from MITI. This project was successful and led to the commercialization of 64K DRAM in 1981, earlier than in the United States.

Since then, the Japanese electronics industry has become competitive internationally and has expanded its share of the world market. The trade friction between Japan and the US that first arose over the US import of Japanese steel in 1969 came to an end with the Japan–US agreement for the self-imposed control of electronics chips in 1986. The above episodes are part of the history of the national system of innovation in Japan. It should be noted that the VLSI project was launched at the same timing on the development trajectory of electronics as various venture businesses took off, as described above. This means that it was the most appropriate timing for businesses to take off.

Collaborative research projects in the US and Europe

Stimulated by the Japanese success with collaborative research on VLSI, the United States and Europe started competing projects in the 1980s. The United States Department of Defense undertook the VHSIC (Very High Speed Integrated Circuits) project between 1980 and 1984, recognizing that VLSI circuits are very important for advanced weapons and that very high speed and highly integrated IC chips should be developed. They spent $350 million, more than twice the budget of the Japanese project, and six companies participated. The European Community started an even bigger electronics project, ESPRIT, in 1984. It lasted ten years and cost 1.5 billion ECU, ten times the budget of the Japanese project. In 1988, the United States organized a civil collaborative research consortium, SEMATECH (Semiconductor Manufacturing Technology Institute), comprising 14 leading electronics companies. In the same year, the European Community launched JESSI (the Joint European Submicron Silicon) project that included Siemens, Philips, SGS Thompson and others and cost $4 billion. In 1990, these two organizations of SEMATECH and JESSI were joined to collaborate through IBM as the mediator. These actions by Western countries, however, do not appear to have been best-timed because they were undertaken during the latter part of the development trajectory of electronics.

National system of innovation in the United States

The United States recognized the importance of basic science early on, and in 1945, the National Science Foundation (NSF) was set up, based on a proposal to President Roosevelt by Vannevar Bush. Following this, the United States enjoyed great prosperity during the 1950s and 1960s. However, they declined in international competitiveness due to the hollowing out of the domestic economy caused by the higher value of the dollar. President Carter sent a message on innovation to Congress in 1979, in which he stated that regaining international competitiveness was a crucial issue. This led to the passage of the Stevenson–Wydler Technology Innovation Act in 1980, which was aimed at strengthening the competitiveness of the United States by activating venture businesses and motivating the private sector by implementing various countermeasures. The Young Report to President Reagan in 1985 was more strictly aimed at promoting competitiveness and knowledge transfer from academia to industry. Many engineering centers sponsored by the NSF were set up in universities. By 1991, 25 science centers had been set up in universities to stimulate basic research for industry. The National Technology Panel was organized in the Office of Science and Technology Policy directly overseen by the President, the first 'Important Technology List' was prepared, and a 'Critical Technology Plan' was proposed to the President based on an evaluation of technology competitiveness in the US, the EC and Japan. The United States has certainly engaged in major efforts to strengthen the international competitiveness of its industries by promoting a national system of innovation.

Development of the electronics industry and the role of government in the Asian region

The innovation paradigm expressed by the three trajectories is useful for estimating the timing of various actions. This procedure was used to analyze the development of the electronics business in the Asian region.

The behavior of firms and the actions of government related to the electronics business in Japan can be described on a map of innovation paradigms. Japanese firms and government took action on the technology trajectory and succeeded in developing 64K DRAM in the early stages of the development trajectory. This timing on the development trajectory is important in connection with the later success of the Japanese electronics business worldwide.

The electronics business in Korea started at the root of the development trajectory, which was ten years behind that in Japan. The development trajectory of electronics lies between 1973 and 1998 in the US and Japan. The Korean government proclaimed a guiding principle of fostering the electronics industry in 1969 and established a policy giving overriding priority to high-tech products including electronics and computers in 1977. The Promotion Act for Electronics and Machinery Industries was enacted in 1981. These government actions stimulated the industry. The timeline for the industry was this. Fairchild and Motorola launched factories for assembling transistors in Korea in 1966. Samsung Electronics Company was established in 1966, introduced technologies for 64K DRAM in 1983, and succeeded in developing the 1M DRAM on their own in 1986. Major contributions to these activities came from people who had worked in Silicon Valley, and their actions contributed to the take-off of the Korean electronics business. Since then, Samsung has become the leading producer of DRAM in the world. In short, the Korean electronics industry took off in the middle of the development trajectory with strong support from the government and grew up to become a leading DRAM producer.

The development of the electronics business in Taiwan is another success story. The Taiwanese electronics industry started with the setting up of a governmental research institute, the Industrial Technology Research Institute (ITRI), in 1973, in which the Electrical Research and Services Organization (ERSO) was founded the following year. This institute played a central role in establishing the Taiwanese electronics industry. ERSO dispatched researchers to RCA in the US to learn about electronics technologies, and later these researchers set up an electronics company, UMC, as a spin-off from ERSO. Another important event was the construction of the Hsinchu Science-based Industrial Park in 1980, which played a key role in the development of the Taiwanese electronics industry. The park functioned as the center of the electronics business and maintained close contact with Silicon Valley. It was a boon to the many venture businesses launched around 1987. Most of the firms were rather small and could not establish their own manufacturing plants for electronics products. A foundry plant to manufacture products was thus set up for common use. This plant worked well, producing even small lots of various products. This is the background for the large growth of the Taiwanese electronics business, which now has a considerable share of the world market for many products. This history of the development of the Taiwanese electronics industry follows the same electronics trajectories as those described above. The transition of the above innovation paradigm can be interpreted based on Vernon's life-cycle theory of innovation (Vernon, 1966), which is described in Chapter 3. That is, the development of the electronics industry proceeded from the United States to Japan and then to Korea and Taiwan.

In other Asian countries and regions, the electronics business has also developed along the same electronics trajectories. The electronics businesses in Singapore and Malaysia developed with the same timing described above, but they were more open to the advances of foreign enterprises, and their governments played an active role in stimulating the domestic business climate. India had no substantial activity during the first half of the development trajectory, but in the latter half, after the 1980s, the government welcomed foreign enterprises. The establishment of information technology parks, especially of the one in Bangalore in 1995, quickly activated the Indian electronics business, which depended mainly on the United States. Though we are now in the last half of the development trajectory, business opportunities for India have been steadily expanding along the trajectory.

It is interesting to remember that although Japan introduced Western technologies 100 years later than the West during the Meiji Restoration, the world businesses are now completely synchronized without any time lag.

International Competitiveness Related to Exchange Rate

In the upswing of a Kondratiev cycle, competitiveness is governed by innovation diffusion, and international competitiveness is technology oriented. In the downswing, however, competitiveness is governed by different factors. In the present downswing, for example, the exchange rate has been an important factor determining international competitiveness.

After the Second World War, the world economy enjoyed a prosperous era during the upswing of the fourth Kondratiev cycle. At the end of the upswing, however, the competitiveness of the United States had been weakened due to the higher value of the dollar, which hollowed out the domestic economy. Japan, on the other hand, enjoyed a cheaper yen, which led to increased exports. This switch in competitive strength led to severe trade friction between the United States and Japan.

After the Plaza Accord in 1985, Japan endured a lengthy period of a higher valued yen, which hollowed out the domestic economy. Japanese export-oriented industries were forced to advance overseas and internationalize. This advancement abroad accelerated the development of the ASEAN (Association of Southeast Asian Nations) countries.

The yen remained at a higher value even after the crash of the bubble economy in Japan in 1990. While the failure to control the liquidity of the money supply was an important factor in this, another important factor was the rise of hedge funds and derivatives, which have distorted the world's monetary system.

Thus, Japan's domestic economy has suffered while the globalization of Japanese businesses has advanced. The recent surge in China's economy has accelerated this gap, and deflation in Japan has been promoted by the large gap between the lower valued Chinese yuan and the higher valued Japanese yen.

Technology Fusion by Existing Industries

Existing industries follow the development of innovations with keen interest and skillfully incorporate these innovations into their activities. This is the fusion phenomenon of innovation technology as described by Kodama (1985) and Freeman (1987). Especially in the case of trunk innovations, innovation technologies pervasively penetrate existing industries and change the structure of societal institutions. The analytical procedure of innovation paradigms using trajectories is useful for estimating the progress of technology fusion. Again, innovation trajectories provide a very useful tool for estimating the timing of fusion and the emergence of business competitiveness.

We examined two innovations, 'computers' and 'electronics', in terms of technology fusion by existing industries, as described in Chapter 9. The trajectories of mainframe computers came a little bit earlier than those of electronics. The introduction of computers into business as their users must occur at least around or after the diffusion trajectory of computers. If the introduction is for the technology development of an industry, the fusion technology can form around the development trajectory of the computer paradigm. The introduction of computers into the steel, cement and chemical industries, as well as into railways and flight simulators, was examined. The chemical industry in Japan introduced computers as just the users but not as the tools for their innovations, so the fusion trajectories were located around the time of the diffusion trajectory of computers. The fusion trajectories for the steel, cement and railway industries to introduce computers as tools for industry were located a little bit earlier than the diffusion trajectory of computers. The cement industry in Japan in particular introduced computers as tools for producing cement, and the introduction occurred near the time of the development trajectory of computers. Flight simulators had been developed much earlier than computer technology, and their incorporation of computers was carried out as a tool of technology development. Thus, its fusion trajectory fully coincided with the development trajectory of the computer itself because of the development of their products using computers.

The introduction of electronics, that is, semiconductor chips, was examined for the machine tool, camera, automobile and jet engine industries. All of these industries adopted electronic technologies as an evolutionary means, and the fusion trajectories developed along the development trajectory of electronics per se. It is surprising to see such early adoption of electronics technologies by existing industries along this trajectory.

It is interesting that the phenomenon of technology fusion by existing industries can be traced in comparison with the trajectories of an innovation paradigm. The introduction of electronic technologies by retail businesses strongly affected their business competitiveness. This is discussed in detail in Chapter 5.

Maturation of Innovation and Evolution of Paradigms

Maturity period of innovation

Innovation paradigms reach maturity, and the growth of related technologies stops. This stage is sometimes unstable because economic euphoria often occurs in spite of market maturation, inducing a bubble economy and the crash that typically follows. In the latter half of the diffusion trajectory, there are certain characteristic patterns of behavior of firms. The products are narrowed down to a dominant design, and the number of firms is reduced. Firms sometimes form strategic alliances to escape the fate of natural selection. An example of a strategic alliance between Japanese and US firms is described in Chapter 9 for the electronics business. These patterns of behavior can be interpreted in terms of their location on the trajectories.

Evolution of innovation paradigms

An innovation paradigm intermittently evolves from one paradigm to another when a new seed for innovation emerges through serendipity. Semiconductor electronics chips evolved apart from vacuum tubes, and now three kinds of paradigms have evolved: computers, electronics and multimedia. Although they developed independently, they are tightly interrelated. The computer and electronics paradigms were described above. The multimedia paradigm is shaping up. This paradigm is tightly supported by the ISDN paradigm, which comprises optical fiber technologies and coherent light emission devices, forming an optical communication system. The diffusion trajectory will enter a mature stage with the introduction of 'fiber to the home' (FTTH) around 2015, and FTTH will become the dominant communication media. The multimedia paradigm will evolve through digitization of all kinds of media; it is now entering a diffusion period based on communication media developed for the internet. While the computer and electronics paradigms are already mature, the multimedia paradigm is still emerging. This gap between paradigms induced a mismatch phenomenon, which brought about an IT recession in the late twentieth century because multimedia had not yet advanced far enough to add sufficient value to the economy. We can, however, expect fairly robust development of multimedia in the near future, accompanying further development of computers and electronics as an extension of the infra-trajectory that will develop along the fifth upswing of the Kondratiev cycle with a peak around 2040.

1.5 PART III: NONLINEAR DYNAMICS OF INNOVATION AND KNOWLEDGE TRANSFER

The third part discusses the above scenario of innovation dynamism and deepens the concept of innovation.

Description of Knowledge Transfer in Innovation

Part III starts with a comprehensive discussion of the sigmoid curve and a verification of logistic expression for the innovation process.

Following Malthus's *Essay on the Principle of Population* in 1798, Sadler (1830) argued that the population growth rate decreases as the population increases, and Verhulst (1838) expressed this in a logistic equation. This logistic equation has since been applied to the expression of biological growth. Schumpeter (1939) recognized that the logistic equation can also be used to describe growth but was reluctant to apply it to innovation processes. A logistic equation was first applied to the diffusion of innovations by Griliches (1957), and several economists examined its functionality. Marchetti (1985) and Modis (1992) described life behavior such as the creation of masterpieces by great music composers and Nobel laureates by a sigmoid curve, and Marchetti (2002) clarified that these sigmoid curves can be expressed by a logistic equation. This implies that knowledge creation and transfer can be expressed by a logistic equation. Mahajan and Peterson (1985) comprehensively analyzed sigmoid-type growth curves using various equations and examined their derivations. Metcalfe (1994a) claimed that the diffusion of innovations is similar to an ecological growth model and is governed by two factors, the intrinsic growth rate (r) of species and the carrying capacity (K), which is the so-called r-K theory, as put forth by Roughgarden (1979) and Slobodkin (1980). Thus, the most general expression of an innovation and a biological growth model should be the product of endogenous factor (g) and exogenous factor (f):

$$dy/dt = g(y)f(y), \tag{1.1}$$

where y is the amount of diffusion at time t.

If the ultimate market is y_0 and $f(y) = y_0 - y$,

$$dy/dt = g\,(y_0 - y).$$

The logistic, Gompertz and Bass equations are derived from this function.

To deal with the lack of a supply side in the logistic model, Metcalfe (1981, 1982) developed a comprehensive description including the market growth rate and the demand and supply sides. He further skillfully clarified the role of economic elasticity and incorporated time-dependent changes in the environment of the industry by introducing the Gompertz model (Metcalfe, 1994a).

Shinozaki (1994) discussed the general trend of biological growth models. Among the various equations for sigmoid curves discussed by Mahajan and Shinozaki, a simple logistic equation appears to be most suitable for describing innovation behavior.

The diffusion of innovations and the creation of new technologies comprise a typical knowledge transfer model and must be discrete phenomena of person-to-person interaction. Such a discrete system should be expressed by a difference equation instead of a differential one. If the system is a huge community like an economic society, however, a logistic equation can be applied on the basis of the mathematical convention of continuity.

Discrete Nature and Complexity in Innovation

Goodwin (1950, 1982, 1990), Peters (1991) and Lorenz (1993) understood that innovation dynamism is a nonlinear discrete system that can be analyzed by logistic mapping and postulated complex systems. Although these approaches are close to that used in his book, there have been few verifications of them using empirical data. We examined complexity in innovation paradigms using survey data and confirmed the existence of several kinds of fractals and self-organization behavior.

Three kinds of fractals

Innovation proceeds in a discrete system of knowledge transfer from person to person. Each innovation product is individually developed along the development trajectory as an element of fractals. Each element is quantumized and has its own small trajectory. Thus, the development trajectory is composed of fractals of products, and each product disparately exists, forming chain fractals along the development trajectory. The development trajectory of electronics comprises a swarm of IC chip fractals, each of which develops along a small trajectory of IC chip development with a time span of 3–4 years. As the fractals evolve, the degree of integration increases in accordance with Moore's law. Adoption of the chips by various industries creates fusion trajectories, most of which emerge along the development trajectory of electronics, forming bundle fractals. Trunk innovations develop across two Kondratiev cycles forming an infra-trajectory and induce various subsequent innovation trajectories, which again form bundle fractals along the original infra-trajectory.

The development trajectory of electronics is composed of various elementary technology systems combined with software such as FMS (flexible manufacturing systems), POS (point-of-sales), CALS (computer aided logistic systems) and EDI (electronic data interchange) systems. The three paradigms of computers, electronics and multimedia are positioned in this order and deeply interrelated. Each development trajectory is composed of various elementary technology systems. These fractals combined with software and other application elements are termed 'system fractals'.

In the chemical field, we can identify many innovation paradigms, which are described in Chapter 6, for the acid/alkali industry, electrochemical industry, synthetic dyestuffs industry, chemotherapy industry, chemical fiber industry, pesticide industry, petrochemical and polymer industries. Each industry is composed of various small fractals. The polymer industry is composed of various plastics, textiles and elastomers. Plastics are composed of fractals exemplified by polyethylene, polypropylene, polystyrene and polycarbonate. The fractals have various sizes, and their time spans vary, for example, from 3–4 years for silicon products to around 10 years for chemical and some other products. These characteristics are consistent with Schumpeter's complex theory of Kondratiev cycles combined with Juglar and Kitchin cycles. The paradigm of innovation, however, is certainly governed by a definite discipline of innovation and extends over a fairly long time period. It is an important point that the discipline of innovation is clearly recognized as being common throughout an innovation paradigm and can easily be distinguished from its fractal components.

Self-organization in innovations

Innovation is a discrete system, and various fractals have been identified, as described above. Another important feature of innovation is self-organization, which can be seen in various phases. The cluster formation of innovations along the upswing of Kondratiev cycles, of bundle fractals of fusion trajectories along the development trajectory, and of infra-trajectories induced by a trunk innovation are ascribed to the self-organization mechanism. The attractor of self-organization and fractal formation can be interpreted as the value-added of innovation.

These phenomena of fractal formation and self-organization indicate that the innovation dynamism is governed by a complex system.

Knowledge Development Dynamics

Given the above consideration, we can term the behavior of discrete and nonlinear innovation systems 'knowledge development dynamics'. That is, innovation systems are composed of four elements and three trajectories. Innovation develops through relevant actors creating innovation products. Innovation develops through knowledge transfer from one person to another, and its products bring about value-added to the economy. Therefore, four elements: actors, products, knowledge and value-added, are involved. As the innovation systems are discrete, they are composed of discrete elements. Product elements are actually inventions to produce patents and products. Patents are intangible, but product elements also include intangible products such as software and systems. Innovations are developed by various actors along three trajectories. 'Actor' is another discrete element, and each trajectory has its own actors. Knowledge is an essential component of innovation and is transferred between actors and trajectories.

An innovation product brings about value-added to the economy, and its pervasiveness and quality is important in determining its impact.

Knowledge development dynamics can be expanded to much wider disciplines beyond innovation dynamics. It covers all human knowledge activities such as knowledge development dynamics of science, knowledge development dynamics of economics, knowledge development dynamics of ideologies, and knowledge development dynamics of arts. These wider activities are discussed in Chapter 11.

Acceptable Difference Equation and its Unique Feature

The discrete nature of innovation is mathematically discussed. May's mapping is not the case of the actual innovation systems to be applied because it has no monotone increasing property. We have found that a difference equation derived by bilinear approach in a condition of gauge invariance has a monotone increasing nature and has the unique feature that any solution of the equation fulfills the solution of the original differential equation irrespective of the span of the difference. This specific feature suggests that fractals with any time span can be included in the parent innovation trajectory. Yamaguchi (1997) also pointed out an interesting phenomenon – any solution of the difference equation derived by mapping the solution of the logistic differential equation fully coincides with the solution of the original logistic equation in any span of difference.

Science Paradigms and their Evolution

The innovation dynamism discussed above indicates that even the technological development period has a logistic nature. In particular, the technology trajectory is often a science trajectory in academia. This implies that the development of science itself can be expressed by a logistic equation. This is examined through various science paradigms, starting with Newtonian dynamics, which can be expressed by a trajectory having a time span of 150 years, from Copernicus' *On the Revolutions of the Celestial Spheres* in 1543 to Newton's *Principia* in 1687. Modern science disciplines – chemistry, biology (embryology), and physics (electromagnetics) – arose during the nineteenth century and these science trajectories have a time span of about 90 years, which matches those of a series of innovations during the Industrial Revolution. This means that modern science and the Industrial Revolution emerged during the same period, building modern civilized society. The science paradigms evolved intermittently. Theoretical physics evolved in a stepwise manner from Newtonian dynamics, relativity theory, elementary particle theory and the ultimate unified theory. Biological evolutionary theory developed from Darwinian theory through genetics to modern molecular biological evolutionary theories. It is interesting to see that schools of economics, from classical economics to Schumpeterian economics, have had the same time span of 90 years, while those of the natural sciences have been steadily shortened.

Technology Foresight and Long-term Perspectives

An innovation paradigm has three trajectories. If an innovation is in the technology trajectory, we can extrapolate future trends of the innovation and predict the timing for business opportunities and the start of actual commercialization, because the time spans of the three trajectories are similar and we have some understanding of the correlation between them. Several innovations currently in the technology trajectory are examined, and future trends are estimated, as shown in Chapter 11.

Technology foresight

Following recent advances in the basic sciences, certain fields will take on more importance in the near future. Since we can clearly identify their technology trajectories, we can estimate their development and diffusion trajectories, in which the diffusion trajectory will start when the technology trajectory ends. The multimedia paradigm has already moved to the diffusion trajectory, triggered by the introduction of digital TV, and the whole scope of the paradigm can be depicted until the end of the diffusion trajectory. The timing for venture business in this paradigm has almost passed. In the field of information technology, the digital age is now starting, and all information media will eventually shift to digitization. Electronic devices will advance into the age of nanodevices, which have a different paradigm than conventional silicon devices and, on a physical scale, could be one-thousandth the size of silicon devices. There are high expectations for quantum computers now that the basic concept has been established. Development is underway for various nanomaterials, which are materials controlled at the nanometer level, and that may have extremely superior performance like higher strength or new functions. In the field of biotechnology, two areas are very promising: genome technology and regeneration technology. Genome technology took off after the completion of human genome decoding in 2000, and it is now entering the age of post-genome and proteomix technologies. Ever since embryonic stem cells were discovered in 1981, various technologies for regenerating organisms have been developed, and various human organs and tissues will one day be reproduced using embryonic or other stem cells. These innovations in biotechnology are now on the development trajectory, and the timing is right for new businesses.

These innovation paradigms are clustered on the upswing of the fifth Kondratiev cycle heading towards 2040.

Long-term scenario

Another important scenario of innovations involves such global issues as energy resources and the environment. In this scenario, various innovation paradigms can be expected, especially the development of new sources of energy and energy-related technologies such as solar energy technologies, fuel cells and superconductors. These innovations have been in development for quite a long time. However, while solar cells have already been commercialized, their diffusion is not so pervasive due to competition from oil and other existing resources. Moreover, the conversion efficiency of solar cells is already in the mature stage, based on the analysis using a logistic procedure. Fuel cells are now receiving much attention. If we could use cheap hydrogen in amounts comparable to those for oil, fuel cells would become a promising energy source. The superconductor paradigm has a long history, starting with the discovery by Onnes in 1911 of the first superconductor, mercury. The development trajectory of superconductors has been in a mature stage, and the diffusion trajectory has taken off in the case of small electromagnets. Magnetic levitation technology is now ready to go, but its economical feasibility is problematic. Heavy electric applications are also anticipated. The diffusion trajectories do not take off simultaneously for all applications; rather, they depend on economic conditions.

A serious problem in the near future will be a shortage of oil. While oil consumption has continued to accelerate since the early 1900s, oil reserves are becoming depleted and no major discoveries of oil reserves have been made. When the change in estimated oil reserves over the past years is plotted, it forms a logistic curve, indicating that we are already in a mature period. The ultimate discoverable reserve of oil is estimated at 2.5 trillion barrels, so the amount of usable oil will remain at 1.5 trillion barrels, which means that oil will run out by 2040. According to the analysis by Hubbert (1969), the actual peak of oil production will occur when exactly half the total reserves have been consumed and after the peak, oil production cannot be increased. Oil depletion will thus soon become a serious problem because we are almost at the peak of world oil production. Campbell (2005), Deffeyes (2001) and Duncan (2000) estimated that the production peak in the world will be between 2003 and 2009 and even in the Middle East it will be in 2009.

Global periodicities and knowledge development trajectories

The longer time-scale of the earth indicates that there have been various periodicities such as forming and splitting apart of super continents, climate change expressed by Milankovitch cycles, and sunspot cycles. Sunspot cycles have a hierarchical structure: 11 years, 60 years and 300 years. The development of the economy could have the same cyclic structure because we have already observed Juglar cycles of 10 years and Kondratiev cycles of 60 years, and modern civilization probably has a 300-year cycle, as described below.

While we have observed that innovations develop and form logistic curves having a time span of about 30 years and form a cluster along the upswing of the Kondratiev business cycle, various knowledge activities in society such as science, economics, arts, scientific thought, ideology and various societal activities can be expressed by the same knowledge development trajectories, and these activities were observed to be synchronized with innovation cycles, forming a cluster in each historical period.

Mega-structure of society

We have experienced a cycle of bubble economy, crash and depression in each of the four Kondratiev cycles. Looking at the longer trend of society, we can predict that the S-curve representing modern civilized society has a time span of about 300 years and consists of five Kondratiev cycles. This curve is now entering a maturity period.

1.6 PART IV: NEW INSIGHT INTO THE ECONOMICS OF TECHNOLOGICAL CHANGE

The final part of this book is a comprehensive discussion of innovation dynamism. In this discussion, typical features of innovations are extracted, and their contributions to the economics of technological change are clarified.

This book is the first attempt to describe the entire picture of the innovation paradigm using three trajectories. It is especially useful to express an abstract matter such as technological development by using visual representation, that is, specific trajectories. An important characteristic of this analysis is that it expresses the progress of innovation by using nonlinear trajectories. While the diffusion of innovation products is known to obey a logistic equation, it is an interesting finding that the trend of technological developments can be described by a logistic equation as well. This comprehensive analysis using logistic equations has revealed various interesting features and has deepened our understanding of innovations.

Analysis of the innovation paradigm using trajectories revealed three features. One is the nonlinear nature of innovation, which means that each trajectory has a finite nature, eventually reaching saturation. This implies that the time of innovation is crucial. The second feature is the exogenous nature of the technology trajectory, which triggers innovation. This implies that innovations often originate outside the economy, especially in academia. The third feature is the importance of knowledge transfer between and within the trajectories. The concept of 'field' for knowledge transfer is of particular importance because knowledge transfer is essential for the progress of innovation.

Nonlinear Nature of Innovation

The nonlinear nature of innovation means that a trajectory reaches saturation within a finite time span. This makes it possible to identify the trajectory simply by determining its time span. It also indicates the timing of various phenomena related to innovations, as shown by the timing of emerging venture businesses, the timing of governmental action, and the timing of technology fusion by existing industries. Thus, innovations should be recognized as a function of time.

Maturation of innovations

The fact that innovations mature indicates that there is a limit to the development of an innovation within a finite time frame. The technology trajectory reaches maturity for core technologies, which is when diffusion begins. The development trajectory is limited in that new products cannot be expected during the maturity period. In the diffusion trajectory, no further growth of the market can be expected within the capacity of the economy at that time.

Inherent potential of innovations and value-added

An innovation has an inherent technological capacity. For example, the steam engine is a general-purpose technology that was adapted for use in various forms of transport, such as locomotives and ships, but not for automobiles and airplanes. That is, the steam engine has a technological capacity which cannot be exceeded. This suggests that there is a time to end research and development of each technology. An innovation reaches its own limits within a certain time period and retains its performance afterwards. This resembles the evolution of living organisms. The coelacanth evolved 350 million years ago, and its shape has remained the same ever since.

In the course of innovation diffusion, the value-added increases and reaches a limit at maturity. The impact of an innovation on the economy depends on its inherent potential and capabilities. It can be measured using an economic impact index, that is, the product of the unit value-added and the population, which is the number of people receiving the benefit.

The time span of innovations will become shorter. With the tremendous increase in the speed of information transmission, the time span depends on the capability of the human brain to process the information. As this capability is unlikely to improve, the time span should not fall below some threshold, say 20 years.

Exogeneity in Innovation – A Wider Concept

The concept of an innovation paradigm having three trajectories has not heretofore been recognized, and most discussions of innovation have been carried out within the region of the trajectories of development and diffusion, focusing on the behavior of firms. Of course, many discussions of innovation have dealt with the basic research activities of universities, but the importance of the technology trajectory has not been explicitly recognized. Basic research at universities is carried out independently of the economy and is exogenous in nature.

According to our analysis, core technologies are often created at universities, and they become fully developed at the beginning of the development trajectory. This means that almost all core technologies had been fully developed before firms became involved with technology transfer from the universities. Recently, there has been a tendency to directly commercialize the scientific knowledge created at universities, and venture businesses are more readily launched. This means that firms must acquire the knowledge at an earlier stage to be successful. Firms should be involved in innovation early on the technology trajectory and should ideally collaborate with universities. Rosenberg (2000) discussed this issue and pointed out the importance of the endogeneity of university research.

Linear model of innovation revisited

The three trajectories of the innovation paradigm are always in the order technology, development and diffusion. This means that basic research, new product development and market formation occur in this order. This is the linear model of innovation, which contradicts the accepted feedback model. This contradiction is ascribed to the introduction of the technology trajectory, an essential element in this work. The feedback model is built around the development and diffusion trajectories. Knowledge transfer between the two trajectories is essential, and the feedback effect is an important factor. This is the feedback model versus the linear model. In our study, we also emphasize the importance of knowledge transfer, which in our case occurs between and within trajectories. Our consideration of the technology trajectory in addition to the conventional development and diffusion trajectories has given new life to the linear model.

Two concepts of evolutionary theory

Nelson and Winter (1982) skillfully described the innovation behavior of industrial firms as the same process as the evolution of life. Using this evolutionary theory of economics, they described the behavior of firms as 'routines' and suggested that firms play the same roles as genes. Routines evolve through regeneration, mutation and survival of the fittest. Firms having the best routines survive in the struggle for existence, which is a form of natural selection.

There are two types of evolution in organisms, micro-evolution and macro-evolution, as described by Hall (1998). Micro-evolution is exemplified by the evolution within the homo family from ape-men and homo erectus to homo sapiens, and macro-evolution corresponds to the bifurcation of humans from apes. A big jump in evolution, as from reptiles to mammals, is called a mega-evolution. If this classification is applied to innovation behavior, micro-evolution corresponds to the progress of innovation that evolves along the innovation trajectories, and macro-innovation corresponds to the jump from one innovation paradigm to another as a result of evolution. This can be termed 'paradigm evolution' or 'paradigm shift', and basic research in academia exogenously induces paradigm evolution. The mega-evolution is the evolution from the ancient civilizations of 5000 years ago to the modern civilization of today.

The Knowledge Transfer Field of Innovation

In the innovation paradigm, it is important that knowledge of the innovations be transferred within a trajectory and especially across trajectories. This phase of transfer is referred to as the 'field of knowledge transfer', and it comprises various actors in the human community participating in activities along the trajectories. Nelson and Winter (1982) defined routines in the course of economic change, and we can identify various divisions of labor to carry out these routines in the course of an innovation. Each trajectory has its own actors, and knowledge transfer between these actors is the decisive factor in innovation. The knowledge transfer fields dynamically shift with time, forming an innovation paradigm. An innovation paradigm evolves to another paradigm if a unique idea or unexpected discovery emerges. The knowledge transfer field of an upcoming innovation is not well enough formed at the beginning to enable someone to notice the upcoming innovation. This is certainly an exogenous state of knowledge in the early stage of technology trajectory. As the technology trajectory gradually takes shape, people start to notice the significance of the innovation and a development trajectory begins.

Uncertainty and the Markov process

Nelson and Winter (1982) claimed that an innovation is a kind of Markov chain system. That is, an innovation evolves through a trial-and-error process, and the process is stochastic, with a high level of uncertainty. While this is reasonable, it is interesting to see that the resulting paradigm has well-defined trajectories in a deterministic system. There is no contradiction because, in the course of innovation development, there is much uncertainty as it is a stochastic process, but the final state of development reaches a deterministic logistic trajectory. Such a framework is interpreted to correspond to the systems of light and electromagnetic waves – the two phases of stochastic probabilities correspond to the wave and deterministic state of a particle.

Inductive Thinking and Entity of Innovation

The insights obtained from our analysis of innovation dynamism summarized above facilitate the discussion of the economics of technological change. The final part touches on the basic stance of the study.

The way of thinking – inductive or deductive – is important for the approach to innovation analysis. Although conventional economics mainly took a deductive approach, Schumpeter used an inductive approach to study innovation dynamism. Neo-Schumpeterians have taken the same approach in examining the economics of technological change using various fact data. In this book, innovation is first analyzed from the viewpoints of science and technology outside of economics, and then its relationship with economic development is examined.

The trajectory approach based on nonlinearity belongs to the punctuated evolutionary concept according to Andersen (1994), and the trajectories become mature within a finite time span. This is in contrast to Darwinian gradualism. This finiteness makes it possible to evaluate the appropriate timing for participating in an innovation and reveals various specific features of innovations.

Another characteristic feature of innovation is that it is a discrete system. The field of knowledge transfer in the innovation paradigm is actually a discrete system because the knowledge is transferred person-to-person. In the innovation system, not only are the actors in charge discrete but also the innovation products are developed stepwise in a discrete process. This leads to fractal formation along the trajectory. Various fractal formations expand the innovation to wider markets, leading to techno-economic paradigm change and institutional change, resulting in a hierarchical structure of the innovation paradigm. This kind of versatile activity is brought about by the self-organization mechanism through the value-added of innovation as the attractor and leads to a complex system.

PART I

Diffusion of Innovation and Economic Growth

During the past 230 years, since the Industrial Revolution, human beings have constructed modern civilized society by applying various technological innovations. Over the last 100 years, innovations have brought about 30-fold growth in gross national products. This book aims to clarify the role of innovation in economic development and the behavior of technological development leading to innovations. It describes them in terms of intrinsic dynamism.

Part I addresses the correlation of innovation diffusion with economic growth. Studies on innovation diffusion have frequently been carried out using a logistic equation. This part first analyzes the applicability of the logistic equation and confirms its validity. It also describes various features of diffusion phenomena.

That the origin of economic growth is innovation is examined on the basis of logistic dynamism. The existence of long business cycles, such as Kondratiev cycles, has been discussed, and while the cause of these cycles has been intensively discussed, no decisive theory has emerged so far. The analysis presented here of their causality focuses on innovation. The emergence and collapse of bubble economies and subsequent depression are discussed in terms of innovation. A new category of 'trunk innovation' is proposed for important innovations constructing economic infrastructures and their significance is clarified. Trunk innovations induce subsequent innovations and trigger evolution of existing industries. Furthermore, it is interesting to find longer diffusion trajectories of innovation beyond Kondratiev cycles. A new concept of innovation diffusion is discussed related to this kind of longer cycle, an 'infra-trajectory', which is deeply related to the institutional changes caused by trunk innovations.

Innovation develops with a kind of dynamism and its analysis results in a well-defined dynamics.

2. Dynamics of product diffusion and economic development

The first part of this chapter reviews how the issue of innovation has been treated to date in economics and gives a historical overview of the economics of technological change, focusing on the correlation between innovation and economic growth. Given this background, the study starts from an analysis of the product diffusion of innovation and describes the innovation diffusion using a logistic equation. In particular, the logistic nature of innovation diffusion is precisely analyzed to identify the intrinsic characteristics, and the tough nature penetrating economic turbulence is clarified through various examples.

2.1 HISTORY OF ECONOMICS OF TECHNOLOGICAL CHANGE

To begin, the history of the economics of technological change is reviewed as the background of this study.

Economic Development and Technological Innovations in Economics

If the origin of modern economics is *The Wealth of Nations*, written by Adam Smith in 1776, it coincides with the Industrial Revolution started by the inventions of the spinning jenny in 1764 and James Watt's steam engine in 1769. Since then, the development of economics has progressed along with the development of modern civilization. In *The Wealth of Nations*, Smith asserted that science and technology play important roles in the division of labor in an economic society – he clearly recognized their significance. Karl Marx pointed out that a capitalist society tends to enhance productivity by replacing laborers with efficient machines. Marx's model of the capitalist economy ascribes a central role to technological innovation in capital goods. Such awareness of the importance of science and technology in economics faded out after the early stages of the development of modern civilization. Later modern economics concentrated on discussing the relationship between labor, money and goods, ignoring the contributions of science and technology.

After the experience of the Great Depression in 1929, reconstruction of economics progressed. J.M. Keynes published *The General Theory of Employment, Interest and Money* (1936) in which he pointed out the importance of an economic policy apart from the principle of laissez-faire. J.A. Schumpeter, who wrote *The Theory of Economic Development* at the age of 29, ambitiously postulated that the driving force of economic development is technological innovation. Schumpeter (1939) reformulated his ideas into the concept of 'Business Cycles', based on Kondratiev's long cycle theory (Kondratiev, 1926), and pointed out that business cycles are caused by innovations. This interpretation is contradictory to Kondratiev's theory. Schumpeter ascribed the cause of three business cycles – Kondratiev long cycles of 50 to 60 years, Juglar's 10-year cycles (Juglar, 1862) and Kitchin's short cycles of 40 months (Kitchin, 1923) – to innovation: 'neue Kombination'. This ascription of three cycles to innovation makes understanding of business cycles rather difficult due to the complicated fluctuations in the economy. In particular, the fact that while the actual peak of the third Kondratiev wave is considered to have occurred in 1920, economic development continued beyond that point, cannot be explained consistently. Kuznets (1940) strongly criticized this complex theory combining three cycles. Hansen (1941) considered that the peak of the third Kondratiev cycle occurred in 1929 and supported Schumpeter's hypothesis of complex cycles. Since then, innovation theory for the Kondratiev cycles has not been sufficiently evaluated.

The economic theory that originated in Kondratiev's business cycle hypothesis was succeeded by Kuznets's business cycle mechanism, and Schumpeter's analysis of business cycles has developed into innovation theory. After the Second World War, business cycle theory was advanced by Hicks (1950), Rostow (1975), Mandel (1975), Mensch (1975), Freeman (1977), Freeman et al. (1982), van Duijn (1979, 1983), Freeman and Louçã (2001) and Perez (2002), and discussion has continued.

After the Second World War, Keynesian economics – especially in terms of economic policy – was highly valued for its contribution to the economic development of advanced countries. Following a period of rapid development of the economy, one of the aims of economics has been focused on discovering the origin of the high growth rate. Though Keynesian economics is based on macroeconomics for the short term, construction of a long-term model for the economy expanded on Keynesian economics and theories of economic development. The economic development model derived by Domer (1946) and Harrod (1949) rested on the basis of a balance between savings and investment.

In neoclassical economics, on the other hand, further discussion on economic growth was carried out in depth. Solow (1957) proposed an economic growth model using a production function in which productivity is expressed as the product of capital *K* and labor *L*, and residual *A* is another factor containing technological progress. According to the analysis by Solow, the predominant portion of residual *A* is shared by the element of technological change. Denison (1962) tried to clarify the five main sources within *A* and analyzed the actual sources of growth. Nelson (1981) examined additional factors in residual *A* and pointed out three: potential factors for enhancing productivity at the level of firms, variables related to the nature of technological advance and a mechanism of interaction between the primary determinants of productivity growth. Pasinetti (1981) outlined three different scenarios under which the expansion and transformation of an economic system can take place. In the first, economic growth takes place by means of population growth, without technical change. In the second scenario, the economy grows because in addition to population growth there is also technical change. The third scenario is discussed in terms of technical change and rates of productivity growth, and expansion of demand is allowed to differ among the different vertically integrated sectors. Stoneman (1983) examined the issue of employment using a production function and pointed out that there is some limitation to compensation theory between capital and labor in technical change. One improvement in the clarification of technological change by neoclassical analysis has been brought about by the endogenous growth model, in which technological change is endogenously treated. This approach to a new growth model was developed by Arrow (1962a), Romer (1986), Lucas (1988), Barro (1991) and Thirlwall (2002).

Analysis by production function reveals various types of industrial behavior, and a productivity index, such as total factor productivity and labor productivity, well expresses the state of economic growth and industrial capability. This approach is well suited to process innovations but not to product innovations. Since the most important technologies in the history of modern industrialization are product innovations, the essence of technological change is not captured by the factor augmentation approach of the neoclassical model, which is constructed on the basis of equilibrium properties.

Innovation Theories of the Neo-Schumpeterian School

Keynesian economic policy did not function well during the period of stagflation caused by the oil crisis, and Schumpeter's innovation theory again received greater attention, together with business cycle theory. Neo-Schumpeterians have accumulated their analyses, and the economics of technological change has become a central theme in explaining the modern economy.

Neo-Schumpeterians have accumulated evidence of how innovations have helped modern economies develop and have carried out extensive surveys to clarify the transition in the development of innovations. The many surveys include studies of 200 companies in the UK by Carter and Williams (1957), the behavior of petroleum refineries by Enos (1962), the SAPPHO project at SPRU by Rothwell et al. (1974), which was a series of surveys of successes and failures in the chemical and scientific instruments industries, the textile industry by Rothwell (1976), and innovations in textiles, ship buildings, coal excavators, fork-lifts, steel-making, electric power plants, nuclear power technologies and semiconductors by Pavitt (1980). One of the most comprehensive studies was carried out by Freeman (1982) in *The Economics of Industrial Innovation* in which he clarified the development of major industries and the essence of innovation. Another important study is described in *Economics and Technological Change* by Coombs et al. (1987) in which comprehensive descriptions of the theories of innovation economics are summarized. Rosenberg (1969, 1976, 1982) and Mowery and Rosenberg (1998) analyzed various innovations in depth, from steam engines to electronics.

Another approach to the economics of technological change is analyzing the product diffusion brought about by innovations. Griliches (1957) first discovered that the diffusion of innovation products obeys a logistic equation in the case of hybrid corn. Many studies of innovation diffusion were subsequently carried out, and they will be described in detail in a later section.

Neo-Schumpeterians aim to clarify a dynamic system that evolves towards more efficient economic organizations. This means moving towards a more dynamic economic system by destructing the conventional economy, a process referred to as a 'gale of creative destruction' by Schumpeter. In this sense, the economics of technological change is incompatible with conventional Keynesian economics and neoclassical economics in which equilibrium is a prerequisite. Apart from neoclassical economics, Nelson and Winter (1974, 1982) proposed an evolutionary model of economic growth that focuses on the behavior of firms and analyzes such dynamic behavior of firms as uncertainty, competition and actions to break the existing equilibrium. This does not mean converging to an equilibrium or maximizing profit. Firms differentiate themselves and grow by seeking a new decision rule in a competitive field. This is a dynamic evolutionary model in which innovation proceeds along an inherent technology locus, a 'natural trajectory'.

Rosenberg and Frischtak (1984) proposed a concept of innovation clusters in which technology (T) clusters and market (M) clusters are identified separately. Utterback and Abernathy (1975) revealed that versatile products produced in the course of innovation converge into a dominant product, a 'dominant design'. Dosi (1982) proposed concepts of 'technology paradigm' and 'technological trajectory' (which is not always the same as 'technology trajectory' defined later in this book). Adopting the concept of a techno-economic system proposed by Perez (1983), Freeman (1984b) proposed the theory of a 'new technology system' to describe a dynamic economy in which technological change not only enhances value-added and profits of firms but also changes the socio-economic system itself. This means there is a pervasive structural change accompanying the institutional changes in industry, the economy and society. This concept coincides with the T cluster concept proposed by Rosenberg and Frischtak. Freeman and Louçã (2001) and Perez (2002) recently discussed innovation cycles based on the techno-economic paradigm and institutional change.

In the course of their innovation studies, the concept of 'path-dependence' was proposed by David (1985, 1988) and Arthur (1987). It refers to those phenomena whose outcomes can only be understood as part of a historical process. Early pressure and decisions can 'lock-in' (David, 1985; Arthur, 1985) a certain technology or market structure for a far longer period of time than may be socially optimal (Mowery and Rosenberg, 1998). This concept is closely related to the concept of 'dominant design' proposed by Abernathy and Utterback (1975). These considerations indicate that as innovations develop, they form complex interrelations and have strong path-dependence.

2.2 LOGISTIC DYNAMICS OF INNOVATION

Since the Industrial Revolution, the economy has developed through innovations creating economic infrastructures. The diffusion of innovation can be described by a logistic equation, as first pointed out by Griliches (1957), and many economists confirmed this relationship; for example, Mansfield (1961, 1963, 1969), Metcalfe (1970), Fisher and Pry (1971), Nakićenović and Grübler (1991), Modis (1992), Marchetti (1979, 1980, 1996). Some authors, for example, David (1975), Davies (1979), Metcalfe (1981, 1994a) and Stoneman (1984), however, have proposed alternative or modified models for describing the diffusion of innovation products. Because the diffusion of new products in the market is quite often retarded by various types of economic turbulence, such as recession and war, evaluating whether the equation fits the diffusion profile well is rather difficult. Hirooka and Hagiwara (1992) and Hirooka (1995c, 1999c, 2002a, 2003a, 2003c) demonstrated that the diffusion of innovation products proceeds according to a logistic equation in a sound economy but is disturbed by economic turbulence. This is clarified when diffusion is analyzed using the Fisher–Pry plot.

The logistic equation is

$$dy/dt = ay\,(y_0 - y) \tag{2.1}$$

where y is product demand at time t, y_0 is the ultimate market size, and a is a constant.

The solution to this nonlinear differential equation is

$$y = y_0/[1 + C\exp(-a\,y_0 t)]. \tag{2.2}$$

If the logistic equation is expressed by the fraction $F = y/y_0$, the equations (2.1), (2.2) can be represented by (2.3) and (2.4):

$$dF/dt = \alpha F\,(1 - F) \tag{2.3}$$

$$F = 1/[1 + C\exp(-\alpha t)]. \tag{2.4}$$

This logistic equation was transformed by Fisher and Pry (1971) to form a linear relation over time t:

$$\ln F/(1 - F) = \alpha t - b. \tag{2.5}$$

The ultimate market size, y_0, is determined by the flex point of the logistic curve, $y_0/2$, which is the secondary differential of (2.1), and the adaptability of the logistic equation can be examined using the linearity of the Fisher–Pry plot. The α is the diffusion coefficient of the product to the market. If $\Delta\tau$ is taken for the time span between $F = 0.1$ and $F = 0.9$, it conveniently expresses the spread of the logistic curve. This is a conventional expression of the time dependence of the product diffusion in the market, as shown in Figure 2.1. This kind of treatment was also used by Marchetti (1979, 1980, 1988).

The diffusion process was examined using the above procedure for 17 products, including five bulk chemicals, four engineering plastics, six electric appliances, crude steel and automobiles (Hirooka and Hagiwara, 1992; Hirooka, 2003c). The linear correlations of the Fisher–Pry plots for 17 products were examined as shown in Figure 2.2–2.5. The intrinsic diffusion coefficients, α, were determined based on the flex points and the results are shown in Table 2.1 together with time spans. 'Intrinsic' means the coefficients calculated by the determination of the slope of Fisher–Pry

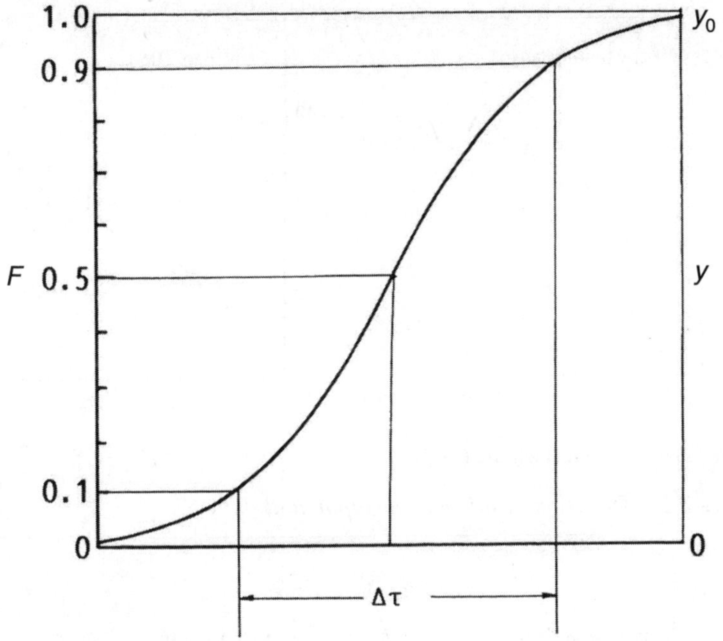

Figure 2.1 Logistic equation and time span Δτ

plot, excluding the delay by economic turbulence. These results clearly indicate that:

1. The diffusion of new products obeys a simple logistic equation during sound economic conditions.
2. The diffusion is easily disturbed by economic turbulence, such as recession and war, so that the demand for products greatly decreases, diverging from the locus of the original logistic equation.
3. It is noteworthy that after a recession, the diffusion of the products resumes, taking the same slope of Fisher–Pry plot as before the recession. This strongly supports the diffusion of a product having an inherent trajectory with a specific diffusion coefficient.

The diffusion coefficients of the products listed in Table 2.1 ranged between 0.23 and 0.94, and the time spans were between 4.5 and 19.0 years. These values of the time spans interestingly correspond approximately to the business cycles of Kitchin of four years and Juglar of ten years as referred to by Schumpeter.

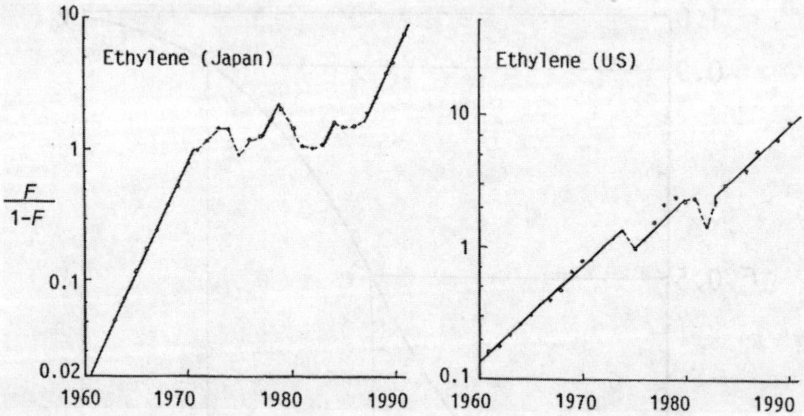

Source: Hirooka and Hagiwara (1992).

Figure 2.2 Diffusion of ethylene in Japan and the US

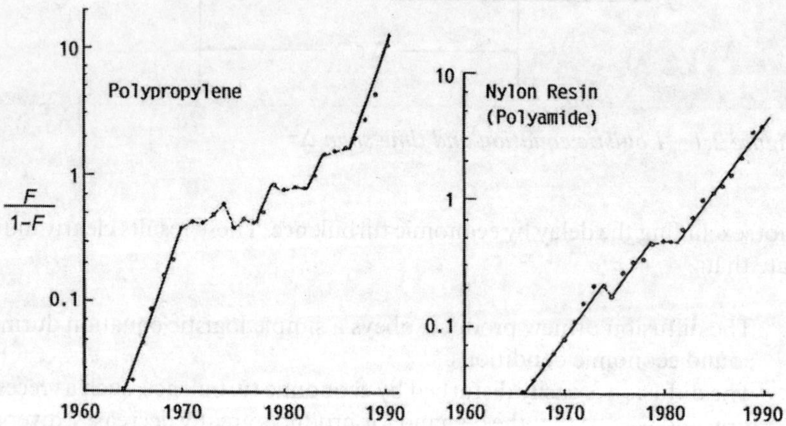

Source: Hirooka and Hagiwara (1992).

Figure 2.3 Diffusion of polypropylene and nylon resin

Although the growth rate of ethylene, the main product of petrochemicals, is slower in the United States than in Japan, the market size is significantly larger in the United States. A possible explanation for these differences in the diffusion rate is the different susceptibilities of the markets to ethylene: the Japanese may have the skills needed to find applications through quicker responses. An alternative explanation is that the Japanese petrochemical industry has lately started to introduce US

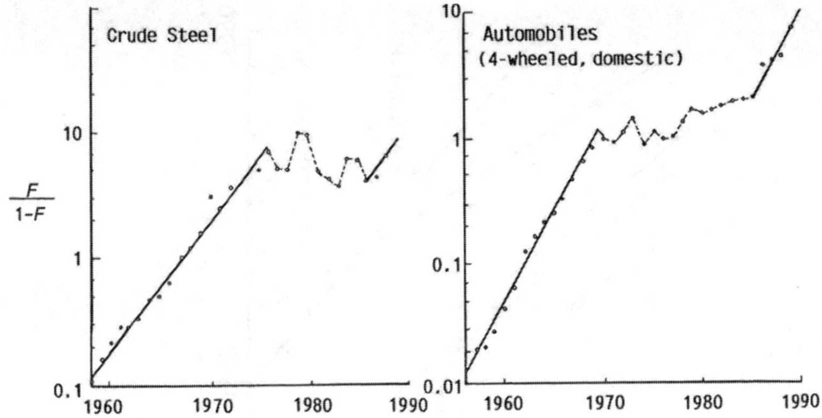

Source: Hirooka and Hagiwara (1992).

Figure 2.4 Diffusion of crude steel and automobiles

Table 2.1 Diffusion coefficients of innovation products (Japan)

Product	Time span (Δt) (years)	Intrinsic diffusion coefficient (a)
Chemicals		
Ethylene	11.3	0.39
Polypropylene	9.1	0.49
Polyvinylchloride	19.0	0.23
Polystyrene	11.9	0.37
Nylon resin	18.1	0.24
Polyacetal	15.6	0.27
Polycarbonate	18.6	0.24
PPE	12.5	0.35
Crude Steel	15.8	0.28
Automobiles	12.5	0.32
Electric Appliances		
Refrigerators	7.0	0.65
Color TVs	9.5	0.46
Microwave ovens	6.5	0.67
VTRs	7.1	0.61
Word processors	4.5	0.94
Facsimiles	8.6	0.51

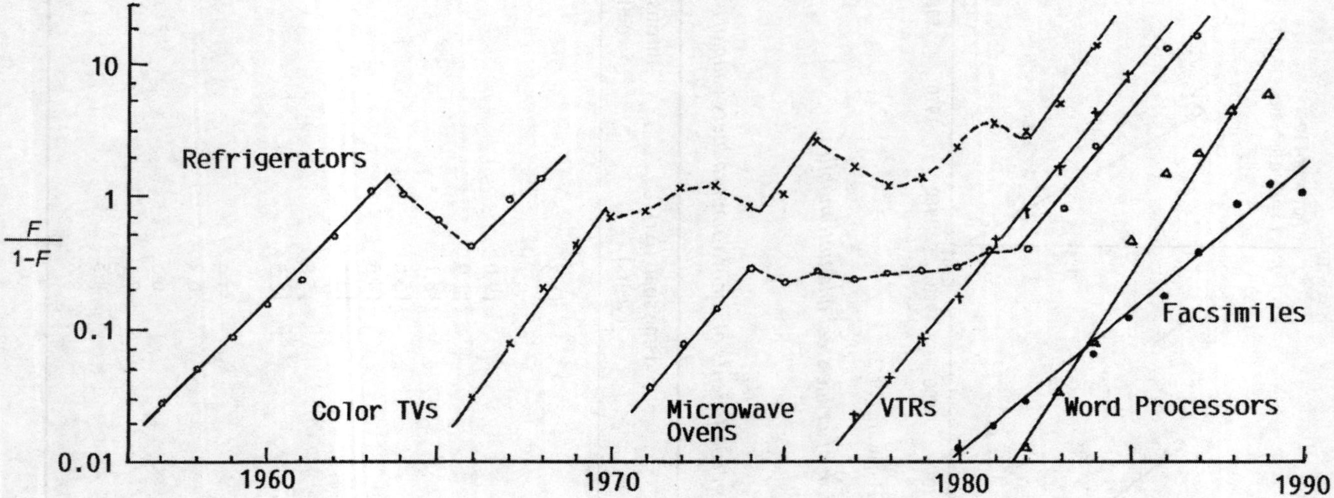

Source: Hirooka and Hagiwara (1992).

Figure 2.5 Diffusion of electric appliances

technologies so as to more easily find applications. Because the United States discovered these applications earlier, Japanese firms have the follower's advantage.

Although there were diffusion delays due to recessions in both countries caused by the oil crises of 1973 and 1979, the delay in the United States was shorter, and the economy recovered more quickly. This could be interpreted as meaning that Japan strongly depended on imported petroleum (naphtha) while the United States was able to use domestic natural gas. The Fisher–Pry plot reflects this kind of situation.

2.3 CHARACTERIZATION OF PRODUCT DIFFUSION

Estimation of the Ultimate Market Size

Logistic analysis shows that the ultimate market for innovation diffusion can be estimated before the ultimate market size has been reached. If we can get the half value of the saturated market size at the flex point, $y_0/2$, the ultimate market size can be estimated to be twice the half value. Table 2.2 shows a comparison of market size estimated from the half value of the market size and the actually realized market size. The flex points were between 1970 and 1985, and the ultimate market size was reached no later than 1995. The comparison was thus carried out using the market size in 1996. The calculated and actual results coincide fairly well, meaning that the ultimate market size and the time at which it will be reached can be roughly estimated at the midpoint of market development.

Analysis of Retardation Caused by Economic Turbulence

We have seen that the retardation for the diffusion of innovation products was caused by an oil crisis and that the market of innovation products is rather fragile. We now analyze this phenomenon in more detail using the depression in ethylene demand in Japan. In Japan, the oil price increased five-fold during the first oil crisis (1973) and three-fold during the second (1979), 15-fold in total. As the petrochemical industry in Japan was using naphtha, a fraction of oil, as feedstock, the soaring price of oil directly impacted the business, causing a serious decline in demand. General-purpose products suffered the most serious damage because ethylene, the major material in such products, has the cost of naphtha feedstock, accounting for 86 per cent of the total cost. Engineering plastics suffered less because they relied less heavily on naphtha.

Table 2.2 Estimated and actual ultimate market size in Japan

Product	Flex point (year)	Ultimate market	
		Estimated (000/yr)	Actual (1996) (000/yr)
Ethylene (Japan)	1971	7 070	7 138
Ethylene (US)	1972	18 900	22 226
Polyvinylchloride	1973	2 630	2 511
Polystyrene	1980	2 260	2 230
Polypropylene	1983	2 120	2 730
Nylon resin	1984	214	203
Polyacetal	1984	150	133
Polyphenylene ether	1984	90	77
Crude steel	1966	108 000	98 801
Automobiles	1968	6 020	7 078
Color TVs	1971	10 700	10 110
VTRs	1983	7 320	6 667

Notes:
Dimension unit from ethylene to crude steel: tons.
Domestic market size from ethylene to crude steel: production in Japan.
Domestic market size from automobiles to VTRs: demand in Japan.

Sources: MITI, Japan Automobile Manufacturers Association, and Japan Electronic and Information Technology Association, Japan.

Figure 2.6 shows the transition in ethylene production in Japan. The trend was already analyzed using the Fisher–Pry plot shown in Figure 2.2. The consumption of ethylene smoothly increased along a typical logistic curve until 1973; then, when the oil crisis struck, ethylene demand dropped, shifting down from the logistic curve. In the figure, the dotted line indicates an extrapolation of the original logistic trend by 1970. As shown in Figure 2.2, ethylene consumption later recovered to the original logistic curve.

Figures 2.7 and 2.8 show the relationship between ethylene demand and the relative price of naphtha in Japan, with the price in 1990 set to unity. They show that the demand for ethylene moved inversely against the price of feedstock. Before the oil crisis, ethylene demand increased smoothly along a logistic curve, but then the demand decreased quickly as the price of naphtha rose. Demand eventually recovered as the price of naphtha returned to a more reasonable value, and the original logistic curve resumed. This trend indicates that product diffusion is retarded by a drop in demand due to rise in product price.

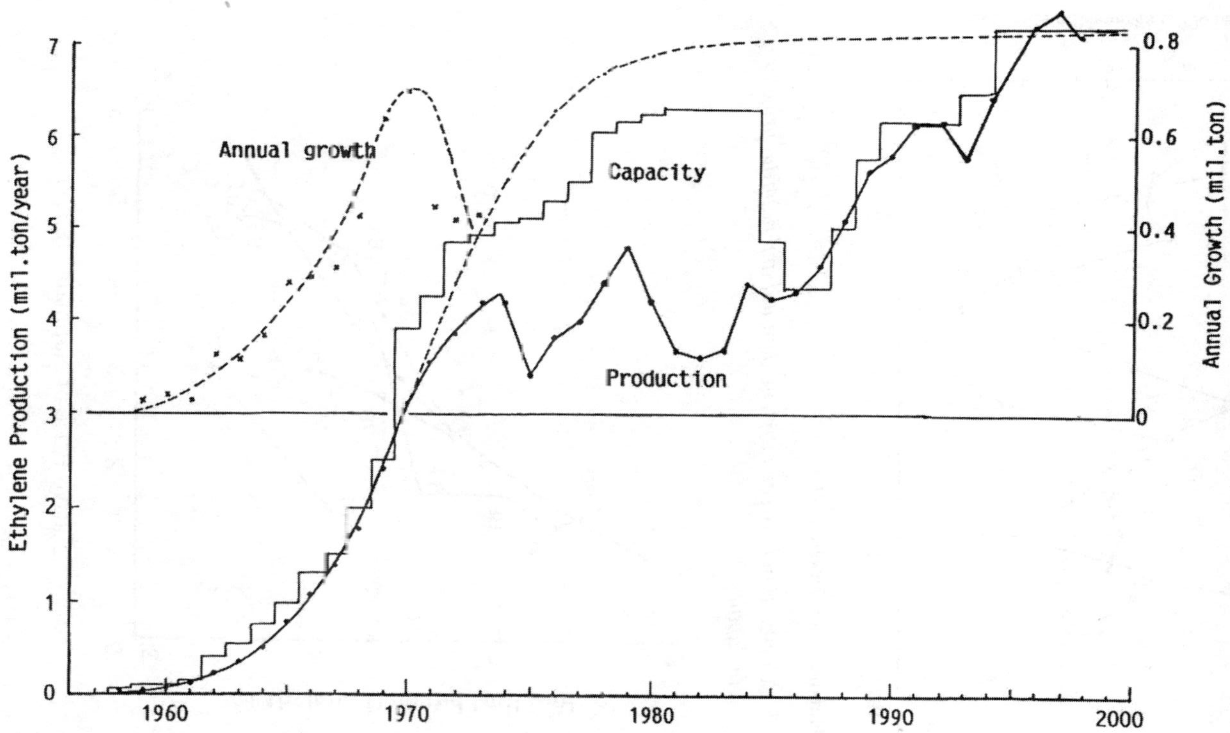

Source: Hirooka and Hagiwara (1992).

Figure 2.6 Ethylene production in Japan

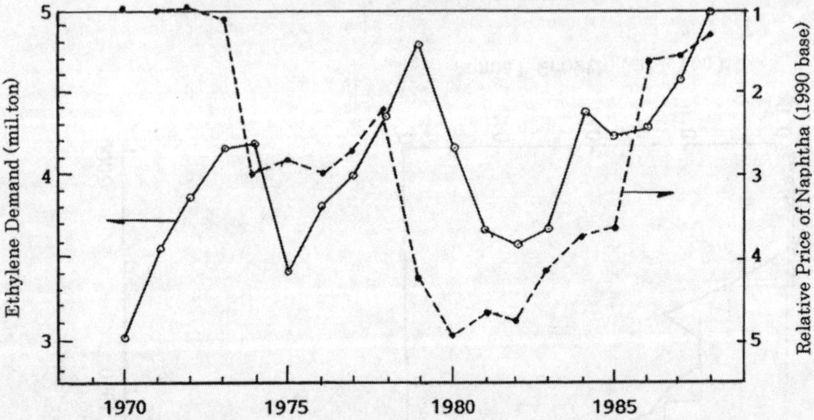

Source: Hirooka and Hagiwara (1992).

Figure 2.7 Correlation between ethylene demand and naphtha price in Japan

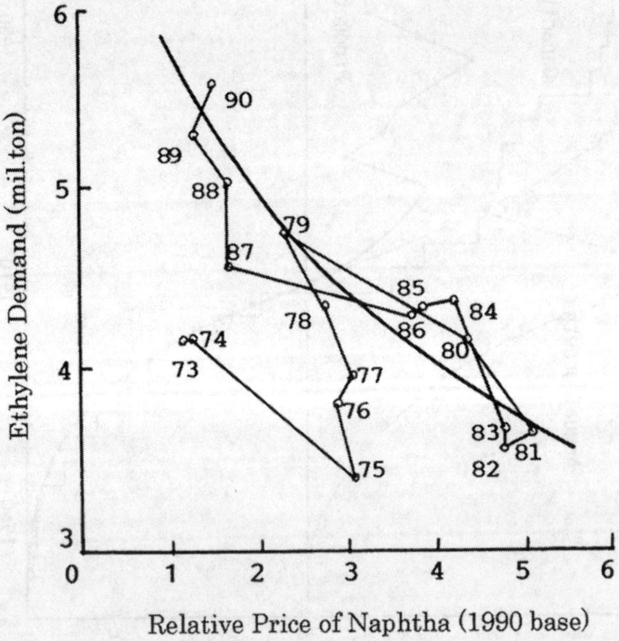

Source: Hirooka and Hagiwara (1992).

Figure 2.8 Ethylene demand and naphtha price in Japan

Market Receptivity and GNP Elasticity of Innovation

Innovation products weather economic turbulence and penetrate the market with an intrinsic diffusion coefficient. Innovations thus have tough permeability as shown in Figures 2.2–2.5. The diffusion of innovation products tends to be fairly retarded during such turbulence as oil crises but resumes its inherent trajectory after the turbulence passes.

During the diffusion process, the gross national product, GNP, is enhanced, especially during periods of economic prosperity. For example, the GNP per capita in Japan increased from ¥154 000 in 1956 to ¥572 000 in 1971 in constant 1965 prices. This corresponds to a 3.7 times increase over 15 years and an annual growth rate of 9 per cent. The fact that the diffusion of innovation products obeys a logistic equation indicates that the penetration proceeds rather independently of the economic conditions, for example, GNP. This infers that innovation products have an intrinsic permeability. Innovation products have their own potential, and market receptivity must be defined based on this potential. For example, when TV sets first came on the market, one TV set was generally enough for each household. This threshold was mostly independent of the economic conditions. Although innovation diffusion is rather sensitive to economic turbulence, innovation products have an inherent ultimate market.

In the early stage of an innovation, diffusion strongly depends on the development of technology, and products are quickly improved. This trend implies that the diffusion of innovation is technology-oriented and can be independent of economic conditions in the first stage of innovation. When a market enters the maturity stage, competitiveness strongly depends on the cost, so demand becomes dependent on economic conditions, for example, GNP. In the later stages of innovation, the demand for products levels off, but it can increase depending on the economic situation.

If the demand for innovation products depends on other factors such as GNP, the diffusion function can be expressed together with these factors. Logistic equation (2.1) in Section 2.2 can be also expressed as

$$1/y\; dy/dt = \alpha(1 - y/y_0). \qquad (2.6)$$

The α is the intrinsic growth rate coefficient, an endogenous factor, and y_0 is an exogenous factor. If these factors are functions of time t, $\alpha = \alpha\,(t)$, $y_0 = y_0\,(t)$, the general logistic equation can be expressed as

$$y = \frac{1}{e^{-\tau}\left[\int_0^{\tau} \frac{e^{\tau}}{y_0} d\tau + K\right]}, \qquad (2.7)$$

where $\tau = \int_0^t \alpha dt$.

If the ultimate market, y_0, increases as a function of GNP,

$$y_0 = y_0(1+b)^t, \tag{2.8}$$

where b is the annual growth rate of GNP, market size y can be expressed by

$$y = \frac{\exp \frac{a y_0}{\ln(1+b)}(1+b)^t}{a \int \exp\left[\frac{a y_0}{\ln(1+b)}(1+b)^t\right] dt + C}. \tag{2.9}$$

While these functions are generally not solvable, the trend of GNP dependence can be ascertained by examining the actual data.

Table 2.3 shows examples of the long-term dependence of product diffusion on GNP. The diffusion of crude steel and automobiles in the United States started around the beginning of the twentieth century, and the markets for both matured just before the crash in 1929. The demand for both picked up again following the Second World War. The GNP elasticity was more than unity during the growth period, showing that the growth rate was much higher than the growth in GNP. The elasticity sharply

Table 2.3 GNP elasticity on steel and automobiles in the United States

	1900	1925	1940	1950	1963	1980	1993
GNP (bil.$) 1958 Price	77	179	222	346	537	914	1778
Growth rate (%/y)		3.4	1.4	4.5	3.4	3.2	5.3
Crude steel							
Demand (mil.tons)	9.2	40	70	88	102	102	92
Growth rate (%/y)		6.0	3.8	2.3	1.1	0	0
GNP elasticity		2.5	3.1	0.5	0.3	0.0	0.0
Automobiles							
Demand (mil.cars)	4.1	3735	3717	6665	7853	8979	8606
Growth rate (%/y)		31.3	0	6.0	1.3	0.8	0
GNP elasticity		>10	0.0	1.4	0.3	0.2	0

Sources: Data from Mitchell (1998); Census, USA (2002); Kelly and Fenton (2003).

decreased towards zero during the maturity period in the subsequent cycle following the war. This indicates that the diffusion of innovation products finally reach their market potential during the maturity period irrespective of GNP. The detail will be discussed in Chapter 5.

Table 2.4 summarizes the diffusion behavior of innovation products during growth periods, comparing it with the GNP growth rate for the same period. The GNP growth rate is sufficiently high during the upswing of Kondratiev cycles. (The discussion on Kondratief business cycles will be carried out in depth in the following Chapters 3 to 5.) The table shows that the growth rates of GNP during the upswing of the third and fourth Kondratiev cycles were 6 per cent and 10 per cent per year, respectively. These values are significantly larger than those in the rest periods, which were usually less than 3 per cent. In the upswing of the third wave, however, the period of high growth rate was rather short, from 1921 to 1929, because a fairly severe depression occurred during 1919 and 1921, after the First World War. There was a similar retardation due to the oil crisis during the fourth cycle. On the other hand, the diffusion of innovation products during the growth period was fairly large, exceeding the GNP growth rate by more than two or three times, as shown in Table 2.4. This suggests a reason for the diffusion of innovation products being rather independent of economic growth during the growth period. That is, innovation products penetrate the market at a rate higher than the growth rate of the economy. Thus, the rate of innovation diffusion can be determined irrespective of economic growth.

Table 2.4 Diffusion rate of innovation products

			United States		
Product	GNP	Crude steel	Oil	Electricity	Automobiles
Interval	1921–29	1885–1900	1900–15	1902–17	1910–25
Time span (years)	8	15	15	15	15
Growth rate (%/y)	6.0	12.0	10.4	13.7	22.4
			Japan		
Product	GNP	Crude steel	Automobiles	Ethylene	ICs
Interval	1960–70	1955–70	1960–75	1963–73	1975–90
Time span (years)	10	15	15	10	15
Growth rate (%/y)	10.0	16.5	18.1	28.3	29.4

Sources: Mitchell (1998); DOC, USA (1940); MITI (1998).

2.4 SUMMARY

The diffusion character of innovation has been studied and the following characteristics are disclosed.

1. Innovation diffusion can be described by a logistic equation. This means that the innovation has a nonlinear nature and becomes mature within a definite time span.
2. The diffusion of innovation is easily retarded by economic turbulence and resumes the original diffusion rate when the economy recovers to a sound condition.
3. This means that the innovation diffusion is tough enough to penetrate through any economic condition and is a kind of physical phenomenon with an intrinsic diffusion coefficient.
4. The ultimate market is independent of GNP and finally converges to its own market potential.

3. Kondratiev business cycles and innovation dynamism

Since the Industrial Revolution, the world economy has repeatedly experienced large fluctuations under the construction of the modern industrial society. Based on Kondratiev's theory of business cycles, many discussions on economic fluctuations have been carried out among economists. For example, Schumpeter (1939) pointed out in his book *Business Cycles* that innovation is the decisive cause of business cycles. This assertion, however, has not been fully accepted as the ultimate theory.

This chapter discusses the issue of business cycles, focusing on their interrelationships with innovation dynamism, and clarifies the process of cycle formation. Various analyses are presented that clarify related topics, including the clustering of innovations, the synchronization of world business cycles, the evolution of innovation paradigms and the international competitiveness of follower countries.

3.1 KONDRATIEV BUSINESS CYCLES AND THEIR INTERPRETATION

Discovery of Business Cycles and Kondratiev Long Waves

Since the Industrial Revolution, the world economy has experienced business fluctuations in which economics alternate between prosperity and recession. A small fluctuation was observed as a Kitchin wave in 1923 (Kitchin 1923), which has a 40-month cycle, that is, 3–4 years, and has been recognized as an 'inventory cycle'. Clarke (1847) first identified long business cycles in which two cycles of 10–11 and 54–55 years are brought about by famine and recession. This interpretation was adopted by Jevons (1878), who proposed the idea of 'solar economics', which says that the cyclical changes in sunspots bring about climate changes, which cause economic fluctuations. In France, Juglar (1862) identified another business cycle of 7–10 years, which was later dubbed the 'investment cycle'. A longer cycle was found by Spiethoff (1923), who identified recession periods of 1822–42 in England and 1874–94 in Germany and prosperity periods of 1843–73 and 1895–1913 in Germany. The first concrete recognition of long business cycles was made by N.D. Kondratiev (1926), a Soviet economist. He discovered and reported in two papers that there were business cycles of 50 to 60 years in a capitalist society.

Kondratiev (1926) gathered data on price indices, interest rates, wages, amounts of foreign trade, amounts of imports and exports, production and consumption of coal, and production of pig iron and lead, and analyzed them using a 9-year moving average in order to compensate for contamination by Juglar cycles. He identified 50-year business cycles and determined that two-and-a-half of these long cycles had occurred during the previous 140 years. He discussed the cause of these cycles and interpreted them comprehensively as the interaction among endogenous factors, strictly excluding such exogenous factors as technological innovations, war, revolution, gold rushes and frontiers. In particular, he focused on clarifying the great circular movement of the relative price level between industrial and agricultural products. Kondratiev held the post of general manager of the Institute of Business Research between 1920 and 1928, which enabled him to gather a huge amount of data through his travels to England and the United States, which he used in reaching his conclusion about long business cycles in capitalist societies. As he had already published his paper in 1922, his work was clearly not triggered by Spiethoff's work on long cycles. Kondratiev belonged to the neo-Narodniki school, his theory was criticized by Trotsky, he fell foul of Stalin and was exiled to Siberia.

Just at that time, the Great Depression struck, and many economists tried to explain the economic disaster, to discover its cause and to develop policies for coping with it. Schumpeter (1939) took Kondratiev's business cycle theory into account when he reconstructed his own economic theory of technological change; although Schumpeter made innovation theory his central theme.

According to Schumpeter's interpretation of Kondratiev business cycles, technological innovations occur intermittently in clusters, activating the economy and bringing about institutional change in society. He recognized three cycles: the first was driven by the rise of the textile industry, the second by steam engines and railways, and the third by electricity, chemicals and automobiles. He predicted that the fourth would start in 1953.

As the third cycle had been considered to have peaked in 1914–20 by Kondratiev (1926), many economists conceived it around that timing: in 1913 by Mandel (1972), in 1920 by Rostow (1952, 1978). These were also based on the emergence of a heavy recession in 1921. Schumpeter (1939) also estimated the peak of the third Kondratiev cycle to be in around 1920 (p. 794) based on the recession from 1919 to 1925. Schumpeter's postulation, however, contradicted the fact that business still continued to prosper even during that period after 1920. Hansen (1941) asserted that the peak of the third cycle was in 1929 and supported Schumpeter by interpreting the depression as the stagnation of innovation and disappearance of frontiers. He further tried to introduce construction cycles of 20 years, but this made the explanation more complicated, and his innovation theory faded away without further support. After the Second World War, advanced countries widely accepted Keynesian economics as a powerful tool for setting economic policy, but they showed little interest in Schumpeter's innovation theory. However, when the oil crisis struck between 1973 and 1979, stagflation could not be sufficiently overcome by following policies based on Keynesian economics, and Kondratiev business cycle theory was again reconsidered.

Schumpeter's Innovation Theory for Business Cycles

Schumpeter viewed 'business cycles' in terms of innovation. Schumpeter first became interested in business cycles around 1904, according to the lecture he delivered at the University of Tokyo in 1931 (Schumpeter, 1931; Kanazashi, 1991). That was 35 years before the publication of *Business Cycles*. In his lecture he pointed out that there are three important factors in controlling the expansion of an economy: external forces, forces that produce change (for example, a population increase), and progress that produces turbulence (for example, an advance that improves a manufacturing process). Among them, forces that produce change should be within the economy itself. An upswing in the business cycle is produced by the rise of new industries, such as the diffusion of railways, chemicals and automobiles. For example, income in the United States increased by 15 per cent due to the spread of cars. While such advances of new industries bring prosperity to the economy, existing industries are pushed aside, and this distortion can cause depression. This idea comes from his concept of 'gale of creative destruction'.

Schumpeter (1931, 1935) postulated that once an innovation leads to the creation of a new market, its neighborhood is activated, and prosperity spreads throughout the economy. Schumpeter discussed this effect in his definition of the second approximation and suggested a kind of institutional change. He had already discussed the importance of Juglar cycles (1931) and proposed a complex theory of three cycles, Kondratiev, Juglar, and Kitchin, in 1935. He identified three clear Kondratiev cycles: the first driven by the invention of spinning machine, the second by the iron industry and railways, and the third by electricity, chemicals and automobiles.

In *Business Cycles*, Schumpeter devoted many of the 1095 pages to a comprehensive discussion of business cycles. Although he did not necessarily devote most pages to a description of innovation itself, he put a high priority on innovation and developed an innovation theory for business cycles. From the viewpoint of innovation, he discussed the first Kondratiev cycle in chapter 6, the second in chapter 7, the first part of the third cycle, 1898–1913, in chapter 7, the third cycle after the First World War, 1919–29, in chapter 14, and the latter half of the third cycle and the Great Depression and subsequent period in chapter 15. The analysis of the third cycle was made separately because the cycle was disturbed by the First World War, 1914–18.

Schumpeter carried out his analysis of business cycles as a complex system combining three kinds of business cycles: Kondratiev long cycles of 50–60 years, Juglar medium cycles of about ten years, and Kitchin's short cycles of 40 months (3–4 years). That is, one Kondratiev cycle consists of six Juglar cycles and one Juglar cycle is composed of three Kitchin cycles. Actual economic conditions are explained by this complex system. Schumpeter put particular emphasis on the Juglar cycle for analyzing business cycles, a most noteworthy point that we will discuss in more detail later.

Another characteristic feature of Schumpeter's analysis is that each Kondratiev cycle is considered to consist of four phases (recovery, prosperity, recession and depression), and each phase corresponds to several Juglar cycles. A Kondratiev cycle initiated by a recovery phase forms a prosperity phase consisting of two Juglar cycles, reaches a recession phase covering the peak of the Kondratiev cycle, and then declines, forming the downswing of a depression phase.

The correlation of Kondratiev cycles with technological innovations was extensively examined by Schumpeter. The first cycle, in which the spinning industry was the main actor, is described in chapter 6. He rather ironically describes how the invention of the spinning machine itself did not directly activate the economy. He discusses many economic circumstances of that age, but simply states that 'The development in the cotton trade alone would be adequate to explain a Kondratiev upswing' (p. 275). The second wave, in which the contributions of railways and iron production were decisive, is discussed in chapter 7. Railway construction in England started in the early 1830s, and full-scale construction began in the 1840s, reaching 10 000 miles by 1860. The mileage increased by roughly 50 per cent over the next decade and again by about 50 per cent to 1910 (p. 342). Railway construction in the United States started six years later and reached 30 000 miles by 1860 (p. 327). The main actors in the third Kondratiev cycle were electric power, crude steel and automobiles. The supply of electric power grew from 3.2 billion kWh in 1899, to 20 billion in 1914, and 97 billion in 1929 (pp. 411 and 768). Automobile production grew from four cars in 1895 (p. 772) to 3.8 million in 1929 (p. 801). The oil supply increased from 60 million barrels in 1897 to 250 millions in 1913 (p. 417).

The summary above covers the main points made by Schumpeter in *Business Cycles* about innovation diffusion along Kondratiev cycles. Although the actual data on innovation diffusion coincides with the timing of upswings of Kondratiev cycles, there is not sufficient evidence of a causal relation between innovation diffusion and upswings, as his analysis is focused on Juglar cycles. Freeman and Soete (1997), in contrast, carefully analyzed the description of business cycles by Schumpeter and illustrated the Schumpeterian concept of long waves based on successive technological transformations in a table, which does not strictly follow his work. They asserted that the most important feature in his view was innovation, which, despite its great variety, he saw as the main engine of capitalist growth and the source of entrepreneurial profit.

Economic Theories for Kondratiev Cycles

After the Second World War, there was much investigation into the origin of Kondratiev cycles. Freeman (1984a) invited opinion leaders to submit papers and edited a comprehensive book reviewing these investigations in which J. Delbeke summarized the major theories as an overview. The main business cycle theories, slightly modified, are represented as innovation theory (Schumpeter, Mensch, Kleinknecht and van Duijn); capital theories (Mandel, Forrester, Wallerstein and Gordon); labor theory (Freeman); raw materials and foodstuffs theory (Rostow); and war induction theory (Dickinson, Goldstein and Modelski). Each theory is fundamentally summarized as a concept of scarcity and abundance of certain factors.

Mandel (1975), drawing on the theory of Marxian economics, defined industrial cycles as the successive acceleration and deceleration of capital accumulation. In the upswing period, the mass and rate of profit and the volume and rhythm of capital accumulation all increase. Conversely, during a crisis and the subsequent period of depression, these factors all decrease. He ascribed the turning point to endogenous factors, especially organic constructions of capital towards downswings and various exogenous shocks towards upswings. Forrester (1977) developed a completely new method of analyzing group dynamics in which the capital goods sector plays a crucial role. His theory is based on two important pillars creating endogenous fluctuations. The first pillar is a multiplier accelerator mechanism, and the second is rational behavior following a capital-stock adjustment model. The economic agents continually exaggerate and expand their capacity to catch up with demand in the upswings and collapse to a level below the long-term average in the downswings. Wallerstein (1979) developed a global system of social structure and discussed it on the basis of a dynamics of the historical evolution of capitalism. This is a theory of expansion and shrinkage of the social system. Gordon (1978) proposed a theory of accumulated social structure and discussed investment cycles, focusing on the production process, especially the movement and structural concept of the labor market.

Rostow (1978) ascribed the basis of long business cycles to the periodic cycles of 50 years caused by the rise and fall of food and energy prices every 20 to 25 years. He emphasized growth dynamics and the importance of three activity factors: forces set in motion by a leading industry in terms of growth, changes in the profitability of producing foods and raw materials, and large waves of international or domestic migration. The second factor especially generates a price revolution, and once an imbalance between demand and supply emerges, a sharp rise in the relative prices of food and raw materials occurs at the expense of the prices of industrial goods.

The war induction theory postulates an exogenous factor. Dickinson (1940) ascribed the long business waves to post-war expense and discussed the interrelation between lengthy economic growth and the history of wars. Goldstein (1985) and Modelski (1987) considered that during the upswing in a Kondratiev wave, a war takes place, resulting in a recession. They examined the causal relationship between cycles of wars and Kondratiev waves.

Innovation Theories for Business Cycles

Among the theories about Kondratiev business cycles, Schumpeter's innovation theory did not initially attract much attention. After the oil crisis, however, it drew more attention since Keynesian economic policy proved to be ineffective. Various investigations have been carried out to clarify the mechanism, especially the relationship between innovation and business cycles. Here we summarize these investigations.

Freeman (1977) focused his argument on the scarcity and abundance of labor; although the argument is actually an innovation theory to interpret the cause of business cycles. During an upswing, major new technologies generate both new investment and new employment on a large scale. As the result of rapid technological change and increasing competitive pressures, the relative prices of the new goods fall and cost advantage increases. The main employment effect during an upswing is a steep increase in employment. After about a quarter of a century, the new branches of industry are firmly established, and their role as generators of additional new employment diminishes and eventually disappears. During a downswing, competitive pressure increases, and labor-saving technical changes become increasingly important. The loss of impetus for the growth of employment in the new industries is reinforced by the acceleration of labor-displacing technical changes and economies of scale. This process is most intense during a Kondratiev downswing.

Mensch (1975) tried to clarify the correlation between innovation and economic development on the basis of Kuznets business cycles (Kuznets, 1953, 1958). His important observation was that the emergence of technological innovations is not a continuous phenomenon but actually involves discontinuous clusters of innovations. He distinguished between inventions and innovations and defined 'lead time' as the time from a basic invention to its commercialization. Its reciprocal value represents the speed of the transfer processes from invention to innovation. He determined various speeds of innovations and suggested that innovations emerge as a group in the economic depression stage of the business cycle. He interpreted this phenomenon as the 'depression trigger effect': the depression exerts pressure for survival, so innovation is attempted as a means of escape. This argument is at least partially based on the fact that innovation accelerates during a depression.

Marchetti (1979, 1988) reprocessed Mensch's data using the idea that inventions and innovations are cultural pulses and as such should be analyzed using the population dynamics of Darwinian origin, coded into Volterra-Lotka equations. He demonstrated that the Fisher–Pry plot of cumulative numbers of inventions and innovations creates straight lines, indicating that these phenomena can be expressed by logistic equations over time. Marchetti also examined the diffusion of many innovations in testing the fitness of the logistic equation under Fisher–Pry notation and found proof for his conjecture. He compared his results with their position on the Kondratiev business cycles, but did not always obtain a clear correlation.

Freeman et al. (1982) investigated innovation systems to clarify their role on the Kondratiev cycles, especially using their concept of new technology systems. Freeman et al. strongly questioned Mensch's 'depression trigger hypothesis' for two basic reasons. First, they questioned the data Mensch used, suggesting the samples were too ad hoc to adequately cover the major industrial fields involved in long wave booms. They noted also that there is no adequate definition of basic innovation. Second, they doubted the plausibility of the depression trigger hypothesis to explain the clustering of innovations. They argued that R&D activity decreases during a depression, and their survey suggests that clusters of patents exists during a long wave boom as well as in recessions. They argue that the clustering is of the diffusion processes for these innovations, not the clustering of the innovations themselves.

Rosenberg and Frischtak (1984) discussed clusters of innovations, distinguishing between technology (T)-clusters and market (M)-clusters. T-clusters are technically related, while M-clusters represent clusters of common stimuli of a generalized increase in demand or represent other favorable macroeconomic conditions. Kleinknecht (1984) further examined the clustering of innovations and empirically proved that it does occur.

Van Duijn (1983) thoroughly studied Mensch's treatment and recognized the importance of the time lag between inventions of technologies and their diffusion in the market, based on his examination of the development course of modern industrialized society. He particularly noted that clusters of key technological innovations led to the building of new industries and became the trigger for long business cycles. He also tried to express the life cycles of new industries using S-shaped curves. New industries mature and the demand eventually levels off.

More recently, Freeman and Louçã (2001) and Perez (2002) extensively analyzed innovation clusters and discussed their relationship with Kondratiev cycles. They clearly identified the existence of innovation clusters in Kondratiev cycles but had doubts about the existence of long business cycles in terms of economic aggregates. They focused on the concept of a techno-economic paradigm and on the institutional change of society by technological revolutions.

Those who ascribed the cause of Kondratiev business cycles to factors other than technological innovation also recognized the existence of industrial development during economic growth and the importance of technologies. For example, Rostow (1978) noticed that there were four clusters of technological innovations in the back of the economy: (1) spinning machines and iron-making using coke and steam engines; (2) railways; (3) steel-making; and (4) internal combustion engines, new chemicals and electric power. Mandel (1972) recognized four stages: (1) steam engines in handcraft manufacturing; (2) steam engines in the machinery industry; (3) electric power and internal combustion engines; and (4) electric and electronic appliances.

3.2 EXAMINATION OF INNOVATION CLUSTERS AND KONDRATIEV CYCLES

Most of the economists discussed above recognized the important impact of technological innovations on activating the economy and suspected some correlation between them. One of the reasons Schumpeter's innovation theory of business cycles has not yet been widely accepted seems to be the lack of direct evidence of a causal relationship. We would like to challenge this issue. Another problem is the correct identification of Kondratiev business cycles. In fact, the Kuznets cycles used by Mensch deviate from the others, and there are various fluctuations within the cycles.

We carefully examined the position of the Kondratiev business cycles and obtained the most probable positioning, as shown in Figure 3.1.

A Kondratiev cycle is identified by determining the degree of economic development, using the GNP and other economic aggregates. Technological innovations add value to the economy, so the diffusion of innovation products corresponds to the amount of value added by the creation of new markets. Therefore, the diffusion curve of an innovation product directly reflects the degree of value added by the innovation per se. More importantly, the diffusion curve has a nonlinear S-shape, which means that the diffusion tends to saturate within a definite time span. This means we can locate the relevant innovation. We tried to determine where the diffusion curves of innovation products are located on the Kondratiev waves. This is the central subject of this study. The locus of the diffusions of various technological innovations is depicted on a map of Kondratiev waves. The locus is normalized so that in a saturated market it is unity. These innovations are those playing a crucial role in the formation of infrastructures in industrialized society. As shown in Figure 3.1, all the innovations are clustered at the upswing of a Kondratiev cycle.

The upswing of the first cycle is that of the Industrial Revolution in the UK, when the textile industry took off. The subsequent economic development can be interpreted as the result of the development of the textile industry, which can be seen through the diffusion of mule spinning machines and cotton consumption along the upswing. The second cycle was caused by the development of railways and iron production. For the duration of the second upswing – from 1846 to 1872 – railways and iron production expanded more than ten-fold.

The diffusion trajectories at the upswing of the third cycle – between 1900 and 1929 – are those of the United States, reflecting the fact that economic leadership had shifted from the UK to the US. Over this period, the United States jumped from being a developing country to an industrialized country by introducing various technological innovations from Europe, such as those for steel-making, oil-drilling, automobile manufacturing and electrical generation. The diffusion of these innovations led to the construction of an advanced state. After World War II, the United States again enjoyed prosperity, accounting for one-third of the world's GNP, by developing various innovations in such areas as aircraft and electric appliances and in petrochemical products such as synthetic plastics, fibers and rubber. Japan jumped from being a developing country to an industrialized one during this upswing, through the diffusion of crude steel, automobiles, petrochemicals and electric appliances. This was similar to what happened in the United States at the beginning of the twentieth century.

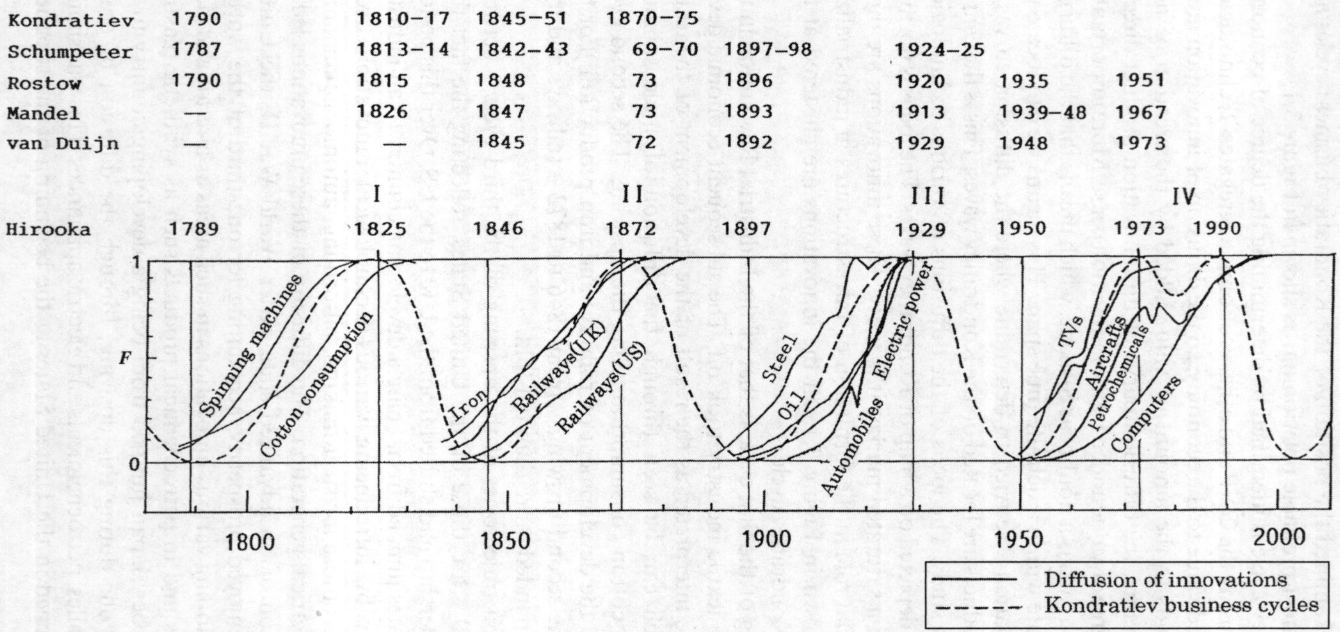

Source: Hirooka (1992a,b, 1994c, 2003c).

Figure 3.1 Kondratiev business cycles and diffusion of innovations

Although this fourth wave was heavily affected by the two oil crises, recovery took place in the latter half of the 1980s with the successive development of the above innovations and the widespread diffusion of information and communication technology and various high-tech innovations.

These results clearly indicate that the diffusion timing of various innovations is always concentrated along the upswings of Kondratiev business cycles, forming a cluster. This strongly supports Schumpeter's hypothesis that innovation is the engine driving the development of the economy. This causal relationship is confirmed by the location of diffusion curves along the upswings, because the diffusion curve of an innovation represents the locus of value added by the new innovation entering the market.

3.3 SYNCHRONIZATION OF WORLD BUSINESS CYCLES

The Industrial Revolution took place regionally in England, specifically Lancashire in the 1770s, and the United Kingdom continued to hold leadership across two business cycles. The big surge of the Industrial Revolution spread over Germany, France and the United States. The United States took the initiative during the third Kondratiev wave and retained it during the fourth. Japan was a latecomer to modern industrialization, joining it at the time of the Meiji Restoration in 1868 after a long closed-door policy and 100 years after the Industrial Revolution in England.

How did Japan catch up with Western countries in terms of industrialization? It is interesting to follow this trend through the Kondratiev long cycles. Figure 3.2 compares developments between the United Kingdom, the United States and Japan on the same map of Kondratiev cycles. During the second cycle, the United States was a little behind England in railway construction and iron production. The success of the United States during the third wave was incredible – it took over leadership in all innovation products such as steel, electricity, automobiles and so on. The production of automobiles in England started in the same cycle, but the actual market was realized during the fourth cycle, after World War II. Japan started her industrialization with the introduction of Western technologies during the third wave, starting with cotton production and railway construction. The industrialization of cotton occurred two waves later than in England, while railway construction was one wave behind. Japan had begun commercial steel production in the Meiji era, but annual production was still less than 7 million tons in 1940 and collapsed to 550 000 tons in 1946 due to the war. The actual industrialization of Japan came about during the high growth period of the 1960s, with the production of steel,

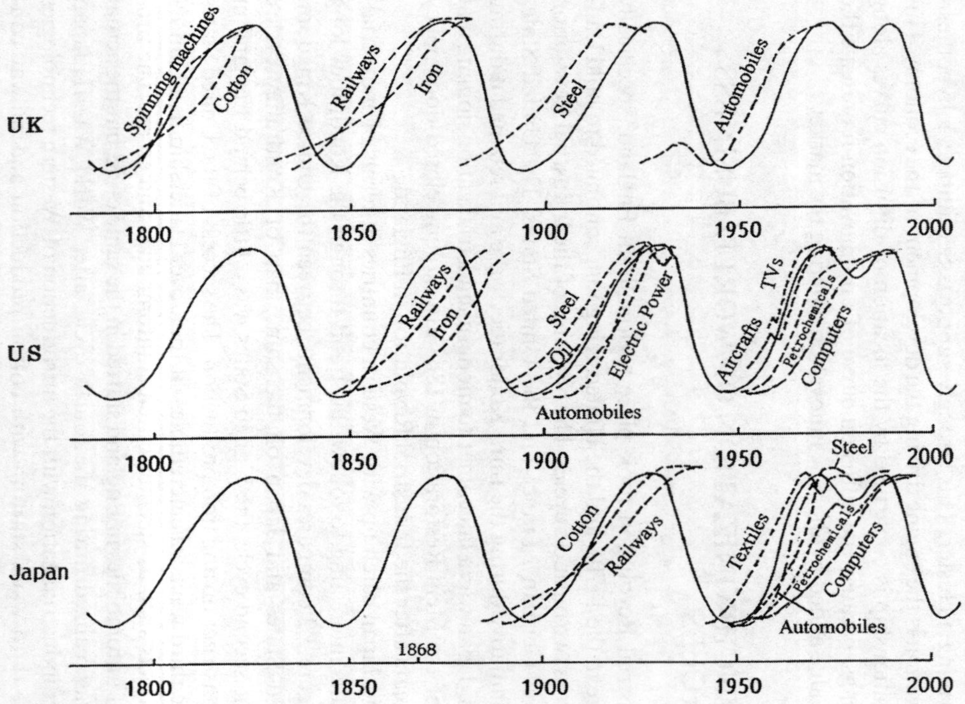

Source: Hirooka (1994c).

Figure 3.2 Synchronization of Kondratiev cycles in world economy

automobiles, textiles, electricity and petrochemicals, and the catch up with the Western countries was achieved around 1970. These events are illustrated in Figure 3.2, where it is interesting to note that the diffusion of innovation products in every country was concentrated in the upswings of the Kondratiev cycles. This means that the business cycles were synchronized worldwide and that the main diffusions of innovation products were concentrated in the upswings. It is interesting to see that latecomers industrialized during the upswings of later cycles. This implies that the diffusion of innovation products slows during periods of recession and resumes during the next upswing. This trend was observed in the diffusion of innovation products in Japan, as illustrated in Figures 2.2 to 2.5. These correlations suggest that innovations induce economic growth and that economic prosperity provides an accessible field for innovations to diffuse. This could be the reason for the gathering of innovations into clusters at the upswings. This could be a kind of chicken-or-egg question as a self-organization mechanism.

3.4 RESONANCE AND EVOLUTION OF INNOVATIONS

Innovations form a cluster at the diffusion stage along the upswings of Kondratiev cycles. Within a cluster, the innovations affect each other, as a kind of resonance phenomenon. Even in the Industrial Revolution, such resonance occurred. The Industrial Revolution started with the invention of spinning machines and steam engines. A spinning industry rapidly developed after the introduction of steam engines as the motive force. The production of steam engines requires iron of good quality, and iron production was promoted by the introduction of steam engines. Along with these innovations, a machine tool industry emerged, supporting the development of these industries. When the spinning industry had been smoothly developed, the bottleneck in the textile industry became bleaching because the process took more than a month. The acid-alkali chemical industry was developed to remove this bottleneck. Such synergy among new industries accelerated the Industrial Revolution, as shown in Figure 3.3.

As the ultimate product of steam engines, locomotives and railways were developed during the second Kondratiev cycle. The development of railways further increased the production of iron. Telegraph lines were constructed along the railways, forming an infrastructure of communications together with telephony. Telegraph technology progressed, leading to the evolution of wireless communication. These communication infrastructures were succeeded by electronics technologies. Likewise, steam engines

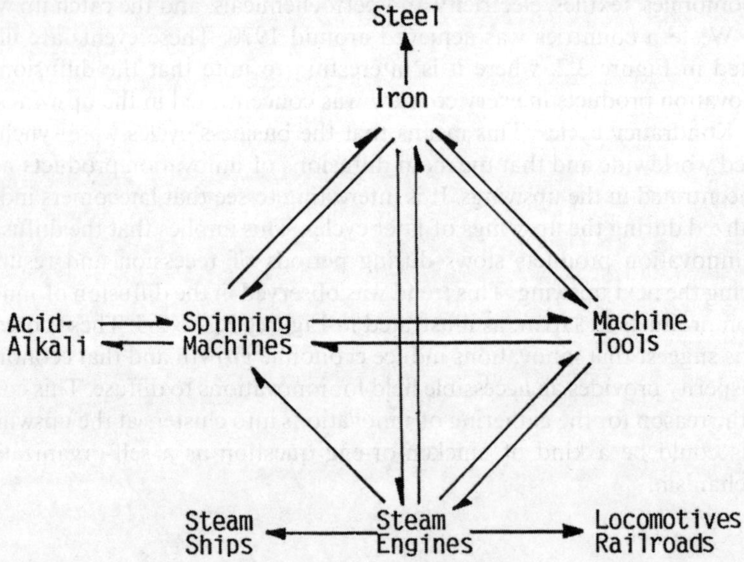

Source: Hirooka (1994c).

Figure 3.3 Resonance effect of innovations in the Industrial Revolution

evolved to internal combustion engines and jet engines. Locomotives were succeeded by automobiles and airplanes. This kind of successive development can be described as a series of evolutions, as shown in Figure 3.4. This figure suggests that an innovation is not an independent phenomenon; rather it affects other innovations and evolves synergistically. Physical distribution developed into various retail businesses, and transportation and communication technologies made retail businesses evolve interactively.

A series of innovation paradigms in a technological field develop intermittently and evolve discontinuously. Table 3.1 illustrates the successive evolution of innovation paradigms by sector. It is interesting to see that almost all paradigms line up with the same timing to form clusters. Each technological field has an increasing number of paradigms and becomes more versatile. This trend brings about an abundantly endowed civilized society. Before the Industrial Revolution, the only energy resource was charcoal, and then coal, oil, electricity and nuclear power were developed later as power revolutions. Now we hope to use hydrogen and fusion as practical energy resources. The same trend is seen in every technological field, and the accumulated results demonstrate the abundancy of modern

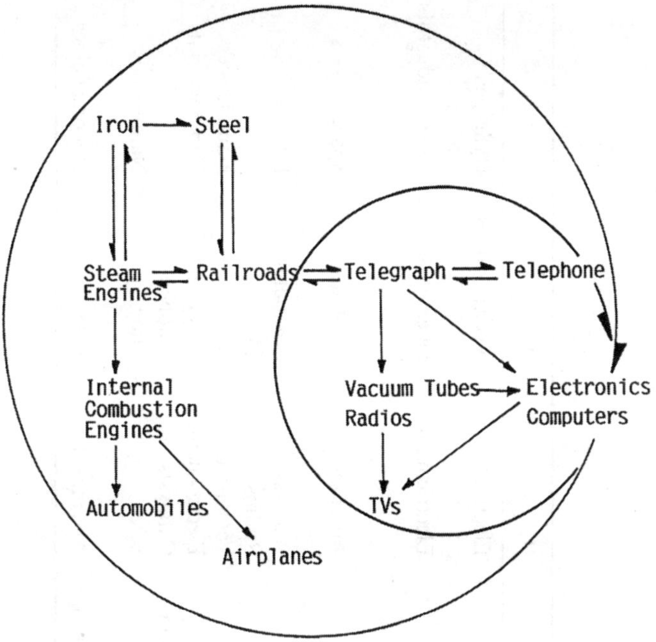

Source: Hirooka (1994c).

Figure 3.4 Clusters of innovations and synergistic development

industrialized society. This accumulation of innovation paradigms is comprehensively illustrated in Figure 3.5.

Such abundant versatility greatly increases energy consumption and productivity, that is, GNP, as shown in Figure 3.6. This figure appeared in the book *The Global 2000 Report to the President – Entering the Twenty First Century*, an extensive survey of global issues by the United States Government (1980). Thanks to the synergistic effect of multiple innovation paradigms, the wealth of nations has been exponentially increased. The GNP of the United States has increased 30 times in the past 100 years, while the corresponding increase in Japan is estimated to be 80 times. Moreover, energy consumption has increased proportionally, as shown in Figure 3.6.

This cumulative build-up of innovations and their synergistic effect is the most important evidence that innovation is the engine of economic development. Other interpretations do not support such cumulative contributions of continuous enhancement of value added to the economy. In short, the intermittent surges of innovation clusters are the cause of Kondratiev business cycles.

Table 3.1 Evolution of innovation paradigms

Phase	0	I	II	III	IV	V
Energy	Charcoal	Coal	Petroleum	Electricity	Nuclear power	Hydrogen Fusion
Motive power	Water mills Wind mills Horse power	Steam engines	Internal combustion engines	Electric motors	Jet engines	Superconductors
Transportation	Stagecoaches Sailing ships	Canals	Railways Steamships	Automobiles	Airplanes	Space rockets
Communications	Letters	Postal mail	Telegram Telephone	Wireless Radios	Computers TVs	Internet
Electronics				Vacuum tubes	ICs	Nanodevices
Chemicals	Alchemy	Acid/alkali	Organic & inorg. chem.	Electrochem. Catalysts	Petrochem. & polymers	Nanomaterials
Iron & steel	Wrought iron	Blast furnace Coke smelting	Pig iron	Steel	Alloy	Nanometals
Manufacturing	Handcrafts	Machine tools	Standardization	Assembly lines	FMS	Nanomachines

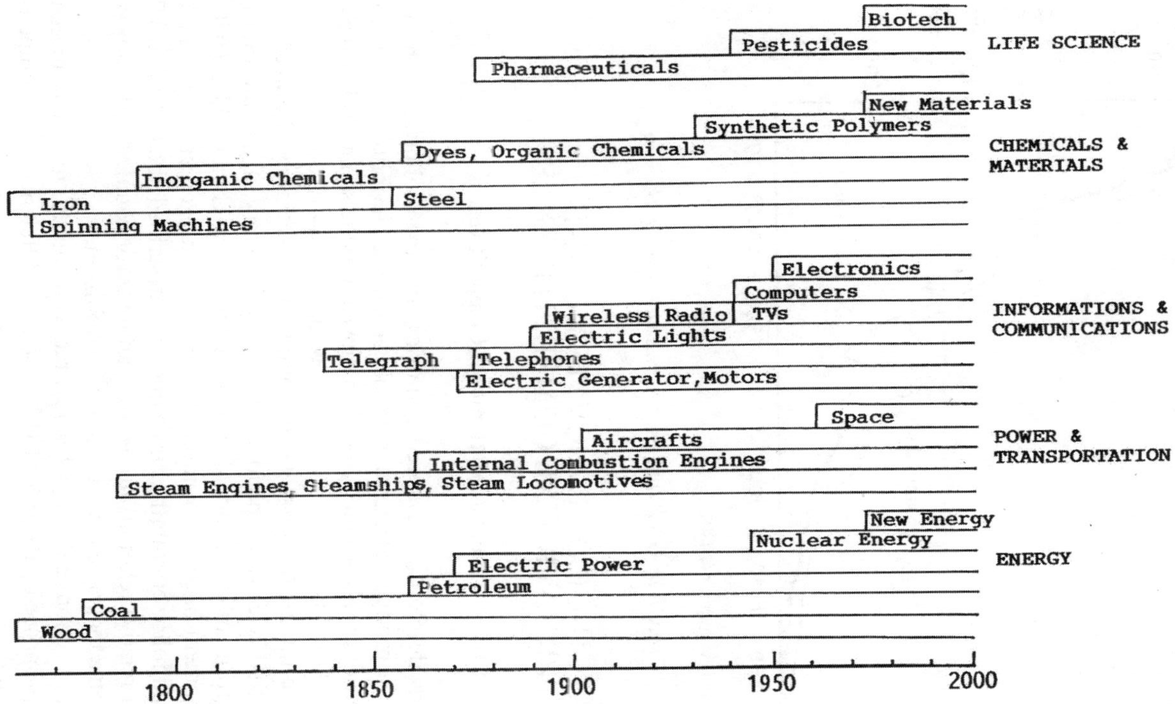

Source: Hirooka (1992a).

Figure 3.5 Formation of modern industrialized society by innovations

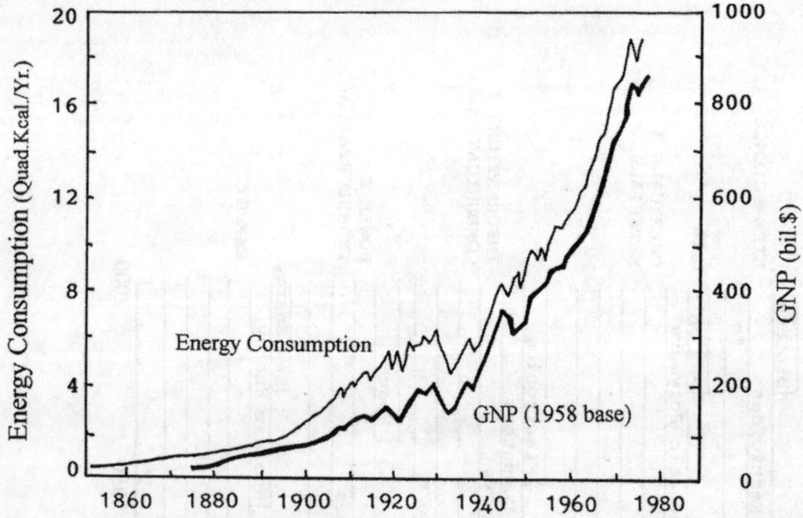

Source: US Government (1980).

Figure 3.6 Growth in GNP and energy consumption in the US

3.5 CATCH-UP OF INNOVATIONS AND INTERNATIONAL COMPETITIVENESS

Life Cycle of Innovation and International Competitiveness

Innovations develop in advanced countries, and then follower countries introduce the products at a lower technological level. The follower countries begin to produce the products and then move up to the same level of technology. Finally, the follower countries begin to export the products to advanced countries. A reverse gap of competitiveness emerges between the two sides, that is, cost competitiveness. This is the story of the innovation paradigm from an international view of technological change. Along with the development of an innovation, the interrelation between the inventor country and follower countries is changing as described above. Figure 3.7 indicates a relative relationship between an inventor country and follower countries on a logistic curve of innovation diffusion.

The basic economic theory of comparative advantage for international trade was proposed by Ricardo (1817). There is a difference in the competitiveness of industries between countries, and every country has more competitive and less competitive ones. For example, while England is more competitive in wool production and high-quality textile production, it is

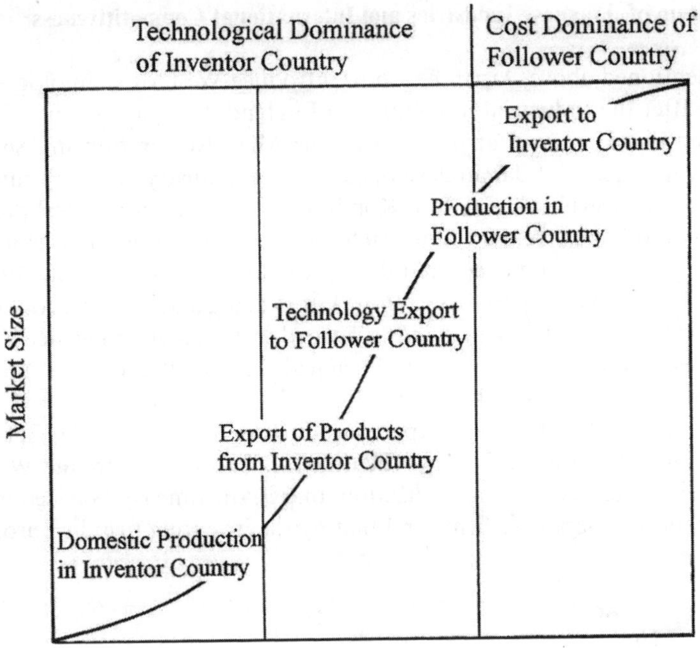

Figure 3.7 International development of innovations and competitiveness

less competitive in raising grapes and producing wine. In contrast, Spain is less competitive in textiles and more competitive in wine. Each country thus has a comparative advantage in different products, a situation that supports mutual trade. Such international trade has functioned for a long time since Ricardo proposed his theory.

However, explaining trade patterns is not always this simple. The United States, as an advanced country with high labor costs, would seem to have an advantage for exporting capital-intensive high-tech goods. The US, however, imports many capital-intensive goods and exports large amounts of labor-intensive agricultural products. Leontiev (1953) discovered this famous paradox, and many studies have addressed it. Vernon (1966) explained it by proposing a life-cycle theory. Innovation products diffuse along a logistic curve, as illustrated in Figures 2.1 and 3.7. In the first stage of innovation, the inventor country develops and produces related products within the country. It later exports them to other countries. Some countries then import the technology required to produce the products and begin to export their version to the original country because their lower labor costs give them a cost advantage. Leontiev's paradox is well explained by this scenario. These processes are clearly illustrated in Figure 3.7.

Catch-up of Japanese Industries and International Competitiveness

As mentioned above, Japan began to introduce Western technologies 100 years after the Industrial Revolution in England. It began constructing an industrialized country at the time of the Meiji Restoration in 1868. As shown in Figure 3.2, Japan developed a cotton industry and built railways during the upswing of the third Kondratiev cycle, but crude steel production was still no more than 7 million tons in 1940. Substantial development of Japanese industries occurred during the fourth upswing from 1950 to 1970. This is when Japan really jumped from being a developing country to an advanced one. Japan was industrialized in terms of crude steel, automobiles, electric appliances, petrochemicals and synthetic fibers and had caught up with the West by 1970.

This development trend of Japanese industries is summarized in Table 3.2 and Figure 3.8. Table 3.2 shows the time lags between Japan and Western countries in terms of industrialization, that is, the time lag between industrialization by Japanese firms and that by the inventing firm in Europe or

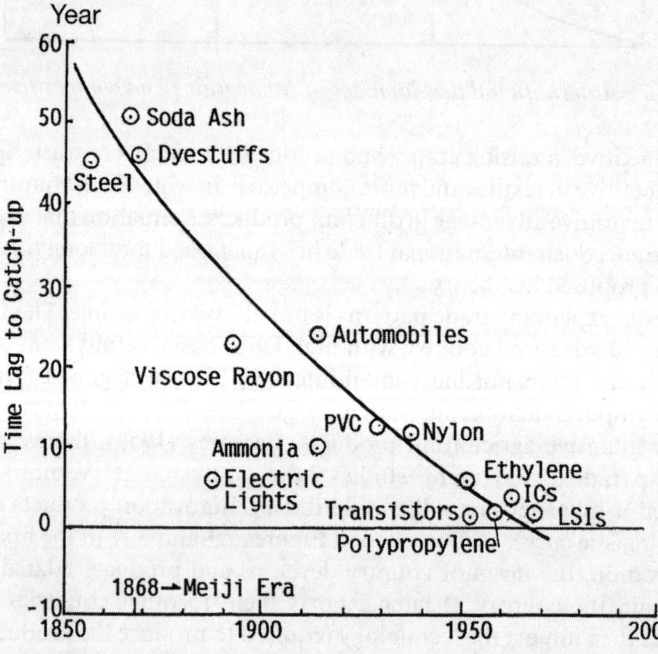

Source: Hirooka (1995b).

Figure 3.8 Innovation time lag in Japanese industries

Table 3.2 Commercialization time lag of innovations in Japan

Technology	Time lag (yr.)	Original commercialization		Commercialization in Japan
2nd generation innovation (1850–1910)				
Steel (Bessemer)	45	1856 Vessemer	UK	1901 Yawata Iron & Steel
Soda, Solvay process	51	1866 Solvay	Belgium	1917 Asahi Glass
Alizarin dyes	45	1868 BASF	Germany	1914 Mitsui Mining
Electric light	6	1881 Edison	US	1887 Tokyo Electric
Viscose rayon	23	1892 Cross, Bevan	US	1915 Azuma Industry
Automobiles	24	1913 Ford	US	1934 Nissan Motors
Ammonia	10	1913 BASF	Germany	1923 Nippon Nitrogen Fertilizers
3rd generation innovation (1910–70)				
Polyvinylchloride	13	1928 Goodrich	US	1941 Nippon Nitrogen Fertilizers
Nylon	12	1937 du Pont	US	1949 Toyo Rayon
Ethylene, Petrochemicals	6	1951	US	1958 Mitsui, Sumitomo Chemical
Transistors	2	1952 TI	US	1984 Sony
Polypropylene	2	1959 Montecatini	Italy	1961 Mitsui, Mitsubishi, Sumitomo
Semiconductor ICs	3	1962 Fairchild	US	1965 NEC
LSIs	2	1968 TI	US	1970 Hitachi

Source: Hirooka (1995b).

the United States. This table clearly shows that the time lag steadily shortened, while Figure 3.8 indicates that Japan had almost fully caught up by 1970.

Figure 3.9 illustrates the production trend of various industrial products in the United States, Japan and Germany (except for chemical fibers, where the production is for Taiwan instead of Germany). Where once Japan was far behind the United States in the production of crude steel and automobiles, it overtook the US during the 1970s. In the case of chemicals, however, while Japan started producing ethylene and chemical fibers at almost the same time as the US, and the industry reached maturity, the scale being much less than in the US. On the other hand, Japan was industrialized much earlier than the United States and Germany in the cases of continuous casting of crude steel, flexible manufacturing systems (FMS) and industrial robots. These Japanese industries were very competitive worldwide, indicating that earlier industrialization is very important for international competitiveness. This also held true for transistor radios, calculators, VTRs and so on, emphasizing that earlier industrialization is crucial.

3.6 BUSINESS CYCLES AND INNOVATION DYNAMISM

We have discussed Kondratiev's long business cycles, focusing especially on clarifying their correlation with the diffusion of innovation products. This section discusses some details regarding the correlation of business cycles with the productivity of innovations to add value; it also clarifies the origin of business cycles.

Recognition and Interpretation of Long Business Cycles

Business cycles have been interpreted in various ways in terms of the economics of technological change.

Schumpeter (1939) argued that the Kondratiev business cycle is composed of two phases, the upswing of prosperity and the downswing of depression, and that the second approximation is described by four phases: recovery, prosperity, recession and depression. In this approach, he uses the Juglar wave with a ten-year cycle as the unit. The upswing starts after the bottom of a depression in the preceding cycle with a recovery that lasts for ten years, followed by two Juglar waves of prosperity. Through a recession phase around the peak, the depression occurs during the downswing of the cycle. Schumpeter stuck tightly to this formulation and described the effect of innovations along Juglar waves. Thus, he did not sufficiently discuss the

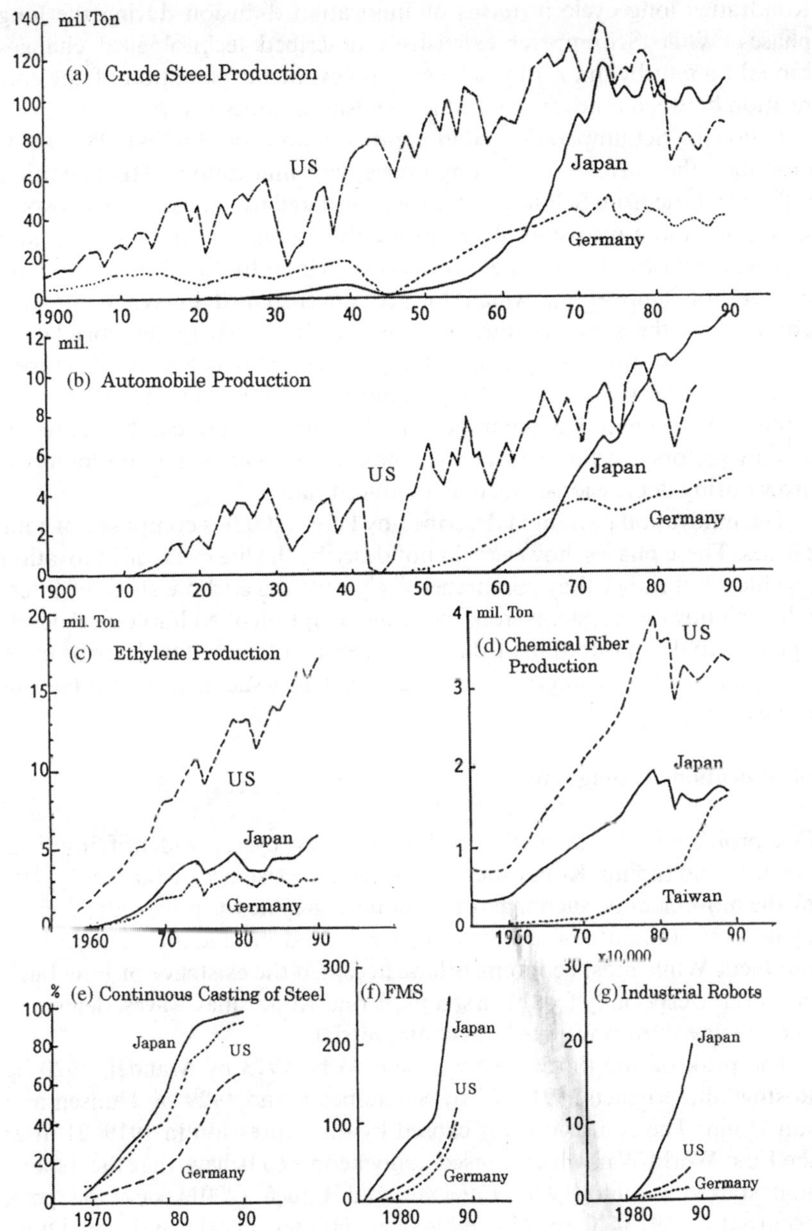

Source: Industrial Statistics of the US, Germany, Japan and UN; Hirooka (1995b).

Figure 3.9 Trends in industrial production by country

Kondratiev long cycle in terms of innovation diffusion during the long phases. While Schumpeter extensively described technological changes brought about by innovations, he gave no explicit explanation of the correlation between innovations and long business cycles per se.

Following Schumpeter's second approximation, van Duijn (1983, 1984) described the correlation of long cycles with innovations. His definition differed a little from Schumpeter's. He considered five phases: two prosperity phases and a recession phase during the upswing, and depression and recovery phases during the downswing. He introduced the life-cycle concept of innovation, which formed an S-shaped curve. While this approach is the same as ours, he was unable to clarify the correlation between innovations and long business cycles. This is because S-shaped innovation life cycles do not always coincide with long business cycles. Van Duijn (1984) stated that 'An important deviating pattern can be observed: leading sectors can survive macro-economic depressions (i.e. continuing to grow during depressions, albeit at a reduced rate).'

The innovation paradigm described by Perez (2002) is composed of four phases. These phases, however, do not describe the life cycle of innovation products but rather they comprehensively describe a whole surge of interrelated innovation systems from the initial irruption of technological developments to the maturation of the market. Perez attempted to describe these four phases using a single S-shaped curve. This issue is discussed later in more detail.

Identification of Long Cycles

One problem in the discussion of long business cycles is identifying their position and timing. Kondratiev confirmed the existence of business cycles by the movement of such macroeconomic aggregates as price indices, rates of interest, amounts of world trade, wages and production of industrial products. While most economists have accepted the existence of long business cycles, especially for the first and second Kondratiev waves, determination of the third wave has been controversial.

The peak of the third wave was said to be 1913 by Mandel, 1920 by Rostow and Kuznets, 1919–25 by Schumpeter, and 1929 by Hansen and van Duijn. The confusion was caused by the depression in 1919–21 after the First World War, which caused many people to believe that the depression started around 1920. Freeman and Louçã (2001) said the peak occurred in 1913–20 (p. 255, table 7.9). Hansen (1941) and van Duijn (1984), however, determined it to be 1929, as did we. Although Perez (2002) described a surge of innovations by using an S-shaped curve for an innovation paradigm, which is differently depicted from ours, the synergy phase she identified coincides well with the upswing of Kondratiev long cycles. She determined the end of the synergy phase to be 1918, which corresponds to the peak of the third wave in our case.

Although Schumpeter (1939) estimated the peak of the third wave to be 1919 (*Business Cycles*, p. 794) or 1925 (p. 907), as shown in his figure 42 (p. 795), actual industrial production in the United States increased steadily until 1929.

Freeman and Louçã (2001) pointed out that the economic growth rate shown in their table 7.9 (p. 255) was abnormally high in the depression period. That is, the economic growth rate during the downswing of 1920–29 was 8.1 per cent while that during the upswing of 1890–1913 was 2.5 per cent in the United States. This contradiction is resolved by considering that the peak was in 1929. Figure 3.10 summarizes the trends of three parameters,

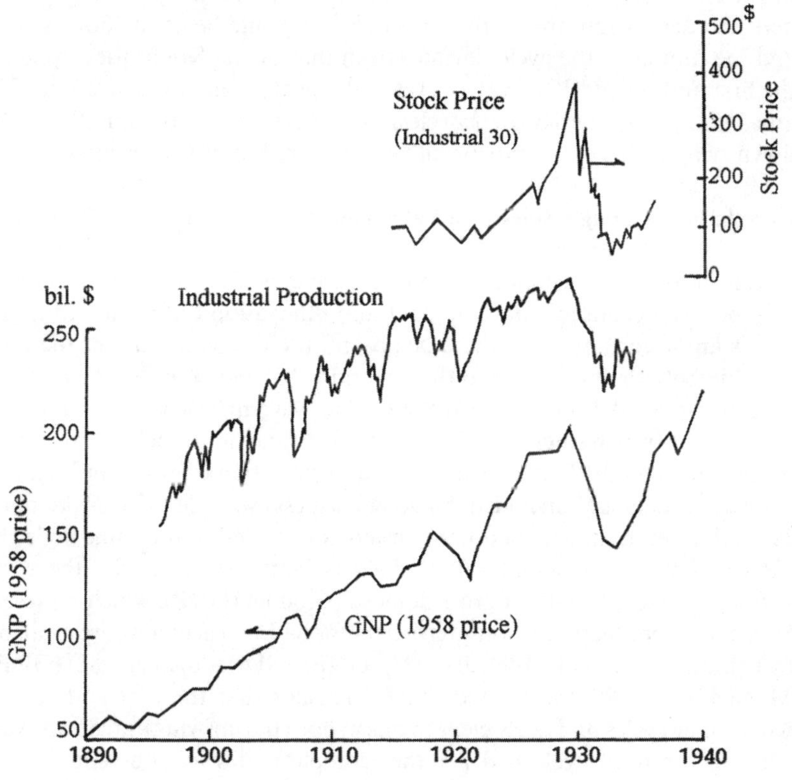

Sources: Industrial production (Schumpeter, 1939); GNP (Mitchell, 1998).

Figure 3.10 Economic trends during 1900 and 1940 in the US

stock prices, industrial production and GNP, in the United States from 1890 to 1940. The data for industrial production were taken from Schumpeter (1939, figure 42), and those for GNP (1958 prices) were obtained from Mitchell (1998). Figure 3.10 clearly indicates that the US economy grew steadily from 1900 to 1929, despite the economic turbulence around 1920. This is confirmed by the continuous growth of industrial production. A rather steep increase in GNP was observed from 1921 to 1929, indicating the existence of a Kondratiev upswing. Stock prices symbolize the economic situation, and they increased sharply from 1921 to 1929. They increased as well in the UK, as shown in figure 56 in Schumpeter (1939). In short, the data for the stock market, GNP and industrial production clearly show quite similar trends in economic development from 1900 to 1930.

Another problem is the Kuznets cycle. Mensch (1975) adopted the Kuznets cycle for his study, but its positioning differs significantly from that of the Kondratiev cycle. In the Kuznets cycle, the turning point from prosperity to depression, that is the 'crash', is in the middle of the downswing, and the timing of the cycle deviates from that of the Kondratiev cycle. In the first and second Kondratiev cycles, the peaks clearly occurred in 1825 and 1872, but in the Kuznets cycles, these peaks are in the middle of the downswings. This discrepancy will be considered in future discussions.

Correlation of Long Cycles with Value-added

If innovation causes economic growth, there must be some causal relationship between economic development and innovation diffusion. Although this is an important issue, there has been little discussion about the relationship between innovation diffusion and value added to the economy.

Kondratiev (1926) recognized the second wave mainly by production of coal and pig iron in England, foreign trade in France and stock market prices in England, and the first wave by wages and stock market prices in England. From these data, he concluded that long business cycles do exist (Nakamura, 1987). However, these data do not always correspond to the value added to the economy – the actual value added is better expressed by the gross national product (GNP) or gross domestic product (GDP), which represent the nation's productivity. Long-term data for national accounts are available from the aggregates of Mitchell (1998) (GNP) and Maddison (1982) (GDP). Mitchell's data for the United States indicates that the first and second Kondratiev cycles had fairly clear trends in both the upswing and downswing whereas the third cycle had a rather complicated structure. This can be ascribed to heavy fluctuation due to the First World War between 1914 and 1918. There is, however, a clear trend of a continuous increase in productivity for 8 to 10 years from 1921 to 1929, creating an upswing. The average annual growth rates corresponding to the upswings were 6.5 per cent for the first wave from 1819 to 1826, 8.1 per cent for the second wave from 1865 to 1873, 6.0 per cent for the third wave from 1921 to 1929, and 4.6 per cent for the fourth wave from 1958 to 1966 in the United States and 10.0 per cent for the fourth wave from 1960 to 1970 in Japan. These growth rates are core components of the upswings, which often included more than 10.0 per cent annual growth. By 1993 the economic growth in the United States for the 100 years from 1893 had seen GNP increase by 30.9 times, and the average annual growth rate was 3.5 per cent. That is, more than 6 per cent growth is significantly high. However, for the United Kingdom, Maddison's GDP data do not show such a clear-cut correlation.

In short, our identification of Kondratiev cycles seems to be appropriate to their use in discussions of the value added to the economy by innovation diffusions. That is, there is a clear correlation between the upswings of Kondratiev cycles and innovation diffusion as the accelerator of value added to the economy. This is an explicit verification of innovation as the engine of economic development.

3.7 SUMMARY

Analysis of the correlation between Kondratiev cycles and innovation diffusion has clarified several points.

1. Various innovations diffuse selectively along the upswing of a Kondratiev cycle, forming a cluster. As the S-curve of innovation diffusion directly reflects the value added to the economy, the formation of innovation clusters during the upswing is evidence that innovation is the origin of economic development.
2. Comparison of the Kondratiev business cycles for the United Kingdom, the United States and Japan revealed that the economies of these countries have become synchronized and share the same cycles. Japan is a latecomer and industrialized by introducing various innovations. It is interesting to note that these innovations also diffused and formed clusters along the upswings of Kondratiev cycles even though they were on later cycles.
3. Innovations within a cluster are interrelated and create a synergetic effect.
4. Innovations belonging to the same field of technology intermittently evolve and cumulatively add value to the economy. This kind of cumulative enhancement of the economy by innovations is strong evidence that innovation is the engine of economic development.

5. GNPs have increased as much as 30 times over the last 100 years. Economic growth rates were found to be higher during the upswings of Kondratiev cycles.
6. Japan industrialized by introducing Western technologies beginning in the Meiji Restoration in 1868. The time lag for Japan to catch up decreased year by year, and Japan fully caught up in the 1970s. Moreover, Japanese industries that commercialized products earlier than Western ones had robust competitiveness.

4. Sequence of bubble, crash and depression

Since the Industrial Revolution, about 230 years ago, the world economy has had four business cycles. The Great Depression during the third cycle in particular has attracted much attention and has been analyzed by many economists using various approaches. There has been, however, no decisive interpretation. Despite such efforts by economists, a serious depression struck Japan during the fourth cycle and continued for over ten years. Previous chapters discussed the innovation paradigms and the four business cycles. This chapter focuses on the Great Depression and the bubble economy that led up to it in an attempt to clarify the causes and the real features.

4.1 ECONOMIC INTERPRETATION OF THE GREAT DEPRESSION

Lengthy recessions have occurred several times since the Industrial Revolution. The one that took place between 1872 and 1897 is called the 'Great Recession', and the more serious one that began in 1929 is called the 'Great Depression'.

The Great Depression motivated economists to re-examine their theories of economics and build new disciplines. John Maynard Keynes stressed the importance of economic policy and developed his new version, subsequently known as Keynesian economics. J.A. Schumpeter advanced his economic theory, combining Kondratiev's theory of long business cycles with his innovation theory of economic growth.

While the Great Depression has been investigated by many economists, there has been no decisive explanation of its causes. Papers on the Great Depression have been written by, among others, Brown (1940), Kuznets (1940), Hansen (1941), Galbraith (1956, 1990), Yoshitomi (1965), Kindleberger (1973), Beckman (1988), Sekioka (1989), Noguchi (1992) and Chancellor (1999).

The predominant cause has been ascribed to the failure of monetary policy and the escalation of speculative activity, especially to the excessive fluidity of the money supply, which caused speculation and a bubble economy.

In contrast, Schumpeter (1939) discussed business cycles and renewed his theory of economic development with the concept of stimulation by innovations. He criticized the business cycle theories of other economists. He described the shortcomings of historical interpretations of crises in his book *Business Cycles*:

> Histories of 'crises' and detailed descriptions of individual crises have been written from the beginning of the nineteenth century. That literature is richer than appears at the first sight because it includes all of the descriptions of particular aspects, most of the attempts to analyze the working of the monetary mechanism and of speculation . . . Historians of crises primarily talk about stock exchange events, banking, price level, failures, unemployment, total production, and so on – all of which are readily recognized as surface phenomena or as compounds which sum up underlying processes in such a way as to hide their real features . . . It did not touch upon the essential things at all. The same objection applies to more recent attempts (pp. 220–21).

Schumpeter ascribed the Great Depression to overlapping downtrends in the Kondratiev, Jugular and Kitchin cycles. He attempted to explain each cycle as having been caused by innovation. Kuznets (1940) strongly criticized Schumpeter's complex theory of innovation cycles.

Regardless of these debates, there are three commonly agreed upon characteristics of the Great Depression.

1. The depression emerged in the United States, the most prosperous country in the world.
2. The depression took place at the peak of prosperity.
3. The depression continued for a long time and affected various areas, including the stock market, manufacturing, and agriculture.

4.2 INDUSTRIALIZATION AND THE GREAT DEPRESSION IN THE US

The first two characteristics above give us a hint about the causes of the Great Depression. Why did such a serious recession suddenly happen at the peak of prosperity? Since Kondratiev cycles can be ascribed to clusters of innovations, it is interesting to analyze the Great Depression from the viewpoint of innovation and industrialization.

Background and Scenario of the Great Depression

The United States introduced various technologies from Europe in the latter half of the nineteenth century and as a result jumped from being a developing country to an industrialized country in the early twentieth century. The United Kingdom enjoyed the second surge of the Industrial Revolution through the diffusion of railways and iron production. Later, the United States developed in such a way that railways reached the West coast and steamships boosted trade with Europe.

The United States introduced Bessemer furnace technology, and steel production increased rapidly at the end of the nineteenth century. Hydraulic power generation was launched at Niagara Falls with the cooperation of Siemens of Germany in 1895. Oil began to be used for lighting in the form of kerosene in the middle of the nineteenth century, and production increased after large oil fields were found in California. Due to the development of the automobile, the demand for gasoline expanded, and the oil industry developed. Though the motorcar was invented by Daimler and Benz in Germany, the United States took over leadership in the car industry after Henry Ford introduced the assembly-line production system. These innovations spread in the United States in the early twentieth century, as shown in Figure 4.1a. At the same time, radio and various consumer durables such as washing machines and refrigerators came into wide use. The development of innovation products in Figure 4.1a is represented by the diffusion trajectories of crude steel, automobiles, electric power and oil consumption. The vertical axis has a normalized scale with the mature market equal to unity. Values higher than unity represent exaggerated demand during a bubble economy, as described later.

Ford's assembly-line process drastically increased the productivity of automobile production, and the price of the Model T car dropped from $850 in 1908 to $250 in 1924. Electric power improved the productivity of manufacturing industries and was also supplied for household use. Trade with Europe was boosted by the diffusion of steamships – shipping costs dropped to one-tenth their former levels. These innovations drastically improved the productivity of industries and led to tremendous economic development. With this boom, stock prices skyrocketed. That of General Motors, for example, rose from $21 in 1925 to $46 in 1928.

The rapid development of the economy brought huge investment demand, which continued for more than ten years. The rise of stock prices stimulated the trend of raising money through the stock market, and the leverage effect of capital gains further accelerated investment. In 1928, dealings on credit increased abnormally, and market prices rose steeply. In 1929, stock prices skyrocketed – the rise for the three summer months equaled the entire rise in 1928. On 3 September, the average stock price for 25 industrial companies reached a peak of $469, and the Dow Jones Industrial Average (DJIA) reached $381. On Black Thursday, 24 October, there was a rush to sell at a loss, cutting the DJIA to $272.

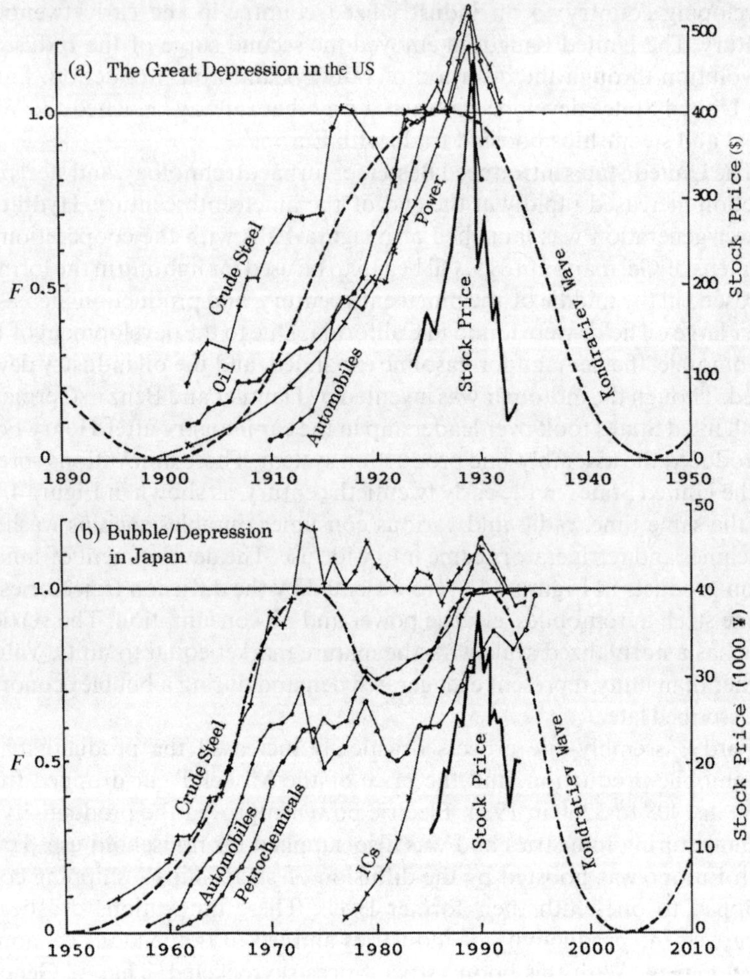

Source: Hirooka (1994b,c, 2000a).

Figure 4.1 Innovation cluster and bubble/depression sequence

Stock prices continued to fall, and the DJIA hit $41 in July 1932, one-tenth the peak level.

Mechanism of the Great Depression

At the beginning of the twentieth century, the United States had become industrialized, and the infrastructures of a modern economy had been laid through various innovations. These innovations activated the economy, and the GDP grew by 2.7 times in the 30 years from 1900 to 1929, corresponding to an annual growth rate of 3.4 per cent on a real price basis, despite the rather severe recession after the First World War. More important is that the annual growth rates were as high as 10–30 per cent for the diffusion of innovation products. For example, the growth rate of crude steel between 1885 and 1900 was 12.5 per cent, that of automobiles between 1910 and 1925 was 22.4 per cent, and that of oil consumption was 10.4 per cent between 1900 and 1915, as shown in Table 2.4. In addition, electricity consumption increased by 31 times and radio receiver production increased by 28 times in the 30 years from 1899 to 1929. These data clearly indicate that the innovation products diffused more rapidly into the market compared with the average growth rate of the economy. This difference triggered deformation of the economy.

This intensified economic development through innovation raised stock prices and induced a bubble economy. The diffusion of innovations, however, reached maturity. The production of automobiles leveled off at 3.7 million cars: 3.62 million in 1923, 3.73 in 1925 and 3.69 in 1926. The production of crude steel was 49.0 million tons in 1923, 49.7 in 1925 and 49.3 in 1927. The consumption of oil and electric power increased throughout the bubble economy, but after the crash the demand dropped drastically. The demand for cars in 1929, however, jumped to 4.45 million, up 18 per cent from the previous year, and the demand for crude steel increased abnormally to 61.7 million tons. These increases were for exaggerated demand, which proliferated like a cancer cell. The exaggerated demand reached more than 20 per cent, as shown by the lines exceeding unity in Figure 4.1a. The actual market, however, shrank drastically (as shown in the figure), with car production of 2.78 million (down 38 per cent) and crude steel production of 44.6 million tons (down 38 per cent) in 1930. The pre-bubble levels were not reached again for another ten years. The crash can be interpreted as having been caused by a failure to recognize the difference between the reality of a nonlinear phenomenon, that is, the market reaching maturation, and the false belief in the linear trend of economic growth.

4.3 ECONOMIC DEVELOPMENT AND THE BURSTING OF A BUBBLE IN JAPAN

Countries can jump from the developing stage to being an advanced country by introducing various innovations and can enjoy a high annual economic growth rate of more than 10 per cent only once. The United States achieved this during the upswing of the third Kondratiev cycle, and Japan achieved it during the upswing of the fourth cycle in the 1960s. Japan's case was quite similar to that of the United States: the Japanese economy grew due to the introduction of various innovations from the United States and Europe and had an annual growth rate of more than 10 per cent in the 1960s. Economic infrastructures were created through the diffusion of crude steel, automobiles, electric appliances and petrochemicals, as shown in Figure 4.1b. At the beginning of the 1970s, Japanese market demand became mature but the economy overheated. Stock prices rose from ¥3000 in March 1972 to ¥5000 at the end of that year, and capital gains grew. Japanese industries began to export their products to Western countries, riding the wave of expanding global trade.

The worldwide prosperity of the 1960s collapsed with the oil crises in the 1970s, then recovered in the mid-1980s. After the Plaza Accord in 1985, the Japanese Yen was highly over-valued, and the Japanese government took steps to stimulate domestic demand to cope with an increasingly difficult export situation. Low interest rates and abundant fluidity of money induced a bubble economy, with the stock market and land prices growing feverishly. This trend was quite similar to the situation before the Great Depression in the United States in 1929, as shown in Figure 4.1a.

The high economic growth in the United States at the beginning of the twentieth century was due to the diffusion of steel, petroleum, electric power and automobiles, and the collapse occurred at the top of the upswing. The Japanese economy followed the same route taken by the US in the early twentieth century; that is, the high growth rate of the economy during the 1960s resulted from the diffusion of such key industries as crude steel, automobiles, electric appliances and additionally petrochemicals. Following the oil crises, the Japanese economy again enjoyed growth due to the later diffusion phase of innovations such as plastics, synthetic fibers and electric appliances together with the contributions of electronics and information technology. Thus, the interval from the 1960s to the 1980s can be seen as an upswing in the Japanese economy in terms of innovation. In the latter half of the 1980s, many innovations reached maturity; only electronics developed further. Thus, there was little room to absorb the flood of money, except through the stock market and land prices. The Japanese national bank, however, continued to issue plenty of money to stimulate domestic demand, but succeeded only in raising the prices of stocks and land, including purchases of American real estate using so-called 'Japan money'. Most firms raised money from equity financing by dealing in stocks and land, and thus the boom escalated.

As seen in Figure 4.1b, the production of crude steel increased during the 1960s, reaching a mature market state of 100 million tons per year by 1973. Automobile production also grew to maturity in the domestic market in the early 1970s. The petrochemical industry, whose development was interrupted by the oil crises, reached a mature level in the latter half of the 1980s. These data clearly indicate that, before the crash of the bubble economy in 1990, the market for various innovations had matured but an exaggerated economy still expanded. This was surely a bubble phenomenon.

4.4 A HISTORICAL EXAMINATION OF THE TRANSITION FROM BUBBLE TO DEPRESSION

The above analysis indicates that there could be a phenomenon common to all the Kondratiev cycles – an alternating sequence of bubble economy, crash and depression. That is, a cluster of innovations brings about a prosperous upswing, which induces a bubble economy based on speculation, followed by a crash and then by a depression. The above section investigated the transitions from a bubble via crash to a depression in the third and fourth cycles, but this alternating sequence could also exist in the other cycles. Certainly, lengthy depressions have occurred since the Industrial Revolution, as described by such economists as Beckman (1988), Sekioka (1989) and Chancellor (1999). Schumpeter (1934b) and Kanazashi (1991) also analyzed the economic crises in the first and second cycles. Here we examine the economic trends around the peaks of the first and second Kondratiev cycles.

Speculation in the Cotton Business during the First Kondratiev Cycle

The Industrial Revolution started in Lancashire in England, driven by the development of the cotton industry. Several innovations – the steam engine, the iron-making process (coke blast furnace), machine tools and so on – were combined into complex systems, and the resulting synergy stimulated the economy. As shown in Figure 3.1, the first upswing was characterized by a rapid increase in cotton demand. The stock market for the cotton business gradually heated up in the middle of the first upswing from 1789 to 1825. Trade with India and America was boosted by newly commissioned steamships, with the result that the stock prices increased. In 1822, economic prosperity was widespread, and speculation in the stock market accelerated. Stock prices skyrocketed from 1824 to 1825, but then, at the end of 1825, the bubble burst – stock prices dropped to one-fourth their value within four years, and a depression began that lasted for 20 years, until 1846.

Speculation in Railways during the Second Kondratiev Cycle

The second Kondratiev cycle was driven by the development of railways and iron production, and the upswing extended from 1846 to 1872. While the first locomotive ran between Stockton and Darlington in 1825, the actual diffusion began in 1845, after the long recession. The success of the first lines and a seal of royal approval eventually swayed public opinion. Railway mania was ignited in 1845, and railways covered 18 000 km by 1855 and 26 800 km by 1875. In the United States, railway construction began in 1830 and substantially took off after 1850. With the switch of the British cotton trade from India to America, the construction of canals and railways escalated in the United States. In the latter half of the upswing of the second Kondratiev cycle, until 1872, there was much speculation in railways, and the bubble collapsed in 1873. A great recession followed in both the UK and the US and continued for 25 years, until 1897. During this 'Great Recession', unemployment in the UK reached 12.5 per cent in 1879. We can see this same pattern in the Great Depression in 1929 in the US.

Alternating Sequence from Bubble to Depression in Kondratiev Cycles

As revealed by the studies described above, every cycle had an upswing with clustered innovations, and the economy developed due to the value added by the innovations. In the middle of the upswing, economic development was boosted, and speculation in the stock market began, igniting a bubble economy. At the top of the upswing, the bubble collapsed, bringing on a long depression. This consistent pattern is illustrated in Figure 4.2.

We may conclude that modern industrialized society is built upon various innovations and that the crashes and depressions which subsequently take place may be explained as the collapse of an overheated economy induced through innovation diffusion. Before the innovation age of the First Industrial Revolution, there were many cases of speculation, for example, speculation in tulips in Holland in 1636 and speculation in the stock of the South Sea Company in the early eighteenth century. Those speculations ended quickly, and their effect on the economy was limited. The later crashes and depressions induced by innovation diffusion, however, were quite different. Their effect was severe and prolonged because the phenomenon induced by innovation dynamism accompanied a long-lasting socioeconomic structural change. Distortion caused by structural change in the socio-institutional framework and overshoot of a bubble economy are more difficult to overcome.

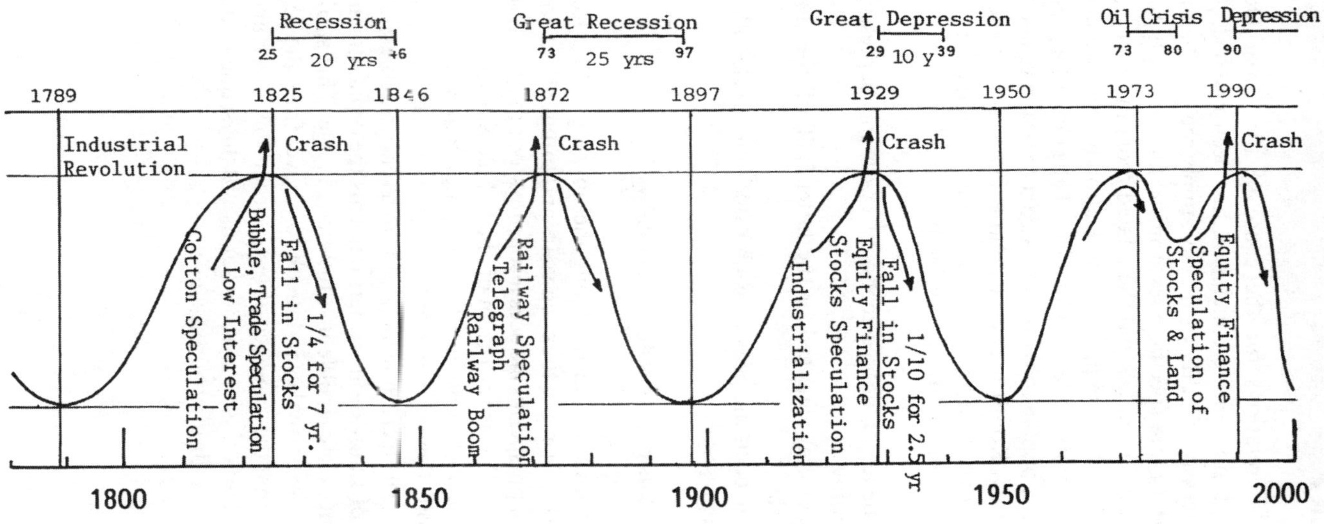

Source: Hirooka (2000a, 2002a, 2003c).

Figure 4.2 Kondratiev cycles and bubble/depression sequence

4.5 SUMMARY

While various interpretations for the cause of the Great Depression have been proposed, no definite conclusion has been reached. Although speculation and excessive fluidity of money supply seem to be direct triggers, there must be a more specific background. The description above of the Great Depression in the US and the related economic changes and of the bubble economy and crash in Japan in relation to innovation diffusion covered several key points. Extensive analysis of the Great Depression and of the collapse of the bubble economy revealed a common phenomenon at the same point in each Kondratiev cycle. Our examination of the first and second Kondratiev cycles also revealed the same phenomenon. Thus, the common phenomena on Kondratiev cycles are summarized:

1. There is a sequence of bubble economy, crash and depression in the capitalist economy.
2. The bubble economy emerged in the latter half of the upswing of Kondratiev cycle.
3. The crash occurred at the peak of each Kondratiev cycle followed by a long lasting depression.

The mechanism of this phenomenon is as follows:

4. A cluster of innovations grows along the upswing of the Kondratiev cycle, which lasts for more than ten years and brings about continuous prosperity. The prosperity induces a soaring stock market and accelerates speculation. This is the origin of the bubble economy.
5. In the latter half of the upswing, the innovation markets reach maturation and level off. This gap between the bubble economy and the leveled-off markets triggers a crash, which leads to a serious depression.
6. The diffusion of innovation clusters deforms the economic infrastructures and induces irreversible institutional change. This strain on the economic framework makes it difficult to recover from the depression.
7. It is concluded that the sequence of bubble economy, crash and depression is triggered by a cluster of innovations and this sequential structure is the cause of Kondratiev business cycles.

5. Trunk innovations and institutional change

The previous chapter suggested that long Kondratiev business cycles are produced by various innovations. It also clarified the causal relationship between business cycles and innovations. This chapter presents more detailed analyses that clarify the substantial features of long business cycles and the pervasive behavior of innovations in terms of the techno-economic paradigm and institutional change in society. New concepts, 'trunk innovation' and 'infra-trajectory' are introduced, and innovation dynamism beyond the realm of technological innovations is discussed.

5.1 NETWORK OF INFRASTRUCTURES AND BUSINESS CYCLES

First, the features of innovations are characterized and the taxonomy of innovations, both technological innovations and other kinds of innovations, is discussed. In particular, the important, pervasive innovations that contribute to economic development – 'trunk innovations' – are discussed.

A Wider Concept of Innovation: Trunk Innovations Forming Infrastructures

Schumpeter (1912, 1934a), in his book, *The Theory of Economic Development*, defined five groups of innovation: those consisting of (1) new products (goods), (2) new processes (production processes), (3) new markets, (4) new sources of supply of raw materials, and (5) new organization. Schumpeter (1939, p. 84) also describes in his book *Business Cycles* that innovation is defined as follows:

> We include the introduction of new commodities which may even serve as the standard case. Technological change in the production of commodities already in use, the opening up of new markets or of new sources of supply, Taylorization of work, improved handling of material, the setting up of new business organizations such as department stores – in short, any 'doing things differently' in the realm of economic life – all these are instances of what we shall refer to by the

term Innovation. It should be noticed at once that that concept is not synonymous with 'invention'.

In general, the discussions here about the economics of technological change have concentrated on technological innovations (1) and (2) above. We have, however, noticed the importance of new resources (4) and new organization (5). Raw materials are not invented as technologies but are simply feedstock for the economic development. When a usage is found for a natural substance, we begin calling it a 'material'. Identifying usage is a kind of innovation. 'New organization' as designated by Schumpeter has been interpreted to include new businesses and new economic systems. This further suggests that a wider influence of innovations is to induce institutional and societal change through construction of infrastructures and networks. This means that the pervasive effect of infrastructure and especially network formation provides a synergistic influence on the economy.

Thus, the innovations leading to new infrastructures and networks in economic development are termed here 'trunk innovations'. A trunk innovation is one that has a pervasive impact on the economy, leading to the formation of infrastructures and networks such as those for energy, motive forces, resources, transportation and communications beyond industries created by themselves. These types of infrastructures and networks enhance economic development, increasing value added to the economy through synergy.

General Purpose Technologies and Trunk Innovations

The existence of clusters of innovations has been recognized, and the interaction of such clusters with the economy has been discussed. Keirstead (1948) introduced the concept of 'constellations of innovations'. Other expressions include 'natural trajectories' by Nelson and Winter (1977), 'technological trajectories and paradigms' by Dosi (1982), 'new technology systems' by Freeman et al. (1982), and 'T-cluster' by Rosenberg and Frischtak (1984). Generic technologies and general purpose technologies have been discussed in these contexts, but these discussions have been limited to the technological innovations and have not included concepts of innovations wider than technologies. That is, the concept of 'trunk innovation' covers a much wider scope and focuses on the value added to the economy.

Bresnahan and Trajtenberg (1995) defined 'general purpose technologies' (GPT) as those technologies having a pervasive influence on the economy. Such technologies include steam engines in the Industrial Revolution, electric power in the third Kondratiev cycle and electronic information technologies nowadays. Not only did these technologies revolutionize their particular industries, they also brought a pervasive benefit to the totality of economic activities. GPTs may not always diffuse quickly, but their applicability sooner or later expands their diverse areas. Information technologies, for example, were expected to greatly impact the economy, but so far their effect has been less than expected. The impact of information technologies has been discussed by, for example, Bresnahan and Trajtenberg (1995), Helpman and Trajtenberg (1994) and David (1991).

Rosenberg (2000) discussed the importance of academic activities in regards to GPTs and pointed out the endogeneity of technology. Though Schumpeter, in his earlier writing, had seen invention as exogenous, he argued in a later work, *Capitalism, Socialism and Democracy* (1942), that modern corporation endogenized inventive activity. Rosenberg (2000) addressed 'engineering' in universities, which played an important role for industry, as an example of 'endogeneity of technology' in universities. Chemical engineering in particular plays an important role in promoting industrialization when a new chemical product or process is invented. Chemical engineering is certainly a general purpose technology contributing to industry. On the other hand, a trunk innovation is not always technology but can be a much wider concept, focusing on the value added to the economy.

A Review of Kondratiev Cycles with Trunk Innovations

The history of modern industrialization is that of the establishment of economic infrastructures. Trunk innovations, as defined above, related to energy, resources, transportation, and information and communications have decisively contributed to the construction of the economic infrastructure over the past 230 years. In the previous chapter, we discussed how Kondratiev business cycles were constructed by various innovations. Many were trunk innovations in the areas of energy, transportation and communications. They include canals, coal and oil. While these are not technological innovations, they made crucial contributions to economic development. Thus, if modern industrialization is discussed in terms of trunk innovations, the contributions of innovation are more readily understood. Trunk innovations enhance economic development synergistically and they pervasively influence the economy through networks. Figure 5.1 shows the correlation between these trunk innovations and business cycles.

Although the main innovations in the Industrial Revolution were spinning machines and steam engines, canals played an important role. Canals were constructed in England during the 1760s and 1820s and helped launch the Industrial Revolution by accelerating transportation, replacing horse-drawn transport. Canals brought about an epoch-making efficiency in the transportation that played a large role in the revolution.

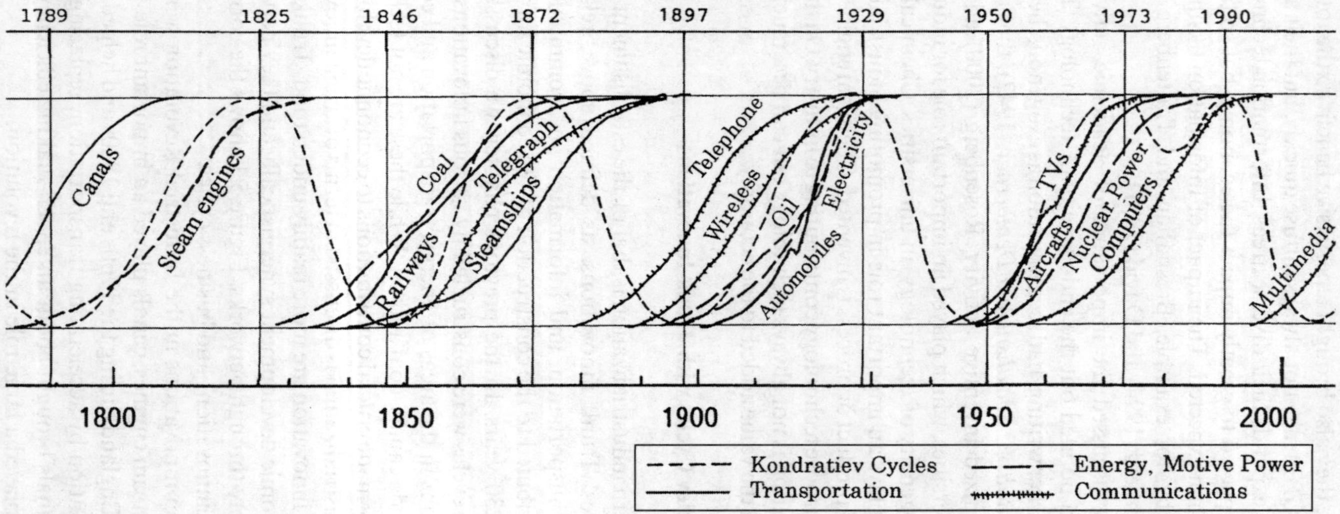

Source: Hirooka (2002a, 2003c).

Figure 5.1 Diffusion of trunk innovations and business cycles

The canal built between Liverpool and Manchester in 1761, served as the main route for carrying Indian cotton from its arrival port to its place of manufacture. In the 1770s, the Grand Trunk Canal between Birmingham and the River Severn, the Grand Junction Canal between London and Birmingham and many others were constructed. Waterways throughout the UK were connected by 1789 (Tylecote, 1992). In Europe and the United States as well, many waterways were constructed during the same period, establishing transportation networks and distribution channels.

In the first and second Kondratiev cycles, energy and motive forces made decisive contributions. Since cotton factories had traditionally used water-wheel power, factories had had to be built near a source of water. The introduction of steam engines in 1789 stimulated the development of the cotton industry because factories were no longer limited to areas with available water. In the iron-making industry, a blast furnace using coke instead of charcoal was invented by Darby in 1709, and the invention of steam-driven bellows made their works more efficient. The Industrial Revolution developed smoothly with the introduction of such energy and motive forces. These interactions, related to energy, motive forces and transportation infrastructures, played an important role in the synergistic effect of innovations.

In the second Kondratiev cycle, the stems of innovation were railways and iron production, and the upswing came between 1846 and 1872. During this period, the consumption of coal skyrocketed due to demands from railways, steam ships and iron-making. Transportation costs decreased to one-tenth of their former price, accelerating the expansion of transportation-related businesses. Telegraph lines were strung along railways, forming a network that provided public communication.

In the third Kondratiev cycle, the United States became the world leader, and trunk innovations built various infrastructures. Oil and electric power played crucial roles in promoting industrial development. Automobiles diffused rapidly into the market with the introduction of the assembly line, leading to the establishment of transportation networks. In the area of communication, telephony and wireless telegraphy became popular, and the diffusion of radio opened up the age of communication.

We are now in the fourth Kondratiev cycle and the world has entered the age of oil; the importance of oil has increased remarkably and it is now a major resource. Transportation has entered a complex age, with widespread use of automobiles, railways, ships and aircraft. The most important development during this cycle is in the areas of information and communication, that is, the establishment of an information technology age based on computers and integrated circuits. In the nascent fifth Kondratiev cycle, the information technology age will continue and further establish a multimedia age through digital systems.

In short, trunk innovations have played crucial roles in each Kondratiev cycle, binding together into clusters. By studying trunk innovations, the construction of a Kondratiev upswing can be more clearly identified.

5.2 THE ROLE OF TRUNK INNOVATIONS IN THE EVOLUTION OF RETAIL BUSINESSES

Another important role of trunk innovations is that they induce subsequent innovations. Pervasive innovations play a big role in inducing various changes in existing industries. Rosenberg (1994) describes how a major innovation provides a framework for a large number of subsequent innovations, each of which is dependent upon the original one. Trunk innovations, as defined above, trigger the evolution of other businesses, even ones not related to technological innovation. This is a kind of techno-economic paradigm change that can lead to an institutional changes in society.

Here we discuss several innovations that induced revolutions in the commerce, retail and distribution businesses. Distribution and retail businesses share a large portion of economic structures and support consumers in many ways. This major component of the economy evolved due to various trunk innovations.

The history of retail businesses has been covered in many published papers and books and is summarized in, for example, Zimmerman (1955), Converse (1959), Lebhar (1963), Inoue (1986), Tedlow (1990), Tsunoda (1992) and Isihara et al. (1995). These documents record a long history showing that several retail business paradigms intermittently emerged based on trunk innovations. B.C. MacCannon (see Tsunoda, 1992), observed that the development history of retail businesses has several paradigms, in which the development time becomes shorter and shorter. The paradigms begin with department stores in which the time span was 100 years, followed by chain stores with a span of 50 years between 1900 and 1950, supermarkets, with 30 years from 1935 to 1965, and discount stores, with 20 years from 1955 to 1975. The history certainly shows that evolution has shortened stepwise.

The diffusion of innovation products can be easily described using a logistic curve, and the development of retail businesses can be described in the same manner. In the case of retail businesses, the development is more effectively expressed by the number of shops instead of by the amount of sales because the number of shops is not affected by price fluctuation. If the ultimate saturated level is normalized to unity, every case can be compared in the same scale. It is difficult to use the measure of the number of stores in the case of department stores, so the development of department stores was described using a single logistic curve, which was determined by the time span of distribution of their establishment.

Evolution of Retail Businesses and Logistic Innovations

Department stores along railways

In the long history of mankind, the first breakthrough in commerce came with the introduction of sailing ships. After the Renaissance, 500 years ago, innovations in shipbuilding and navigation, for example, the invention of the compass, made it possible to cross oceans. The age of Great Voyages began in the fifteenth century, and in 1522 Magellan succeeded in circumnavigating the world. This was a turning point; world trade developed with, for example, cotton and spices being imported from India. Sailing ships can thus be considered the first innovation in commerce.

The economy developed along the upswing of the second Kondratiev cycle, driven by the invention of locomotives and the extension of railways. In England, the railway was first built in 1825, and 20000 miles of track had been laid by the second half of the nineteenth century. Alongside the railways, telegraph network lines were strung.

The first evolution in retail business was the development of department stores, starting with Bon Marché in Paris in 1852 and followed by Au Printemps, Louvre, Samaritaine and Lafayette by 1890. Harrods and Marks & Spencer in London and many department stores in Germany and Italy were built in the same period. In the United States, Macy's opened in New York in 1858, Wanamaker in Philadelphia in 1862, and Marshall Field in Chicago in 1868. The timing of these openings was along the upswing of the second Kondratiev cycle, immediately after the diffusion of railways in the UK and the US. It also coincided with the diffusion of steamships. This correlation clearly indicates that the revolution in transportation made it easier to set up department stores because distribution costs fell to one-tenth of their previous value, making it possible to assemble a wide variety of goods from far-flung places. The construction of a telegraph network, following the railway lines, promoted the development of the mail-order business, starting with Montgomery Ward in 1872 and Sears Roebuck in 1886. These development processes are illustrated in Figure 5.2

Chain stores and supermarkets in the age of motorization

In the early twentieth century, the United States moved from being a developing country to an advanced country. This happened during the upswing of the third Kondratiev cycle, driven by a cluster of innovations including oil, electric power, automobiles and durable consumer goods.

The revolution in energy and transportation in particular drastically changed the economy in the US.

Chain stores first appeared at the beginning of the upswing of the third Kondratiev cycle between 1900 and 1929. A typical chain store was the Great Atlantic and Pacific Tea Company (A&P Co.), which was founded in 1825. At the beginning of 1900, A&P adopted a new business style called 'economy store' through which it achieved economies of scale and rationalization by integration. This kind of business was born out of the changed social background strongly affected by the emergence of various infrastructures including automobiles, electric power and oil. The development of full-scale road systems changed distribution systems. Urbanization was also promoted. Infrastructure development enhanced the feasibility of chain stores. The actual development of chain stores closely followed the diffusion trajectory of automobiles. Chain stores, however, faced difficulties in the Great Depression.

In 1930, just when the crash occurred, a unique shop called 'King Kullen Grocery' opened. This was the first supermarket, a big self-service shop with a large parking area, located in the suburbs. This retailing model contrasted sharply with that of chain stores. Supermarkets became popular and developed rapidly all over the United States along with the extension of highways and the spread of refrigerators. The development of supermarkets expanded from 1200 shops in 1936 to 8175 in 1941. The expansion continued after the Second World War and the 20-year period between 1945 and 1965 was the peak period for openings. Supermarket sales reached 48 per cent of total food sales in the US and the variety of goods sold expanded to include clothes and sundries. During this period, the amount of freight traffic increased by 2.5 times and the age of full-scale motorization began. These trends are depicted in Figure 5.3.

Discount stores during the age of mass production

After the Second World War, the fourth Kondratiev cycle began, with the upswing occurring between 1950 and 1973. The most prosperous age started early in the 1960s as various innovation products flooded the markets. Electric appliances and a variety of synthetic plastics, textiles and other durable consumer goods greatly expanded the markets. This was the age of mass production. 'Discount stores' sold various discount goods in large lots; the first shops, such as Kmart and Wal-Mart, appeared in 1962. Kmart successfully expanded its business to more than 2000 stores in the latter half of the 1970s. A key factor in the success of discount stores lay in the low warehouse costs achieved by introducing forklifts and pallets, which reduced labor costs by 25 per cent. Innovations in packaging through the use of synthetic plastics was another factor. The development of discount stores is illustrated in Figure 5.4.

First convenience stores

Even before discount stores appeared, Southland Company started a unique shop, a 'convenience store', in 1960. A convenience store is defined as one which has a small sales space (less than 300 m^2) sells various goods and is open for a long time each day. Southland Co. began in 1927 as a small shop selling ice for refrigerators. It had long opening hours and began to sell a variety of goods. After the Second World War, the name was changed to '7-Eleven' and the number of shops was expanded by introducing a franchise system, which led to the opening of more than 7000 shops by the early 1980s. The trend in business expansion is shown in Figure 5.4 These stores belong in the same category as discount stores, although they were later distinguished by their introduction of information technology, especially POS systems.

Dawn of retail business in Japan

The Kondratiev wave is now synchronized worldwide, and Japan enjoyed high economic growth during the fourth cycle in the 1960s. This upswing in Japan corresponded to its jump from a developing country to an advanced country. It was the time when various infrastructures were constructed simultaneously, and steel, automobiles, heavy chemical industries and electric appliances were rapidly developed. This was the same situation as in the United States in the early part of the twentieth century. Figure 5.4 plots the growth curve of the automobile industry in Japan and the development of retail businesses, exemplified by Daiei Company, the leading supermarket in Japan. The development of supermarkets fits well with the S-curve of automobile trajectory in Japan. This suggests that transportation facilities could play an important role as the background infrastructure. The first Daiei supermarket was opened in Osaka in 1960, coinciding with the take off of Wal-Mart and 7-Eleven of Southland Co. in the United States on the third upswing of the Kondratiev cycle. The establishment of Daiei was well timed, as the consumer society was just taking off and distribution facilities were being networked.

Evolution of Retail Businesses through Information Technologies

The evolution of retail business described above was brought about mainly by innovations in transportation. The more recent development of retail businesses has come about due to advances in computer and information technologies. The diffusion of computers progressed for 25 years, from the 1960s to the early 1990s. Various industries have increased their value-added

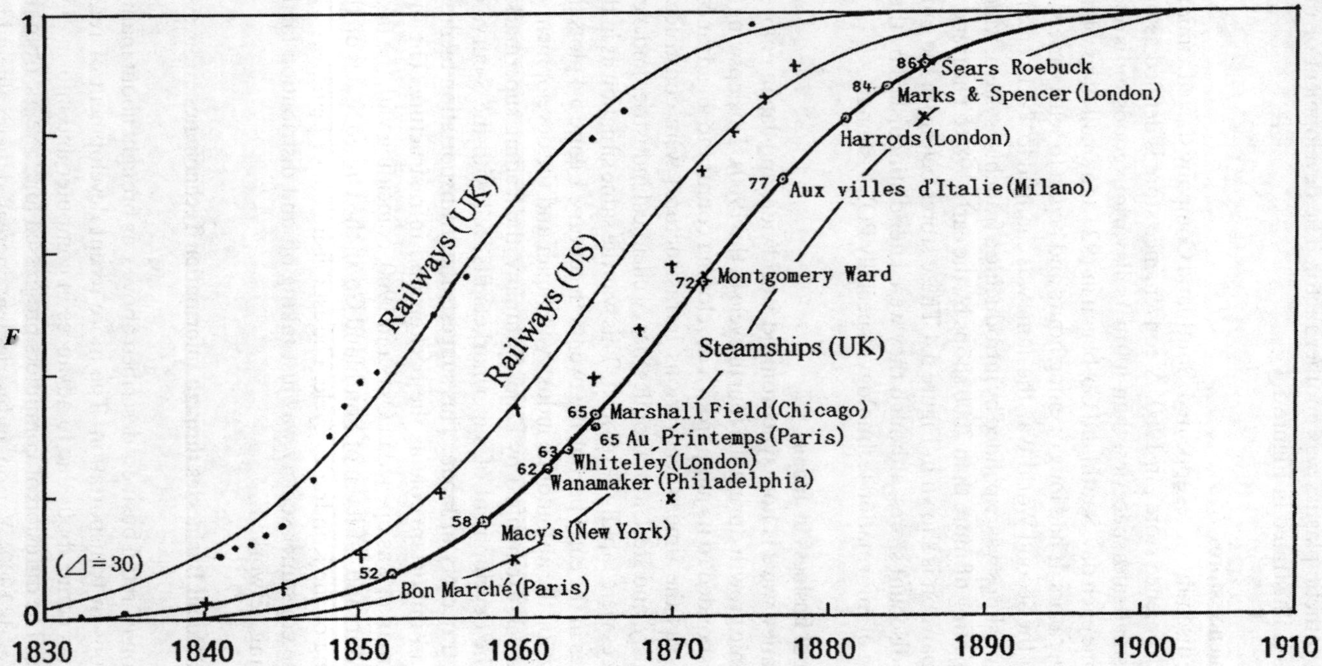

Source: Hirooka (1999a).

Figure 5.2 Development of department stores

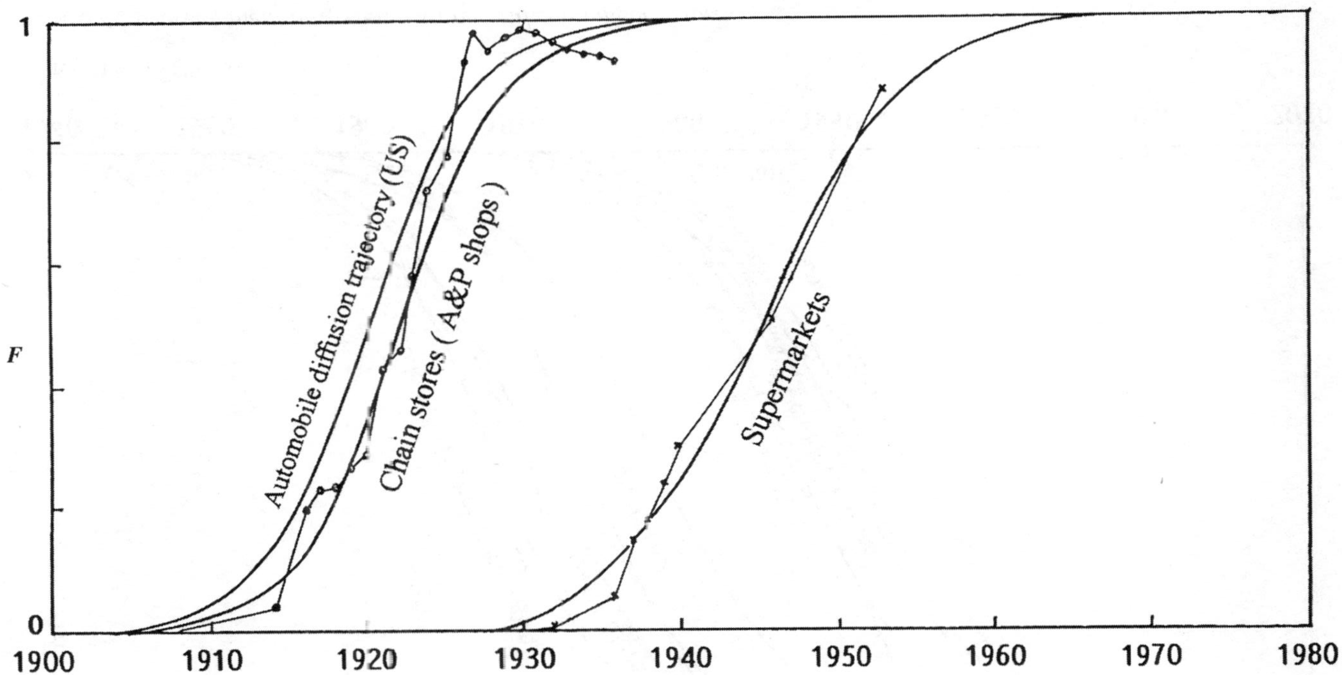

Source: Hirooka (1999a).

Figure 5.3 Development of chain stores and supermarkets

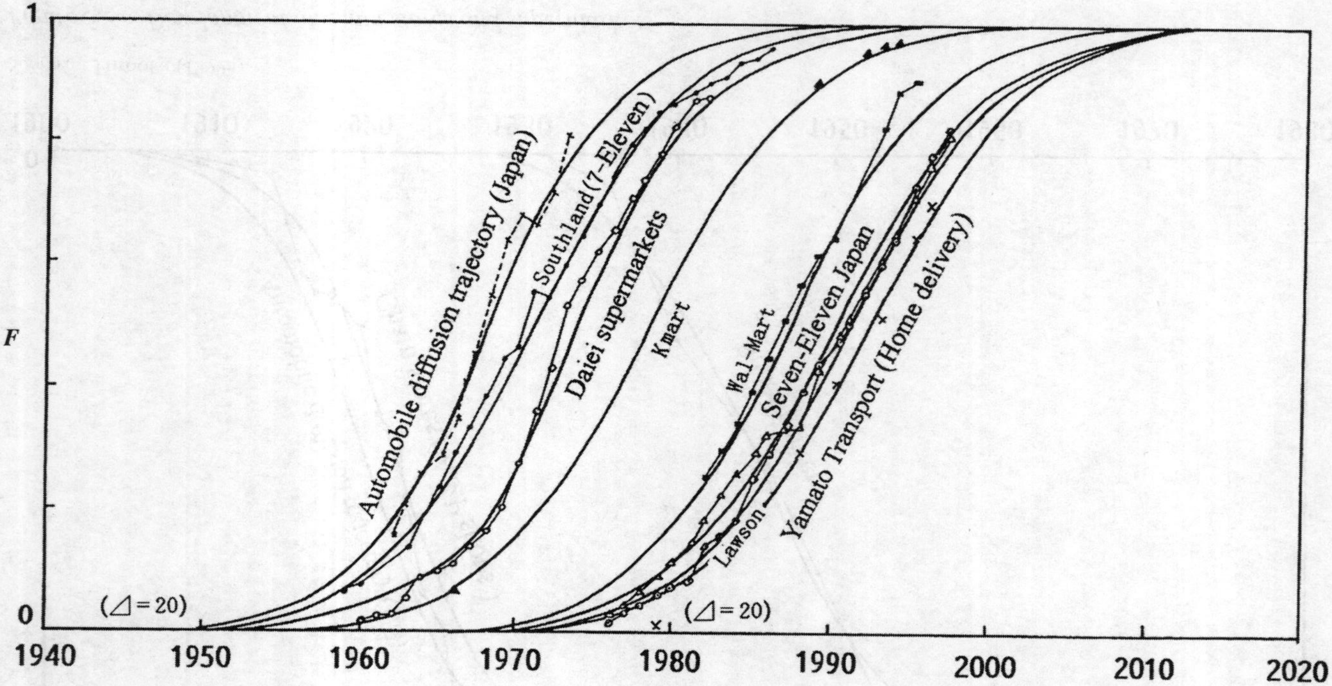

Source: Hirooka (1999a).

Figure 5.4 Development of discount stores and convenience stores

and evolved by making use of information technologies. Retail businesses introduced information technologies about five to ten years behind the diffusion of computers. The YamatoTransport Company in Japan was one of the earliest firms to introduce computers and thereby gained a strong competitive position. Wal-Mart recovered from a setback by introducing computers and information technologies.

Convenience stores with smaller sales floors compared with retail businesses became more competitive by the introduction of POS (point of sales) systems using barcodes, which enabled better-selling goods to be sorted out by single-article management and made logistics more efficient. Other types of retail businesses have also improved their efficiency by introducing POS, EDI (Electronic Data Interchange) and other advanced information technologies. The earlier introduction of such information technologies resulted in better competitiveness, and there have been various success stories.

Counterattack by Wal-Mart

While both Wal-Mart and Kmart began operations in 1962, ten years later, Wal-Mart had only 200 stores, far behind Kmart's more than 2000 stores. In an effort to fight back, Wal-Mart began to introduce computers for inventory control in 1974, and decided to introduce POS systems in 1979. It started experiments with EDI in 1981, expanded its POS systems in 1983, and introduced satellite network systems in 1985. POS systems had been installed in all stores by 1988, and automatic delivery systems were in place by 1991. As a consequence, Wal-Mart exceeded Kmart in sales in 1990 and has developed smoothly since. This is a typical example of acquiring competitiveness by introducing information technologies. The introduction of information technologies by Wal-Mart and Kmart are compared against the diffusion trajectory of computers in Figure 5.5.

Convenience stores in Japan

Seven-Eleven Japan was set up in 1973, introducing the know-how of the American 7-Eleven, a division of Southland Co. Seven-Eleven Japan introduced a computer ordering system and constructed diversified network systems for cooperative distribution in 1978. As a secondary stage facility, POS and EOS (Electronic Ordering) systems were introduced in 1982. Single-article management and logistics systems were introduced using interactive POS and information management computers in 1985. Strategic logistic operations were controlled by introducing an ISDN network in 1991. These advanced operating systems using information technologies provided Seven-Eleven Japan with a strong competitive edge.

Lawson Company is another convenience store, comparable in size to Seven-Eleven Japan. They introduced POS systems in 1988 and completed their first information system. In 1991, they adopted the second comprehensive information system equipped with store computers and POS registers. In 1997, Lawson began installing multimedia terminals, called 'Loppi', in every shop for customers to purchase goods and event tickets, and to access services such as paying public utility bills via machine. These were the first net service facilities to use multimedia.

The trajectories of Seven-Eleven Japan and Lawson are shown in Figure 5.6 in which the trajectories of both companies are prepared according to items of information technology introduced by them in comparison with the diffusion trajectory of computers.

Home delivery service of Yamato Transport Co.

In Japan, parcel delivery had been handled exclusively by the government post offices until Yamato Transport Company launched a private home delivery system in 1968. They equipped computers for the delivery systems and constructed a network of the systems all over the country. This was seven years before the introduction of computers by Wal-Mart and Seven-Eleven Japan. It is interesting to note that the introduction of computers by Yamato Transport developed along the diffusion trajectory of computers themselves as shown in Figure 5.7. The diffusion trajectory of Yamato Transport represents the number of deliveries. They developed Yamato System as the basic software for delivery services and created the first NEKO system for tracking parcels. The second NEKO system for POS barcode control system was adopted in 1980. A wireless system was introduced for collecting parcels in 1984. The third NEKO system was for POS using PDA (personal digital assistants: mobile terminals) in 1986. The fourth NEKO system was introduced in 1993 for customer management system using IC cards. Thus, Yamato Transport Co. has maintained a market share of more than 40 per cent. The strong competitiveness of Yamato Transport is ascribed to early adoption of computer systems.

The Evolution of Retail Businesses and Trunk Innovations

The above description is an analysis of retail businesses in comparison with the emergence of trunk innovations. The evolution of retail businesses takes place intermittently, with the emergence of trunk innovations as the trigger. This interrelation between the emergence of new retail businesses and trunk innovations is clear in all cases. This is a kind of techno-economic paradigm change from old retail businesses to new retail businesses, through a change in the circumstances of business resulting from trunk innovations, leading to an institutional change in society.

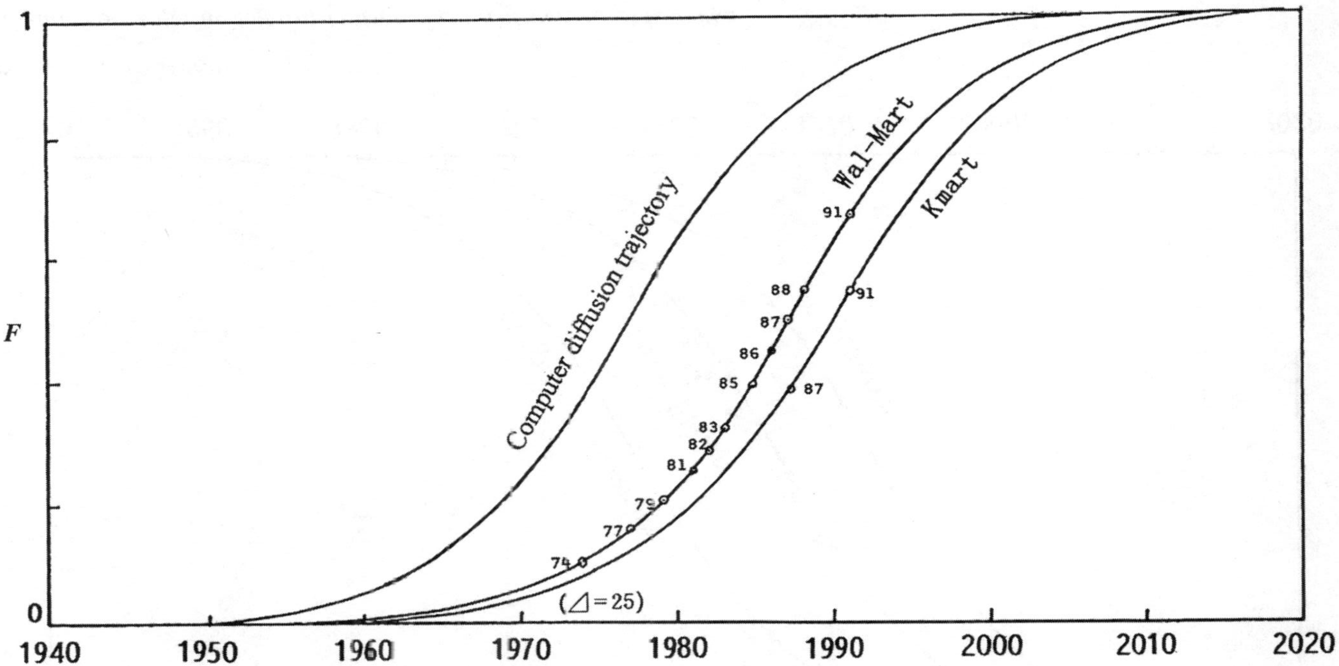

Source: Hirooka (1999a).

Figure 5.5 *Information trajectories of discount stores*

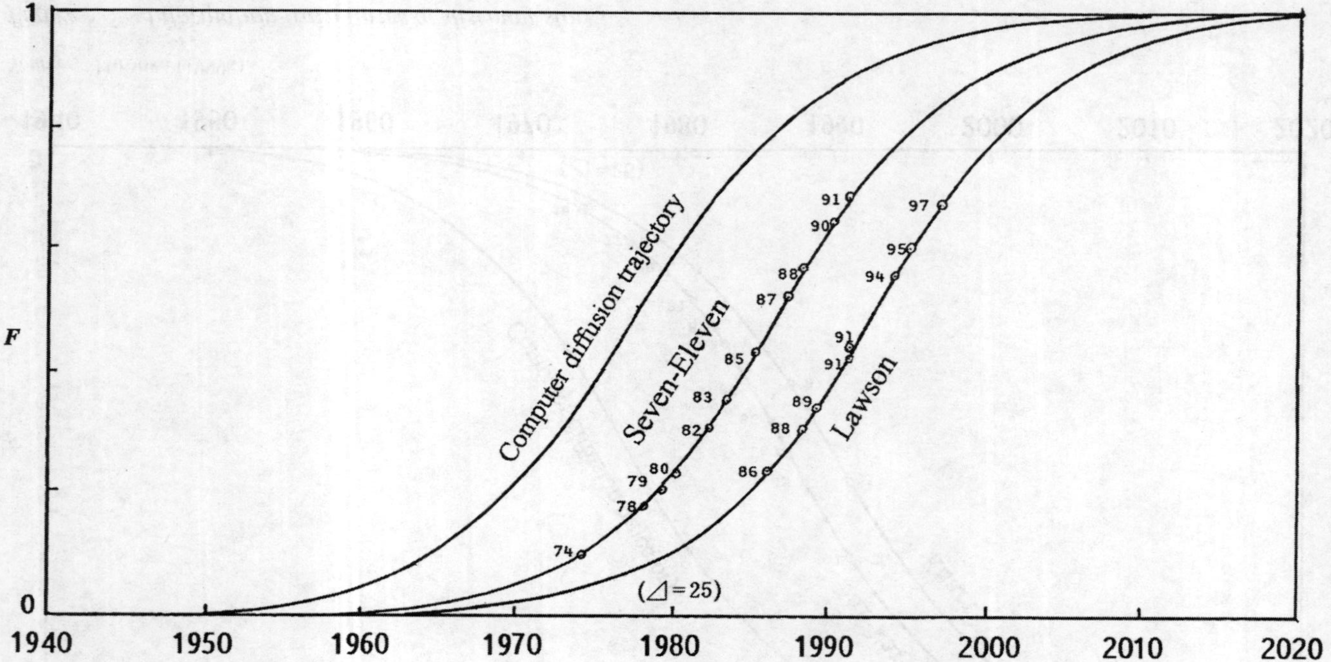

Source: Hirooka (1999a).

Figure 5.6 Information trajectories of convenience stores in Japan

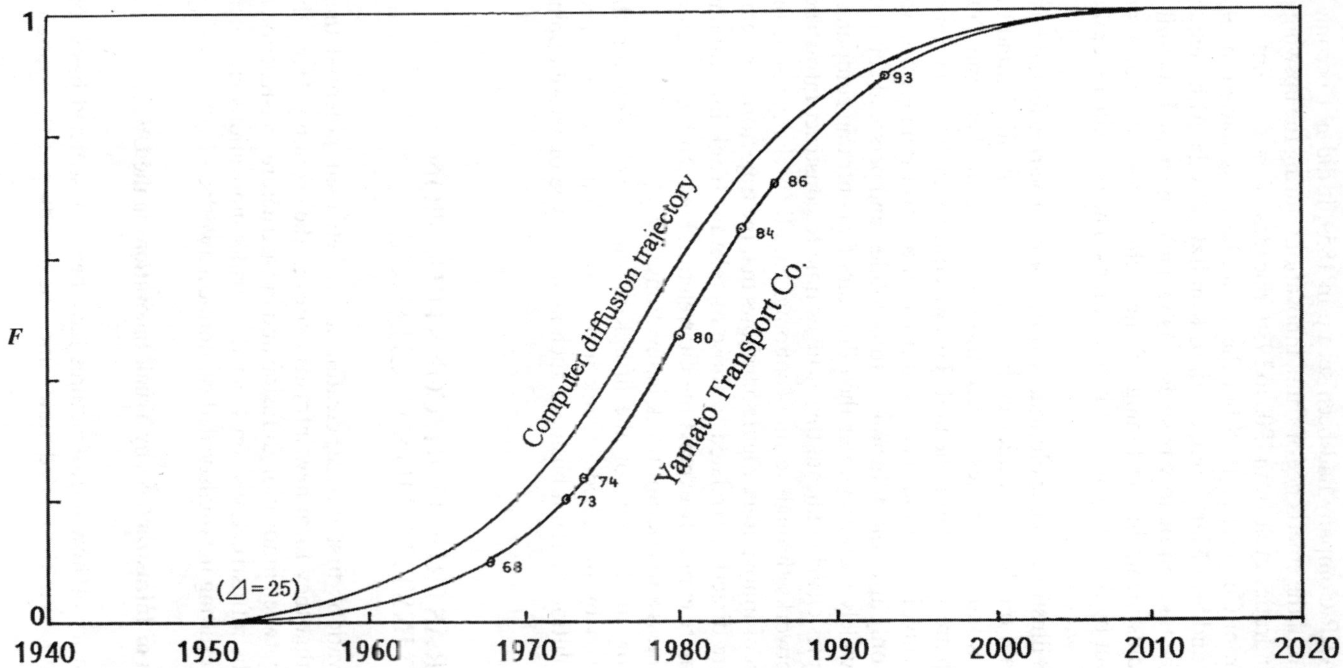

Source: Hirooka (1999a).

Figure 5.7 Information trajectory of home delivery services in Japan

Though A&P Company had been set up in 1859, it did not become a chain store until the emergence of trunk innovations during the upswing of the third Kondratiev cycle from 1900 to 1930. The idea of the discount store originated from Corvette in 1948 but the actual business started in 1961 with Kmart and Wal-Mart along the third Kondratiev cycle. Although the trigger for the take-off of new businesses is variable, the actual take-off is based on adjustments in the timing of the techno-economic paradigm brought about by the emerging trunk innovations. The best kite never flies without a blowing wind.

Figure 5.8 illustrates the evolutionary trajectories of various retail businesses overlaid on the Kondratiev cycles. They are all slightly behind the upswing of Kondratiev cycles, except those for supermarkets and POS stores, which are significantly behind. This was due to the late occurrence of relevant trunk innovations because a truck transportation network was constructed only after the diffusion of automobiles, and personal computers became widely used only after the diffusion of mainframe computers. These examples clearly indicate that it takes time to construct infrastructures and networks through techno-economic paradigm change after the direct impact of innovations. Figure 5.9 shows that the trajectories of retail businesses are directly correlated to those of relevant trunk innovations. This figure more explicitly express the interrelation between the evolution of retail business and relevant trunk innovations.

We are now at the start of the fifth Kondratiev cycle, entering the upswing driven by multimedia trunk innovations. Thus, we can expect another revolution of retail business, such as some type of tele-shopping perhaps.

5.3 INFRASTRUCTURE CONSTRUCTION AND INSTITUTIONAL CHANGE

Trunk innovations first diffuse, creating a market; their potential then expands so that they form new infrastructures in the economy. Here, the correlation between innovation diffusion and infrastructure construction is analyzed. The infrastructures constructed by trunk innovations are interconnected, resulting in sophisticated economic networks.

Construction of Infrastructures by Trunk Innovations in the US

This is the story of how infrastructures have been constructed based on trunk innovations in the United States since the founding of the country.

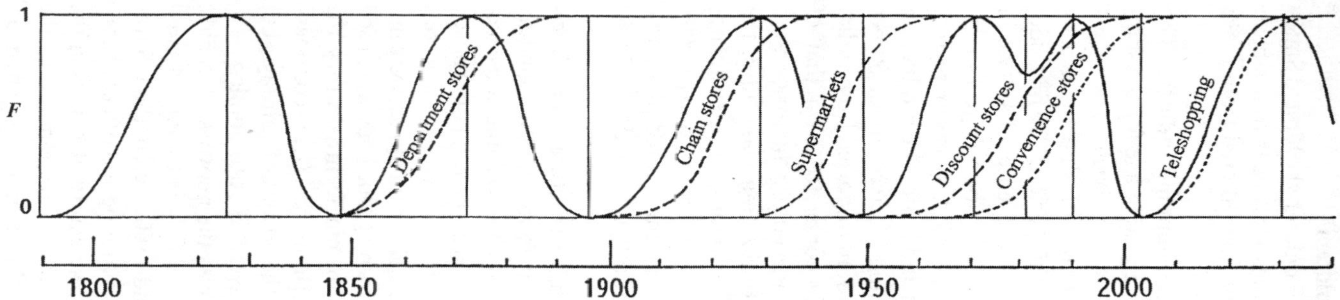

Source: Hirooka (1999a).

Figure 5.8 Evolution of retail businesses vs. Kondratiev business cycles

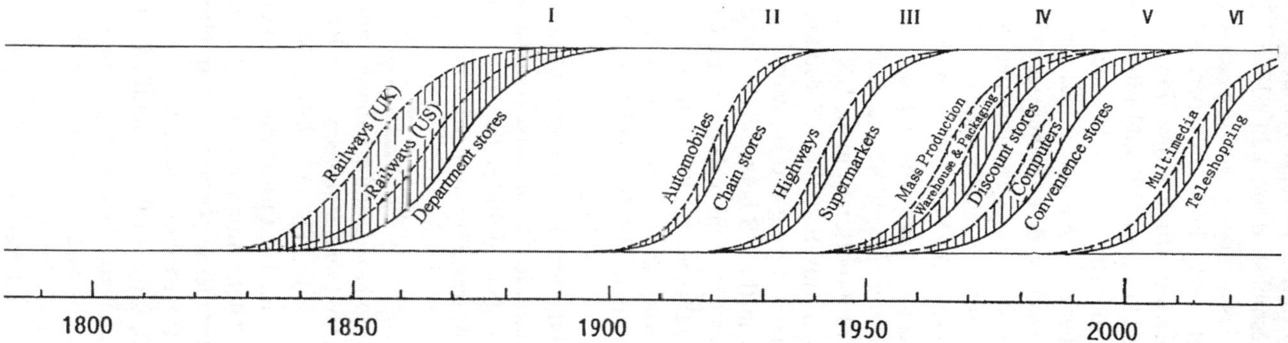

Source: Hirooka (1999a).

Figure 5.9 Evolution of retail businesses induced by trunk innovations

Infrastructures at the dawn of the Industrial Revolution

While the Industrial Revolution was unfolding in England, the United States was established, on 4 July 1776. The United States was eager to establish itself as a country, and thus took various actions that differed from those taken during the colonial period.

The first trunk innovation in the US was the shipping industry. The period between 1793 and 1807 in particular was a golden age for the shipping industry in the US, which grew up quickly on the basis of low-cost shipbuilding and low wages for sailors. The number of ships owned by American companies increased three-fold during this period. The first side-wheel steamboat was invented by Fulton in 1807 and entered service in 1811. International trade expanded, with ships sailing not only to Europe but also to the West Indies and even to India and China. Exports increased, especially of agricultural products such as wheat and flour. The production of cotton rapidly increased, and by 1807 England received about 60 per cent of its cotton from the US, exceeding the amount imported from India. This means that the US economy was tightly connected with that of England during the Industrial Revolution. While cotton factories were built in the US between 1813 and 1840, industrialization was still in the early stages. Full-scale development of the cotton spinning industry did not come about in the US until the second Kondratiev cycle, from 1840 to 1880, following the introduction of the steam engine as the source of motive power in the 1840s. The US was almost one Kondratiev cycle, that is 50–60 years, behind England.

In the first Kondratiev cycle, two important infrastructures were developed in the US: turnpikes and canals. Immediately following the founding of the United States, the most important political issue was the construction of transportation networks for both commercial and military purposes. The period from 1790 to 1830 was the most active period of turnpike construction, and the total reached 400 miles by 1822. These turnpikes were initially toll roads and gradually changed to freeways by 1830. They were used by postal service stagecoaches. By the middle of the nineteenth century, the mail road reached California, becoming the overland mail (transcontinental) road. The turnpikes fell into disrepair with the advent of railways, but were later restored as common roads with the advent of automobiles.

Another important infrastructure in the US was the network of canals, which were constructed at a cost of $31 million, mainly between 1815 and 1843, and a further $66 million between 1844 and 1860. Again, the US was about 50 to 60 years behind England, where the major period of canal construction had been from 1760 to 1820. The infrastructure of canals in the US, built fairly early in the first Kondratiev cycle, corresponding to the Industrial Revolution in England, supported the later development of various technological revolutions. In the US, the construction of the Erie Canal, 363 miles long and connecting the Hudson River with Lake Erie, was completed in 1825. Following its completion, the cost of moving freight dropped to one-fourth of that of horse-drawn transportation. It stimulated development of the western agricultural area and enhanced the activity of New York City as the hub for east and west overland transport and international trade.

Further development of infrastructures

The construction of railways in the US began in 1828, only three years after the first train ran from Stockton to Darlington in England in 1825. Construction activities were well under way during the 1830s and accelerated in the 1850s, with the length of track reaching 30 000 miles. The developing railway network competed with the canals. Following the end of the Civil War in 1865, the construction of railways progressed and the transcontinental railroad arrived at the west coast in 1869. The peak in railway construction was between 1868 and 1892 – a total of 180 000 miles of track had been laid by 1895. The iron industry was fed by the construction of railways. Thus, the iron industry and railways were the main sectors during the upswing of the second Kondratiev cycle. Iron production increased by 25 times, and the amount of freight carried by railways increased almost 100-fold from 1860 to 1910, even though the total length of railway was only ten times greater than at the start of the period.

The oil industry also began on the upswing of the second Kondratiev cycle. The first oil well was drilled in Pennsylvania in 1859, and Rockefeller established the first oil company, Standard Oil, in 1867. The greatest development of the oil industry, however, occurred during the upswing of the third Kondratiev cycle.

These trunk innovations constructed fairly firm infrastructures that played major roles in the industrial revolutions. The historical descriptions above are mainly drawn from Akimoto (1999); some are from Okuno (1991).

Construction of highways in the age of motorization

As described above, turnpike construction happened comparatively early in the United States, between 1790 and 1830. The main roads were developed for stagecoach travel and overland mail. During this period, the stagecoach played an important role, forming a transportation network and connecting production areas to market areas. Turnpikes were then converted into common roads, and more roads were constructed, mostly for farming villages. The total length of such roads was 1.6 million miles in 1900. Along with the extension of railways west across the Mississippi River, many villages were built, and roadways were constructed, reaching 2.9 million miles in 1920 and 3.32 million in 1945. This is remarkable considering that the total reached only 3.9 million miles by 1985.

The roads were initially unpaved, making them unsuitable for automobiles. The Rural Post Roads Act of 1912 funded construction and maintenance of roads for mail service. Technology for paving road surfaces with coal-tar pitch and gravel had been developed in the 1900s, and its application remarkably improved travel by automobile. The Federal Aid Road Act (FARA) was passed by Congress in 1916. The first highway, the Bronx River Parkway, was constructed in the northern part of New York City in 1917. During the First World War, the railways became overburdened, and transportation by motorcars was strongly encouraged. The production of trucks reached 20 000 vehicles per month. This promoted a policy of road development, and the Federal Highway Act (FHA) was passed in 1921. This act was significant because it put the Federal government in charge of road administration with responsibility for establishing a high-level national road network. Its passage fostered development of technologies for road design and construction that would improve safety. The Pennsylvania Turnpike, 160 miles long and finished in 1940, incorporated highly sophisticated designs and technologies; it shortened the drive from Philadelphia to Pittsburgh by truck by five or six hours.

To combat the Great Depression and the effects of unemployment, which exceeded 25 per cent in 1933, President Roosevelt launched the New Deal in 1933. After a further recession in 1937–38, the government decided to revitalize the economy by using public investment, in line with Keynesian financial policy. The construction of a national interregional highway network was made the first priority. The trigger for this was the appointment of the US National Interregional Highway Committee by President Roosevelt in 1941. This committee proposed that the Federal government support the construction of an efficient highway network, leading to the passage of the Federal Aid Highway Act of 1944. The FAHA was renewed in 1956 and again in 1970, 1976 and 1981. After World War II, the highways were further improved. In 1945, 52 per cent of all highway roads were well-paved; by 1985 the proportion was 83 per cent.

This is a brief history of the institutionalization of the automobile brought about by the development of a motorization infrastructure in the United States. Along with the development of a transportation infrastructure, there were various ripple effects. The diffusion of automobiles in the twentieth century accelerated the need for road maintenance, and various acts were passed. An important development was the formation of a mail service network and the Rural Post Roads Act was passed in 1912.

Another impact of the development of transportation infrastructure was urbanization. A prerequisite of urbanization is the establishment of supply lines for food and daily necessities. Urbanization in the US progressed between 1860 and 1930. The number of cities (towns with a population of more than 2500) increased eightfold, from 392 in 1860 to 3165 in 1930; the annual growth rate averaged about 30 per cent. Urbanization led to changes in retail systems as well. The revolution in retail business was described in Section 5.2 above. The rise of department stores, chain stores and supermarkets is well interpreted on the basis of the construction of infrastructures and subsequent institutional change (see Okuno, 1991; Akimoto, 1999).

Long Trajectories of Infrastructures: Infra-trajectories

Grübler's comprehensive study on infrastructures, *The Rise and Fall of Infrastructures* (1999), analyzed the construction of various infrastructures since the Industrial Revolution by determining the time span for the diffusion of innovations using logistic equations. These data indicate that the diffusions of various trunk innovations had rather long trajectories of more than 50 years. Perez (2002) also revealed long surges which extend across two Kondratiev cycles. The present analysis of automobiles and steel productions, given in Table 2.3 and Figure 3.9, also indicates that demand for products resulting from trunk innovations continues to grow beyond one Kondratiev cycle. To clarify this kind of long development of innovations in more detail, various trunk innovations are further examined.

The progress of the various main infrastructures in the modern industrialized economy of the United States are summarized in Table 5.1 and Figure 5.10. The notation in the table is defined in Figure 2.1. The situation in the United Kingdom can be examined by using the data in Figure 5.3 of Freeman and Louçã (2001); we estimated the trajectories for turnpikes (t_1 = 1750, t_9 = 1820) and canals (t_1 = 1755, t_9 = 1830). The data actually locate as long trajectories between 1750 and 1830, which corresponds to the period from the bottom of an assumed zero-order Kondratiev cycle towards the top of the first Kondratiev cycle. The zero-order Kondratiev is assumed to have an upswing from the bottom in 1735 to the peak in 1765.

It is interesting to see that all of these long trajectories developed from the root of the upswing of one Kondratiev cycle to the peak of the subsequent cycle as shown in Figure 5.10. We call this longer trajectory an 'infra-trajectory'. Also interesting is that several infra-trajectories form a cluster, starting from the root of the upswing of the initial Kondratiev cycle to the peak of the subsequent cycle. In each Kondratiev cycle, infra-trajectories form a well-defined cluster. The future trend towards the fifth cycle is estimated by an extrapolation from the past trend of long surges.

The first cluster of infra-trajectories in the United states is composed of turnpikes, shipping and canals, which all start at the root of the upswing of

Table 5.1 Infra-trajectories of trunk innovations in the United States

Trunk innovations	t_1	t_5	t_9	$\Delta\tau$ (Year)	Y_0 2)	Source
1st surge						
Turnpikes	1790	1820	1850	60		1)
Shipping	1790	1825	1860	70		1)
Canals	1815	1845	1875	60	7010 km	1), 2)
2nd surge						
Steam engines	1850	1882	1915	65		3)
Railways	1855	1885	1915	60	482 700 km	1), 2)
Steamships	1848	1885	1923	75		2)
Urbanization	1860	1895	1930	75		1), 4)
Telegraph	1875	1906	1931	56		9)
3rd surge						
Steel production	1900	1940	1980	80		5)
Automobiles	1915	1945	1975	60		6)
Oil consumption	1920	1950	1980	60		6)
Highways	1920	1955	1985	65	6.04 mil. km	2)
Supermarkets	1930	1960	1990	60		7)
4th surge						
Electric power	1950	1980	2010	60		8)
Air transport	1960	1980	2000	40	289.4 bil. km	8), 2)
Computers	1960	1985	2010	50		10), 11)
POS stores	1970	1995	2020	50		10), 11)

Notes: t_1, t_5, t_9: time at $F = 0.1$, 0.5 and 0.9; $\Delta\tau$: time span between t_1 and t_9, and Y_0: saturated market size.

Sources: 1) Akimoto (1999); 2) Grübler (1999); 3) Rosenberg and Trajtenberg (2004); 4) David (1972); 5) Kelley and Fenton (2003); 6) US Historical Statitics (1989); 7) Zimmerman (1955); 8) UN Statistics (1990); 9) US Census Bureau (1924); 10) US Census Bureau (2002); 11) Hirooka (2003b).

the first Kondratiev cycle. This is just before the launching of innovation industries like spinning and steam engines, which appeared in the upswing of the second Kondratiev cycle in the case of the United States. This indicates that the Industrial Revolution had started earlier, with the construction of infrastructures of transportation such as roads and shipping. The second cluster of infra-trajectories is a bundle of steam engines, railways, steamships, urbanization and telegraphic communication networks. These are ripple effects starting from steam engines and the construction of railways. This cluster corresponds to the 'railroadization' of Schumpeter (1939, for example, pp. 168, 325–35).

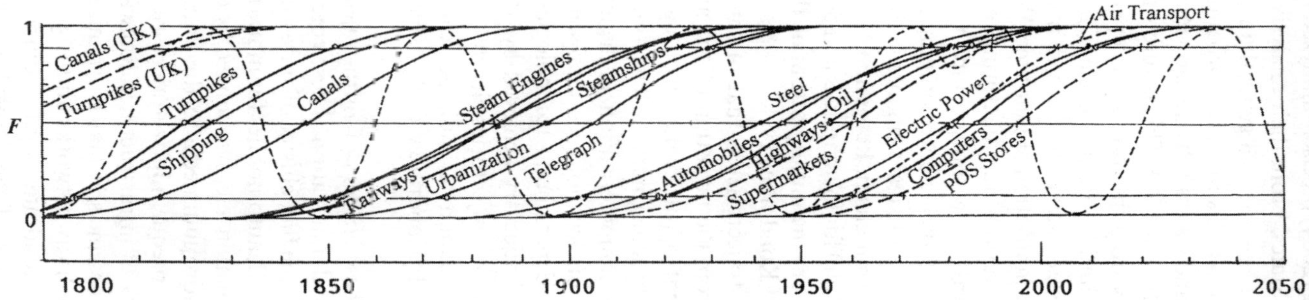

Source: Hirooka (2004a).

Figure 5.10 Infra-trajectories in the United States on Kondratiev cycles

Rosenberg and Trajtenberg (2004) precisely analyzed the development of steam engines in the United States. From their analysis, it is clear that the Corliss steam engine played a key role as a general purpose technology in stimulating the industrialization and subsequent urbanization. The Corliss engine was patented in 1847 and was rapidly developed after strengthening Corliss's patent through petition in 1869. This is surely a trunk innovation and makes an infra-trajectory from the root of the upswing of the second Kondratiev to the top of the third Kondratiev cycle. This is situated slightly earlier than those of railways and steamships which means that the steam engine itself is the core trunk innovation at that time. While steam engines were used as motive forces in various manufacturing industries, it must be more important to have provided new transportation infrastructures such as railways and steamships, and their networks induced a drastic institutional changes in society. The use of steam engines in the United Kingdom spread between 1800 (63 in Birmingham, Leeds and Manchester) and 1880 (110 000 for motive forces, 13 000 for locomotives and 5000 for steamships) to form an infra-trajectory from the initial bottom of the first Kondratiev cycle to the top of the second Kondratiev cycle.

The third cluster of infra-trajectories is that of 'motorization'. These infra-trajectories originate from the invention of the automobile and are carried along by the subsequent logistic evolution of highway networks and retail businesses represented by supermarkets. The fourth cluster of infra-trajectories is symbolized by 'computerization' in the information age: the core innovation is information technologies represented by computers, integrated circuits and related peripheral technologies. The fifth cluster could be a multimedia infra-trajectory achieved by 'digitization' via the internet. This long trend of infra-trajectories clearly indicates that further robust development of information technologies can still be expected, reaching the peak of the fifth Kondratiev cycle, which will be attained around 2035 to 2040.

An important aspect of these clusters of infra-trajectories is that each cluster has a specific core trunk innovation. These core trunk innovations are turnpikes and shipping in the first cluster of infra-trajectories, railways for railroadization in the second, automobiles for motorization in the third, and computers for the information age, symbolized by computerization in the fourth and digitization for multimedia in the fifth. These core trunk innovations induce various subsequent infra-trajectories. This ripple effect is crucial to forming a network of infrastructures. Railways led to the laying of telegraph lines along rail lines and stimulated urbanization. Automobiles led to the development of highway networks and triggered the logistics of supermarkets. In the fourth Kondratiev cycle, computers play a decisive role, and various applications have been derived, for example, POS retail stores, as illustrated in Table 5.1 and Figure 5.10. Thus, such synergistic interrelations can result in a rather well agglomerated cluster. Perez (2002) skillfully named this phase of innovation the 'synergy phase', which adequately explains the above situation.

Furthermore, other infra-trajectories, such as steamships in the first Kondratiev cycle, steel and oil in the third, and electricity and air transport in the fourth, are inclusively formulated within the respective cluster of infra-trajectories. That these infra-trajectories share the same cluster in the relevant Kondratiev cycle can be ascribed to the cluster formation of trunk innovations having the same timing in the upswing of a Kondratiev cycle.

Infra-trajectories sometimes become obsolete and are replaced by an alternative infra-trajectory. This is a kind of struggle for existence. A general scheme of struggle for existence in biological systems is described by the Lotka and Volterra equation. Grübler (1999) comprehensively summarized such substitution models, and Marchetti and Nakićenović (1979) first proposed a multiple substitution model expressed by a Fisher–Pry plot. According to this treatment, the primary energy substitution was analyzed in a sequence of wood, coal, oil, gas, nuclear and new energy. Such substitution also occurs in transportation systems, and we are now in multiple equilibrium, with railways, automobiles, air transport and shipping.

Institutional Change through Trunk Innovations and 'Exaltation' of the Economy

We have seen that every diffusion of a trunk innovation primarily reaches a mature stage on the upswing of a Kondratiev cycle, as shown in Figure 5.1. What is the gap between the initial maturation and the following enhancement in the subsequent Kondratiev cycle? The finding that trunk innovations have a longer infra-trajectory is verified by examining two cases: the changes in domestic demand for steel and automobiles in the United States and Japan, as shown in Figures 5.11 and 5.12.

The demand for steel and automobiles in the United States increased along the upswing of the third Kondratiev cycle and saturated the market just before the Great Depression, as described in Chapter 4. The demand for both products decreased during the Great Depression and recovered to the 1929 level just before the start of the Second World War. As shown above, during this downswing of the third Kondratiev cycle, various infrastructures started to be constructed and affected society through institutional change as described above. Thus the demand for automobiles was enhanced by the expansion and improvement of highways which also facilitated the logistics of industries and created new styles of business, such as supermarkets. The utilities of trunk innovations

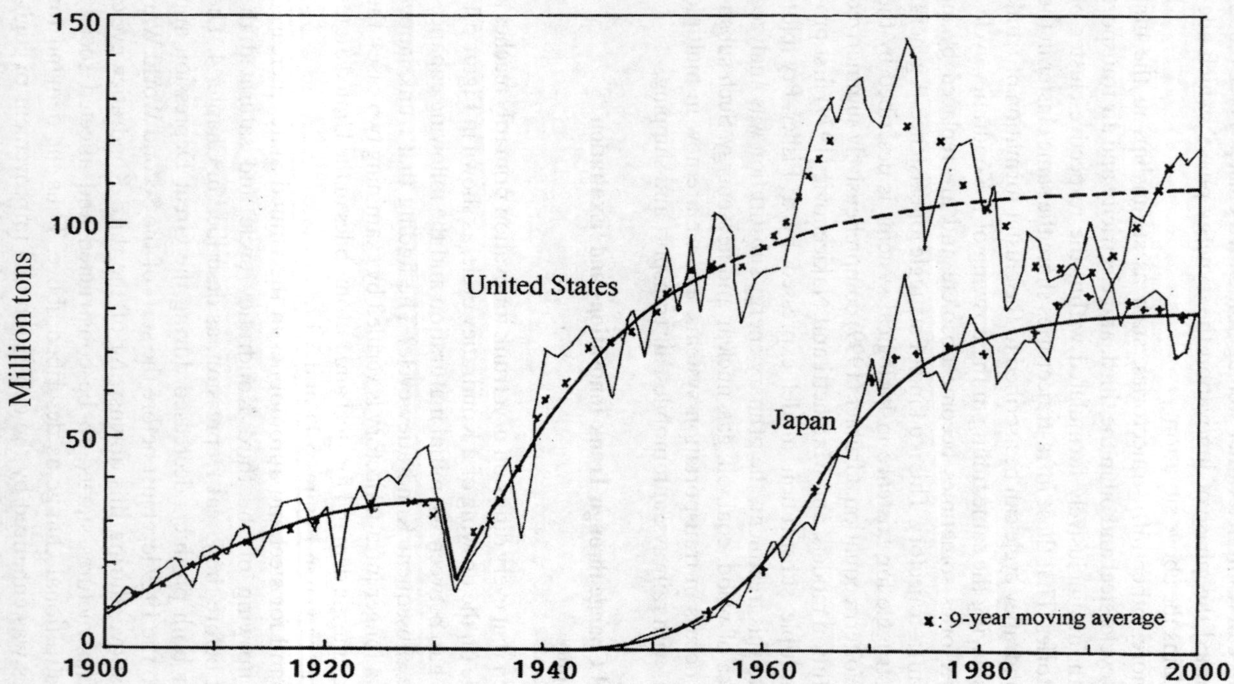

Source: Hirooka (2004a).

Figure 5.11 Domestic demand for crude steel in the US and Japan

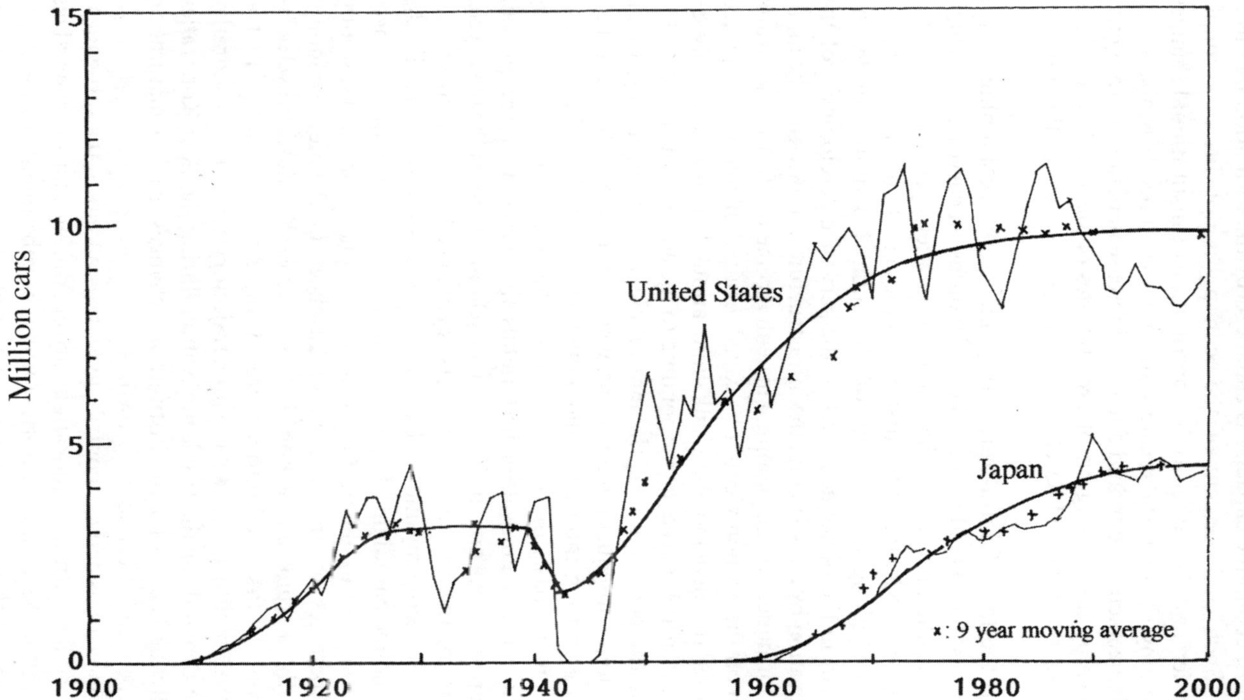

Source: Hirooka (2004a).

Figure 5.12 Domestic demand for automobiles in the US and Japan

are enhanced as they penetrate the economy by creating new infrastructures. That is, economic capacity is enhanced by the formation of infrastructures. Such enhancement of the economy is not due to a simple increase in demand but rather to the contribution of institutional changes in the economy through the creation of infrastructures. It is also not a simple enhancement of demand due to an increase in national productivity, as shown by the leveling off of growth rates of steel and automobile demand after the 1980s while GNP continued to grow, as illustrated in Table 2.3.

In short, the loci of infra-trajectories indicate the positioning of the infrastructures that extend beyond the initial diffusion curves of the trunk innovations in the upswing of the initial Kondratiev cycle and reach the peak of the next cycle. While the first part is the actual diffusion of innovation products along the upswing of the initial Kondratiev cycle, the extended part in the subsequent cycle is the pervasive development of the trunk innovation by constructing the infrastructure and bringing the innovation performance to full bloom. This enhancement effect of the trunk innovation on the economy could be called 'exaltation' and is interpreted to mean that the demand for trunk innovations is driven not only by enhancement of the market corresponding to the advance of the economy but also by the penetration of the innovations into the economy, which induces various subsequent innovations, which are brought about through adoption by existing industries and sometimes by the creation of new industries on the basis of the potential of the trunk innovations. This means that a trunk innovation first fulfills the demand of the present economy up to the peak of the Kondratiev cycle, and then, in the next cycle, the demand for the innovation products is enhanced by the institutional changes in the economy induced by the trunk innovation. That is, the techno-economic paradigm of a trunk innovation induces more advances in the economy during the next cycle, and the demand for the innovation products increases in line with those advances in the new economy. Subsequently, a trunk innovation develops across Kondratiev cycles by molting from the first shell into a more advanced institutional shell. The total innovation diffusion thus has a long trajectory across two Kondratiev cycles. The demand for the trunk innovation during the next Kondratiev cycle finally levels off after the institutional changes reach a maturation stage, independent of economic growth.

Trunk innovations actively contribute to the economy in the subsequent upswing together with the innovations arising during that upswing. This enhancement by the infra-trajectories during the subsequent cycle is a very important role of trunk innovations in the development of the economy. It is a kind of evolutionary behavior.

Interestingly, the steel and automobile industries in Japan substantially started after the Second World War, and the domestic demand for their products is now in the leveling-off stage because the corresponding infrastructures developed quickly. The demand for these products in Japan increased at the same time as the infrastructures were being developed. Production before the war in Japan was almost negligible; although Japan started steel production in 1901, at the beginning of the Meiji era in 1901, production was only 4.7 million tons in 1935. The production of passenger cars was only 1600 in 1940, in a total vehicle production of 46 000, including buses and trucks.

Rostow (1960) analyzed the long-range diffusion of automobiles by country and identified four stages of growth: preconditions for take-off, take-off, the drive to maturity, and the age of high mass-consumption. It is noteworthy that he recognized the existence of another mass-consumption stage after the maturation of diffusion.

Perez (1983) proposed the concept of a techno-economic paradigm, which is a best-practice model made up of a set of all-pervasive generic technologies and organizational principles for guiding the successive innovations of a technological revolution and applying it for modernizing and rejuvenating an entire economy. The notion of paradigm is from the Kuhnian concept, which was first used as 'technological paradigm' by Dosi (1982). The penetration of technological revolution across the economy involves the construction of new infrastructures, and the resulting change of paradigm also brings about widespread institutional change in society.

Grübler (1999) comprehensively studied the formation of infrastructures and found long trajectories. Long-term infrastructure development proceeds through dichotomy of expansion periods. The first growth period reaches the point of saturation and is locked in. The infrastructure then tunnels through the period of saturation and becomes part of the main growth sectors in the next upswing period in line with the above notion of institutional change. These two step developments of innovations occurred in specific pervasive innovations, such as transportation, accompanied by institutional change. The present study has reconfirmed this trend and more importantly, clarified the relative structures of infra-trajectories with Kondratiev cycles.

Perez (2002) comprehensively described how the cluster of innovations has a rather long surge across a 'turning point', which corresponds to the transition from a bubble via a crash to a depression, and how it is fully deployed during the subsequent phase. As an example, she described how automobile consumption began to grow from 1908, exploded during the 1920s but, beyond the crash in 1929, really grew to a maximum in the 1950s and 1960s in what she calls the 'Deployment Period'. In our words, it was unexpectedly enhanced during the fourth Kondratiev cycle. Perez implicitly recognized that the diffusion of innovation products increases from one Kondratiev cycle to the next.

Perez (2002) also described innovation surges and identified their locations in time. Although she doubts the existence of long business cycles as such, the deployment periods of her innovation surges coincide relatively well with our estimation of the upswings in the Kondratiev cycles. That is, the deployment periods described by Perez and our estimated upswings are, respectively, 1798–1829 and 1769–1825 for the first wave, 1850–73 and 1846–72 for the second wave, 1895–1918 and 1897–1929 for the third wave, and 1943–74 and 1950–73 for the fourth wave. This means that the deployment period should correspond to our diffusion period of innovation products and that it substantially corresponds to the upswing of a Kondratiev cycle. In this respect, there is no discrepancy between Perez's findings and those presented here. Her analysis using a long S-curve across the turning point is reasonable because only this treatment can explain the enhancement effect of an innovation beyond the initial Kondratiev cycle forming an infra-trajectory.

According to the analysis presented here, the infra-trajectories bridging the third and fourth Kondratiev cycles coincide well with Perez's description of the fourth surge. The empirical data for the infra-trajectories indicate that every surge must bridge two Kondratiev cycles for it to correspond to an infra-trajectory.

From Business Cycles to Innovation Cycles

To summarize the above discussion, diffusion of a trunk innovation is based on a logistic curve along an upswing of a Kondratiev cycle, gradually bringing about related infrastructures, and forming an infra-trajectory that extends across the downswing, reaching the peak of the next cycle. This pervasive impact of trunk innovations enriches the economy as an 'exaltation' effect. This is the scenario for a comprehensive innovation paradigm describing the interaction of innovations with economic activities.

Perez (1983, 2002) first identified the role of innovations in the formation of economic infrastructures, which she described as a techno-economic paradigm and institutional change in society. This is a brilliant insight, ingeniously expressing the interrelationship of innovations and the economy. The concept of infra-trajectory is closely related to this.

Perez (2002), however, expressed doubt about the existence of long business cycles and thus examined successive long surges of techno-economic paradigms from only the standpoint of technological innovations. Her analysis of S-shaped curves in innovation diffusion is an issue of nonlinear dynamism, as discussed above. The approach using life-cycle theory by van Duijn is the same as ours. All of us recognized the nonlinear property of innovation dynamism reaching a definite saturated level. This seems to be a very important observation. Perez's descriptions strongly suggest that 'innovation' clusters result in a successive dynamism resulting in a pulse of surges. We agree with this scenario and postulate that Kondratiev business cycles are formulated by repeated pulses of innovation clusters, which have intermittently appeared since the Industrial Revolution. That is, we should discuss 'innovation cycles' instead of 'business cycles' because business cycles are simply the result of innovation surges. This approach makes it easier to interpret business cycles. That is, no other origin than innovations has been evidenced for the long business cycles.

Since the start of the Industrial Revolution about 230 years ago, modern industrial society has been built upon various innovations, and only innovations can add value to the economy, as pointed out by Schumpeter. The GDP of advanced countries has been dramatically enhanced by various innovations, and the accumulation of innovations has brought about high economic growth, that is, 30 times greater GDP over the last 100 years. This kind of cumulative integration of innovations is simply the origin of the development of the modern civilized economy, and the accumulation of value-added is not supported in other cycle-formation mechanisms. This is important evidence that innovation plays a crucial role in economic development. As shown by the economic growth rate, exceeding 6–10 per cent a year when a country changes from a developing country to an advanced one through innovations, the impact of innovation diffusion is extremely large, inducing institutional change, and thus the effect of a bubble and subsequent crash is so severe that it takes a long time to recover.

Identifying the Kondratiev cycles is challenging because there is often much turbulence within a cycle. In the fourth wave, the oil crisis greatly affected the economic situation, deforming the shape of the cycle. If we focus on innovation cycles, however, the diffusion of innovation products can be easily identified even across recessive retardations. The conclusion is that the innovation per se is the source of economic development, and the course of economic development can be more clearly understood by studying the behaviors of innovations through the techno-economic paradigm and societal institutional change. Innovations produce economic development, and their diffusion is affected by economic conditions. This indicates that innovations diffuse as a cluster and form an upswing of Kondratiev cycles. This causal relation must be ascribed to a self-organization for cluster formation.

5.4 SUMMARY

The definition of 'innovation' is reconsidered in this chapter, and a wider concept of 'trunk innovation' was proposed, including other resources innovations and various organizational activities as well as technological innovations, which is to revisit Schumpeter's original definition. Trunk innovations play a crucial role in economic development and are especially important in building economic infrastructures. This is the origin of institutional change in an economy.

I. Trunk innovations

1. Trunk innovations are a wider concept than general purpose technologies (GPTs). Trunk innovations are not necessarily technological innovations but they are pervasive, adding much value to the economy.
2. Trunk innovations mainly occur in the areas of energy, resources, motive forces, transportation, and information and communications. The diffusion of trunk innovations selectively forms a cluster along the upswing of Kondratiev cycle and becomes the main force driving economic development.

II. Inducement of subsequent innovations by trunk innovation

3. Trunk innovations induce various subsequent innovations, which stimulate institutional change in the economy.
4. Retail businesses have been induced by various trunk innovations. Department stores were launched along the trajectories of railways and steamships in the latter half of the nineteenth century. Chain stores and supermarkets developed along with the construction of highways and advances in motorization. Discount stores were brought about by innovations in mass production. Convenience stores were made competitive by information technologies, especially POS (point of sales) systems. This kind of interrelationship is well interpreted using the inducement effect of trunk innovations.

III. Construction of infrastructures and institutional change

5. A trunk innovation diffuses along the upswing of a Kondratiev cycle, reaching a mature market at the peak of the cycle. After that, it is interesting to find that it extends across the downswing and reaches the peak of the next Kondratiev cycle, creating a much enhanced market. This longer trajectory is called an 'infra-trajectory'. These two-stage

developments of innovations were also recognized by Perez and Grübler.

6. Infra-trajectories of trunk innovations form a cluster from the root of an upswing of one Kondratiev cycle to the peak of the subsequent Kondratiev cycle. A core trunk innovation induces various subsequent innovations, producing a cluster of infra-trajectories. Core innovations include railways in the second cycle, automobiles in the third and computers in the fourth.
7. The pervasive enhancement of economic development by trunk innovations is ascribed to institutional change in the economy. This pervasive effect not only enhances the market but also induces various subsequent innovations, leading to an 'exaltation' of the whole economy.
8. Long business cycles are ascribed to the intermittent clustering of trunk innovations and a kind of innovation cycle.

PART II

Dynamics of Technological Development and Emergence of New Industries

In Part I we analyzed the development process of an economy through increased value-added due to the diffusion of innovations. It is clear that Kondratiev long business cycles are caused by clusters of innovations, and the emergence of a bubble economy followed by a depression is well explained by the dynamism of innovation diffusion. The trunk innovations form infrastructures, and longer infra-trajectories play a crucial role in the institutional changes in the economy.

There is a long period of technological development before the diffusion of a technological innovation. While almost everyone in this field may be vaguely aware of the existence of such latent periods of technological development, no one has been able to describe it explicitly. While basic research findings at universities are fully open, so that anyone can access them, it is difficult to recognize in advance the feasibility of a future technological innovation emerging on the basis of the on-going science and technology. Even to those specializing in the economics of technological change, that subject remains a black box

The most important aim of the present work is to throw light on this issue and to present a comprehensive scheme describing the innovation paradigm by clarifying the interrelation between the periods of technological development and the diffusion of innovation products. This interrelation reveals various interesting phenomena and gives new insights into the economics of technological change. As the analysis of the innovation paradigm has revealed the nonlinear nature of innovation, it can reveal various sophisticated characteristics of innovations. Thus, the present study clarifies the time profile of innovations and addresses the role of academia and science as the origin of innovations, the relationship between academia and industry, the timing for venture business launching, national systems of innovation related to international competitiveness, the adoption of innovations by existing industries, the evolution of innovation paradigms and so on.

6. Technological development and the innovation paradigm

Although it is easy to describe the transition of product diffusion in a figure, there is a problem of how to deal with the period of technological development. The challenge is how to describe the period of technological development in common terms. A comprehensive paradigm including the periods of technological development and product diffusion is needed. This is the main subject of this book. Also needed is a discussion of how science can be introduced into innovation and of the roles that science and invention play as the basis of innovation.

6.1 LOGISTIC NATURE OF THE INNOVATION PARADIGM

Innovation products appear in the market and begin to diffuse after a long period of development of core technologies. Long before an innovation appears in a market, there is a basic discovery or key invention that eventually becomes the foundation of the innovation and then a long period of technological development. We inclusively call all of these processes an 'innovation paradigm'. As an innovation product diffuses, it forms a logistic curve, which is called the 'trajectory'. Preceding the diffusion trajectory is a period of technological development, and an innovation paradigm consists of these two periods. This chapter discusses how to describe the technology development period.

Description of the Technological Development Period

While there is no common measure for explicitly describing the degree of development of an innovation process, there is a common feature in the development processes of science and technology – we can determine the timing and location of an innovation by examining the distribution of major inventions and discoveries. A clue is the shape of the distribution of core technologies, which we can use to estimate the trend of a series of inventions. One approach is to examine such trends and analyze the innovation paradigm by distinguishing the two phases of diffusion and technological development.

Looking at the course of technological development, there is some indication that, at the technology level, it is hard to make rapid progress in the early stages. After reaching a certain level, technology develops sharply, and in the final stage, the development of technology again slows down. This pattern suggests some kind of sigmoid curve.

The output of an industry is products, which can be described quantitatively. The added value is determined by subtracting the cost of input from the total sales. Although the first stage of an innovation is a latent period of technological development before market formation, the output of technological development is patents and the development of new products. Thus, the trend of patents and new products can be used to describe the transition of the technological development period.

Bibliometric Analysis of the Technology Development Period

Since the output of technological development is patents and new products, we can describe the trends in the annual number of papers and patent applications and in the annual development of new products in the course of the development of a new technology. There have been several case studies of the trends in patents and product development. Kaku (1986) intensively analyzed the number of new products commercialized in the dyestuff industry in Germany and found a steep mountain-shaped distribution of newly developed products across the years 1850 to 1900. Achilladelis (1990) investigated the trend in the number of patents in the plastics and pesticide industries after World War II and also found a steep yearly distribution of the patents in both cases.

We analyzed the data of Kaku and Achilladelis by applying the Fisher–Pry plot. As shown by the lower graphs in Figure 6.1, there were sharp transitions in the distributions. The corresponding Fisher–Pry plots (the upper graphs) are clearly straight lines, direct evidence that the distributions obey a logistic equation. Thus, a technology development period can be represented by a logistic S-curve having a definite time span.

Andersen (2001) used regression analysis to analyze the trend in the number of US patents issued, by field, for the 100 years from 1890 to 1990, and found that eight fields of chemicals, nine fields of electric and electronics, 21 fields of mechanical engineering, two fields of transportation and four fields of non-industrials exhibited trends that can be expressed using a logistic equation. She concluded that the innovation process can be expressed using a logistic equation based on her bibliometric analysis of patents.

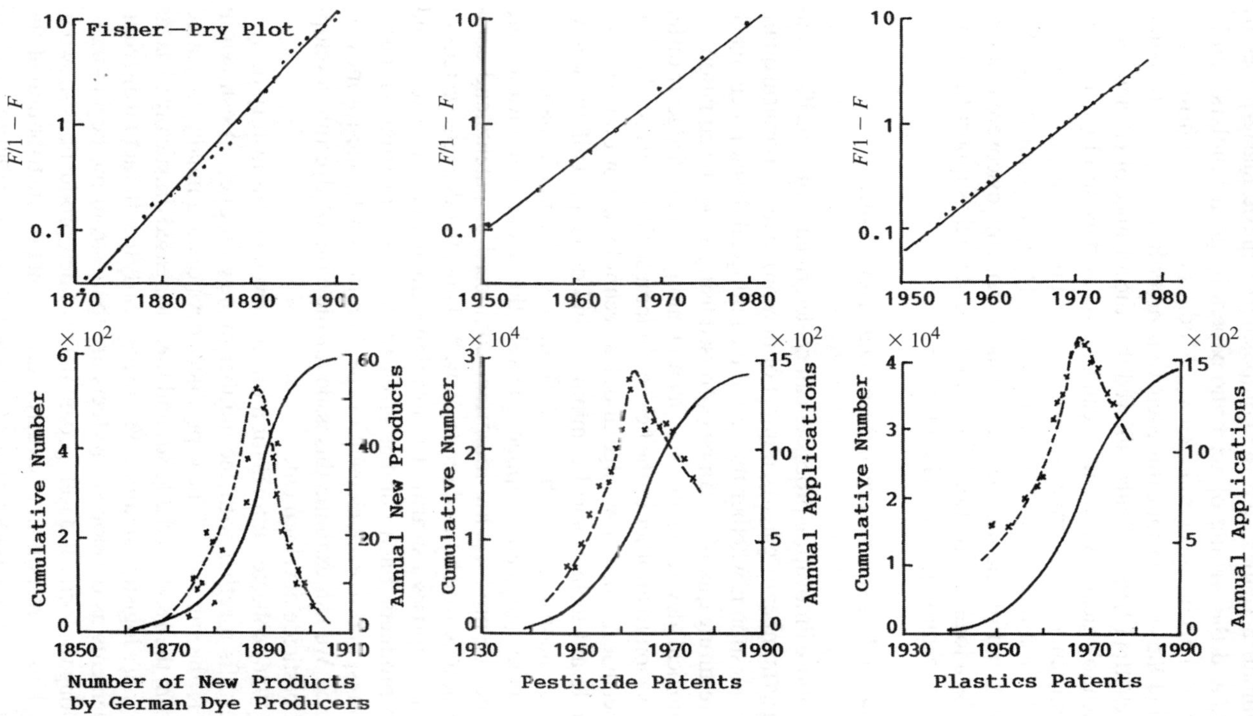

Source: Hirooka (1998b, 2000b); data from Kaku (1986) and Achilladelis (1990).

Figure 6.1 Logistic nature of technology development

From the point of knowledge development, Marchetti (1985) found that various human activities, like the composing of music masterpieces by composers and the writing of scientific papers by great scholars, can be well expressed using a sigmoid curve. Modis (1992) reconfirmed this. Marchetti (1985) proved that this S-curve is described by a logistic equation. Marchetti (1979, 1980) also showed that the advance of efficiencies in various technological developments can be expressed using a logistic equation; the efficiencies of steam engines, generation of electric power, electric lamps and the ammonia production process. He also pointed out that the annual trend of the number of chemical elements discovered can be described by a logistic equation. These findings strongly support the logistic nature of technological development.

Evidence of the Logistic Nature of Technology Development

All of the above innovation processes can be described using specific quantities such as number of patents, number of new products, or physical quantities. The question is whether the actual technological development itself, not some resulting quantity, can be expressed using a logistic equation.

Technological development is a phenomenon of knowledge transfer from one person to another in the field of human society. This is a kind of discrete behavior. The knowledge about a technology, however, spreads over the relevant technological community and develops as if the knowledge homogeneously covers the whole field, to which a mathematical convention of continuity can be applied. That is, the actors of innovation are discrete and the products are inventions composed of patents and commercial products which are discrete as well. Thus, the discrete system of innovation comprises a swarm of innovation elements such as patents and developed products. Each number of elements on the distribution curve should correspond to each element of the inventions with one-to-one correspondence. We will examine this issue for the cases of dyestuff development and pesticide development.

Kaku (1986) collected information on the development history of synthetic dyestuffs together with the number of new products. As shown in Figure 6.1, the number of new products developed annually increased sharply from the 1850s to the 1880s and then decreased drastically towards 1900, obeying a logistic equation. We can see clearly that the actual distribution of elements as developed products are found from the beginning to the end along the locus of developed numbers of new products, as shown in Table 6.1. This correspondence clearly indicates that the distribution of the number of products developed can be replaced by the distribution of innovation elements.

Table 6.1 Development of new products in the dyestuff industry in Germany

Year	Developed products (elements)	Number of new products
1856	Mauve	1
1859	Fuchsine	1
1863	Bismarck brown	1
1868	Alizalin	1
1871		5
1973	Eosin	2
1875		12
1877	Methylene blue	10
1878		20
1879		18
1880	Indigo	5
1883		23
1884	Congo red	10
1886		33
1895		22
1897	Indigo by naphthalene	8
1899		8
1900		3
1901	Indigo by aniline Indanthrene	2

Source: Hirooka (2003a); data from Kaku (1986).

The development times of individual pesticide products are marked on the logistic S-curve of the number of patents for pesticides, as shown in Figure 6.2. As a matter of course, the development time of the new pesticide products, as the actual elements of innovation, are distributed all around the S-curve of the number of patents for pesticides.

These data shown in Table 6.1 and Figure 6.2 indicate that the elements of technological developments themselves are distributed in the same locus as those developments described by specific quantities of innovations and can thus be considered to have a logistic nature. That is, the distribution of the discrete elements of developed technologies directly corresponds to the actual locus of technological developments. Thus, the logistic curve of a technological development can be determined by the concrete distribution of a series of elements of developed technologies, and the size and positioning of the curve is determined by the time span of the element distribution.

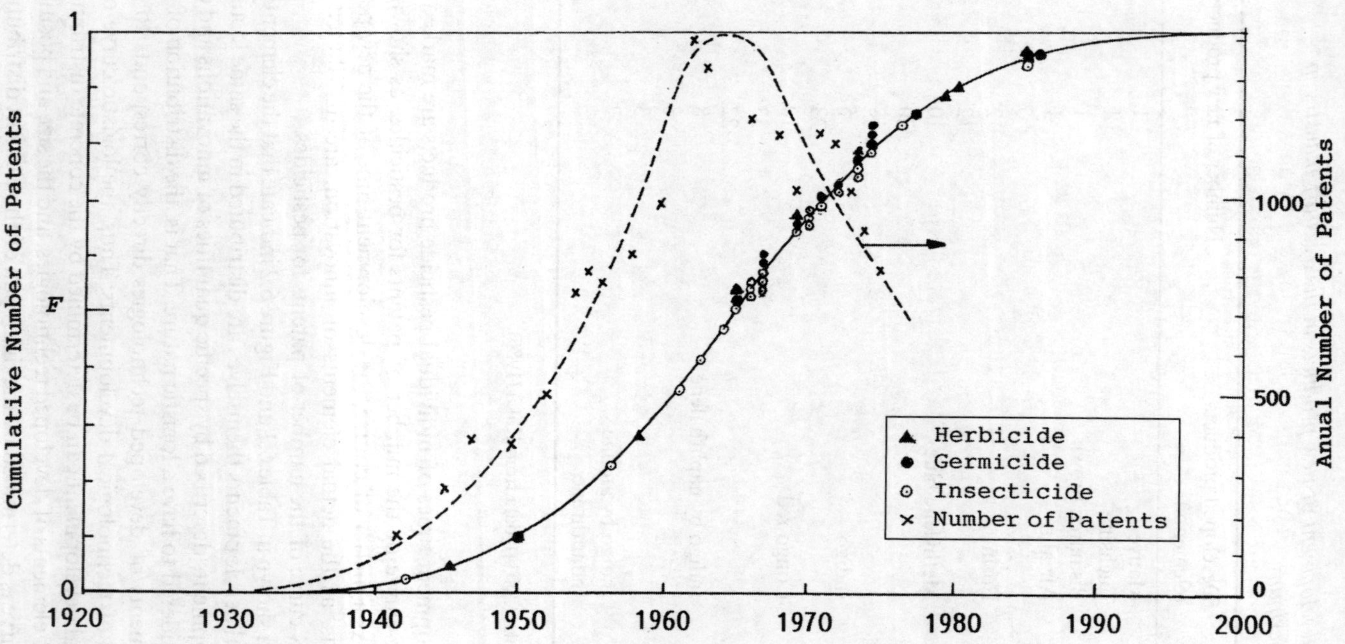

Sources: Hirooka (2003a); data from Sumitomo Chemical Co. (2002).

Figure 6.2 Development of pesticides on patent distribution curve

Of course, the vertical axis corresponds strictly to the level of the technology or degree of maturity. The nonlinear logistic nature of innovation indicates that more elements are concentrated around the middle point of the logistic curve and there is no substantial technology progress before or after the development time span because the origin of an innovation comes at the beginning of the span and no development of the technology occurs after the time span. It is a very interesting and important implication that all elements (that is, inventions) of innovation key technologies occur within a specific time span, forming a swarm. Thus, a trajectory can be easily identified by determining how inventions are gathered during a specific time span, and the time span of the trajectory can be determined by identifying the first and last core inventions in the swarm, without measuring any quantitative mass. Of course, this procedure is feasible when the locus of the trajectory is a logistic curve. This simple procedure for determining a logistic curve provides a crucially important tool for analyzing innovation. This dynamics develops further beyond innovation, expanding to much wider applications on knowledge development, which will be described in later chapters.

How to Determine the Trajectory

It is important to define the actual method of determining the trajectory. As described above, if the measurement of the trajectory is carried out using a Fisher–Pry plot, the slope of the straight line corresponds to the time span in the expression of the normalized scale of the vertical axis. Because the logistic curve goes from minus infinity to plus infinity, we have to define the time span. Thus, we define the time span as the interval from $F = 0.1$ to $F = 0.9$ in which F is the fraction of the vertical axis as the full scale is unity. Elements of innovation which are represented by technologies and new products emerge discretely and this discrete phenomenon makes it possible to determine the time span. Though the time span can be optionally determined, their distribution along the trajectory is seldom beyond $F = 0.1$ or $F = 0.9$, and the swarm is substantially located within this time span. Thus, if we look directly at the swarm of technology elements arranged in order of time developed along the trajectory, we can define the first technology element to be developed at $F = 0.1$ and the last one at $F = 0.9$. This form of time span expression was previously used by Marchetti (1979, 1980, 1988, 1996). Although this procedure seems to be rather vague, the actual determination was successfully achieved for more than 40 innovations, as shown in the next section. These results clearly reflect the nonlinear nature of the trajectories.

An actual determination is illustrated in Table 6.2 and Figure 6.3 for the innovation paradigm for electronics. Considering the development of electronics technologies from the nineteenth century to today

Table 6.2 Elements of electronics technology trajectory

Year	Inventor	Invention	Trajectory
1948	W.H. Brattain, J. Bardeen	Function of point contact transistor	F = 0.1
1948	W.B. Shockley	Patent of p-n junction transistor	
1949	W.B. Shockley, G.L. Pearson, M. Sparks	Establishment of p-n junction concept	
1949	W.B. Shockley	Theory of p-n junction transistor	
1951	W.B. Shockley, M. Sparks, G.K. Teal	Completion of p-n junction transistor	
1952	W.B. Shockley	Concept of unipolar transistor	
1953	G.C. Dacey, I.M. Ross	Unipolar field effect transistor	
1954	H. Krömer	Drift type transistor	
1956	Charles A. Lee	Diffused base transistor	Trajectory
1959	Jack S. Kilby	Invention of solid state circuit	1948–1973
1959	Jean A. Hoerni	Silicon planar transistor	$\Delta\tau = 25$
1960	Dawon Kahng, M.M. Atalla	MOS transistor	
1961	R.W. Noyce	Silicon planar monolithic IC	
1963	F.M. Wanlass, C.T. Sah	CMOS, compensation MOS transistor	
1965	TI	Schottky-cramped transistor	
1967	John T. Wallmark	MNOS, Si-nitride-oxide semiconductor	
1971	Daglass L. Benzer	ISO planar technology	
1972	T.H. Philip Chang	Electron beam stepper	
1973	IBM	Submicron lithography	F = 0.9

Sources: Hirooka (2001b, 2003a); data from Kisaka (2001) and others.

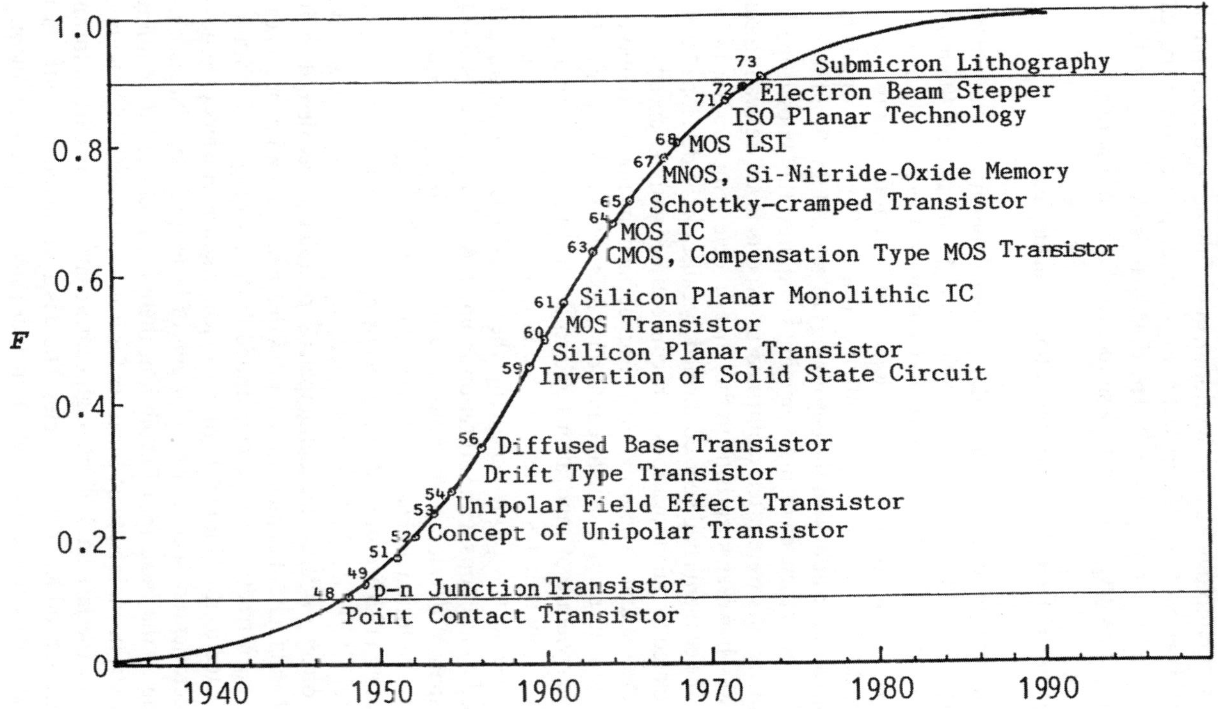

Sources: Hirooka (2001b, 2003a); data from T. Kisaka (2001) and others.

Figure 6.3 *S-curve of electronics technology trajectory*

(Kisaka, 2001), we can easily recognize that the origin of the technology trajectory was the discovery by Shockley et al. of the transistor in 1948 and the last core technology was the development of submicron lithography by IBM in 1973, with a series of core technologies developed almost every year in between, as shown in Table 6.2. The time span is clearly seen to be 25 years, and the S-curve is instantly described as shown in Figure 6.3.

Determination of Time Spans for Innovation Technologies

To confirm this finiteness of technological development, we determined the time spans of the technological development for more than 40 innovations from the age of the Industrial Revolution to the present time (see Figure 6.4). These swarms of technologies were identified using various data from the work of McNeil (1990) and Yuasa (1989) and from other handbooks and textbooks on the chronology of technology. Because identifying the trajectory elements sometimes requires the knowledge of experts, the results were often confirmed by experts in the relevant disciplines. It was, however, surprising to find that the determination was rather easy once the elements had been confirmed, because all the developed technologies related to the innovations were concentrated within a specific time span. We are the first to describe such abstract phenomenon as technological developments using a specific equation. This success can be ascribed to the nonlinear nature of innovations and to the fact that determining the trajectory simply requires measuring the time span of the spread of the innovation elements without considering the vertical axis, which has a scale normalized to fractions with saturation at unity. As shown in Figure 6.4, the time span of developed technologies was around 60 years during the Industrial Revolution and is now shortened to 25 to 30 years.

Kondratiev (1926) proposed the concept of long business cycles and Schumpeter (1939) ascribed business cycles to clusters of innovations. We provided direct evidence for Schumpeter's postulation as described in Chapter 4 and found that most innovations that had a major impact on the economic infrastructure were concentrated in the upswing of Kondratiev waves. Related to this finding, Figure 6.4 clearly shows clusters of innovation trajectories that seem to be the origins of business cycles. That is, corresponding to the Industrial Revolution, there is the cluster of spinning machines, steam engines, blast furnaces for iron-making and acid/alkaline chemicals in the last half of the eighteenth century. This cluster forms the first and second Kondratiev cycles. There is a second cluster in the last half of the nineteenth century, which includes steel-making technologies, telephony, internal combustion engines, electric power and dyestuffs, that is, at the origin of the third Kondratiev cycle. The third cluster is that of the

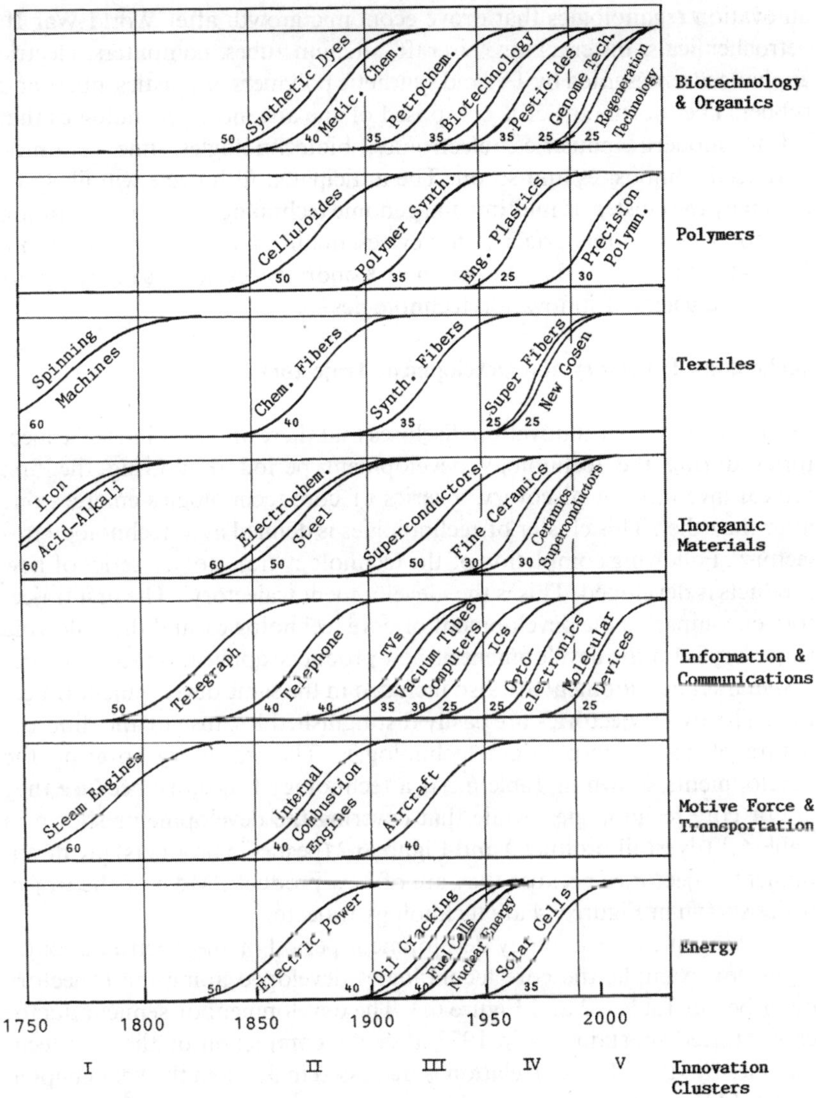

Sources: Hirooka (1994b, c, 2003a).

Figure 6.4 Time spans of technology development periods

innovation technologies that drove economic growth after World War II: petrochemicals, oil cracking, aircraft, vacuum tubes, computers, electric appliances represented by TV, and synthetic polymers of plastics, fibers and rubber. The fourth cluster is composed of the advanced technologies that led to modern industries: electronics, biotechnologies, fine ceramics, advanced composites and so on. The anticipated fifth cycle will likely be driven by the cluster of multimedia, genome technologies, cell regeneration technologies, nanomaterials, nanodevices, quantum computers and so on. This is the first report that the origin of Kondratiev cycles lies in the occurrence of clusters of innovation technologies

Technology Trajectory and Development Trajectory

Detailed analysis of innovation discloses that there are two kinds of trajectories during the technology development period. Following the first radical invention or discovery, a series of core technologies emerges one after the other. This cluster of technologies is defined as a 'technology trajectory'. Following completion of the technology trajectory, a series of new products is developed. This is the 'development trajectory'. The first trajectory encompasses the development of core technologies, and the following trajectory is for the development of new products; application technologies and market development are also included in the same development trajectory. The two trajectories are easily distinguished because of the different nature of the corresponding technologies. The trajectory covering the developments shown in Table 6.2 is a technology trajectory because they are of core technologies, while that covering the developments shown in Table 6.1 (dyestuff products) and Figure 6.2 (pesticide products) are development trajectories because they are of new products. Most of the trajectories shown in Figure 6.4 are technology trajectories.

Looking at the technology development period of the electronics paradigm, for example, the core technologies developed along the trajectory described in Table 6.2 and Figure 6.3. The development of semiconductor chips started substantially in 1973 after the completion of the core technologies by IBM. This correlation is discussed in detail in the next chapter. It should be noted that the development of patents corresponds closely to the development trajectory, but not to the technology one. Few patents are applied for during the technology trajectory and then, at the beginning of the development trajectory, the number of applications abruptly increases due to the rapid increase in the number of firms developing related products, which provides much incentive to acquire patents. Utterback (1994) showed that the number of firms increases quickly as the industrialization of innovation products begins. Actually, the trend in the number of relevant patents corresponds to the development trajectory but not the technology trajectory in general. This will be discussed later in more detail.

6.2 THE CONFORMATION OF THE INNOVATION PARADIGM

Interrelation of Technology Trajectory and Diffusion Trajectory

Thus, an innovation paradigm is composed of three trajectories: technology, development and diffusion, in that order. Interestingly, the diffusion trajectory begins just as the technology trajectory is completing, making a cascade. Although this relationship is an empirical observation, it is important to describing the innovation paradigm and especially to gaining technology foresight. Typical examples of the interrelation between the technology and diffusion trajectories are shown in Figures 6.5 and 6.6 for energy development, motive power, and information and communications technologies. The plots clearly show that the technology trajectory was immediately followed by the diffusion trajectory in a cascade fashion. The source of data is the same as that for the data in Figure 6.4, together with other industrial statistics.

A Comprehensive Description of the Innovation Paradigm

Thus, the relative locations of the technology and diffusion trajectories in an innovation paradigm are such that the diffusion trajectory starts just as the core technologies are being completed. The development trajectory is located just between the technology and diffusion trajectories, beginning when a new product is first developed using the core technologies. Therefore, the development trajectory starts slightly before the completion of the technology trajectory. Thus, the comprehensive paradigm of innovation consists of the technology, development and diffusion trajectories as shown in Figure 6.7.

The first half of the innovation paradigm comprises the technology trajectory, starting from a basic invention or discovery and followed by various relevant core inventions. The first patent application is submitted and after patenting, the inventing firm enjoys the benefits of a monopoly. There are few patents during the period of the technology trajectory, but the number begins to increase as the development trajectory takes over, reflected in an increase in the number of new products and firms. Therefore, the trend in patent applications strongly reflects the changes that occur during the period of the development trajectory. The first new product is developed at the beginning of the development trajectory; when it is introduced into the market sometime later, it marks the beginning of the diffusion trajectory.

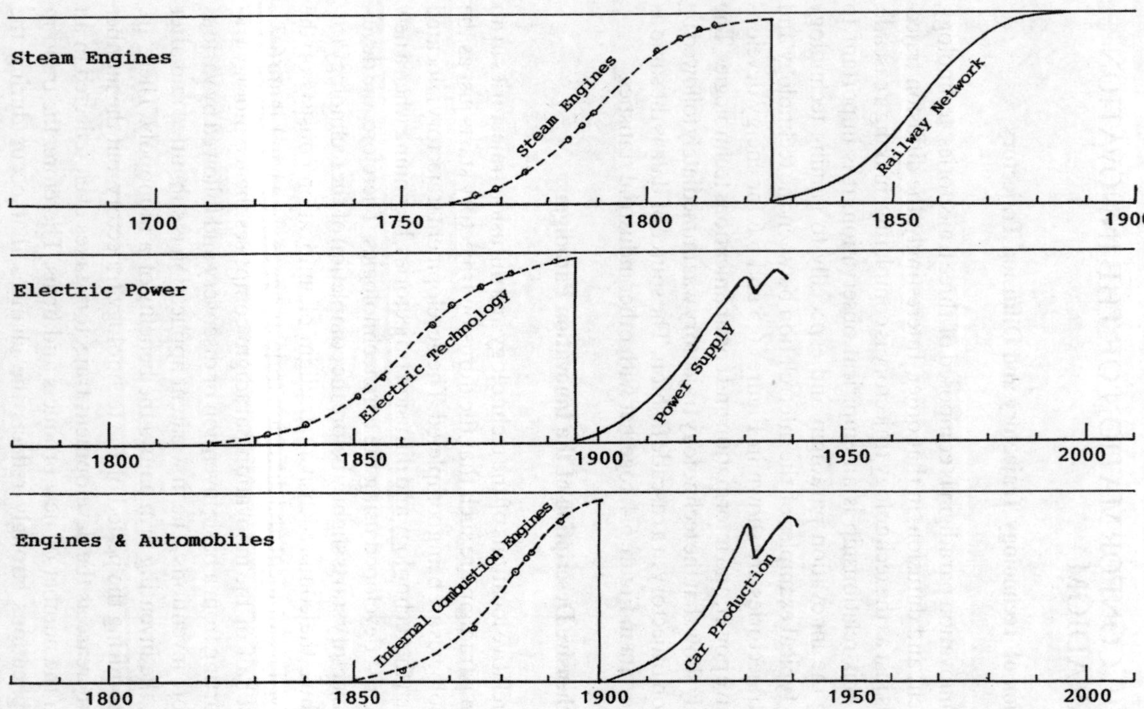

Sources: Hirooka (1998b, 1999c, 2003a, 2005b).

Figure 6.5 Cascades of technology and diffusion trajectories: energy and motive powers

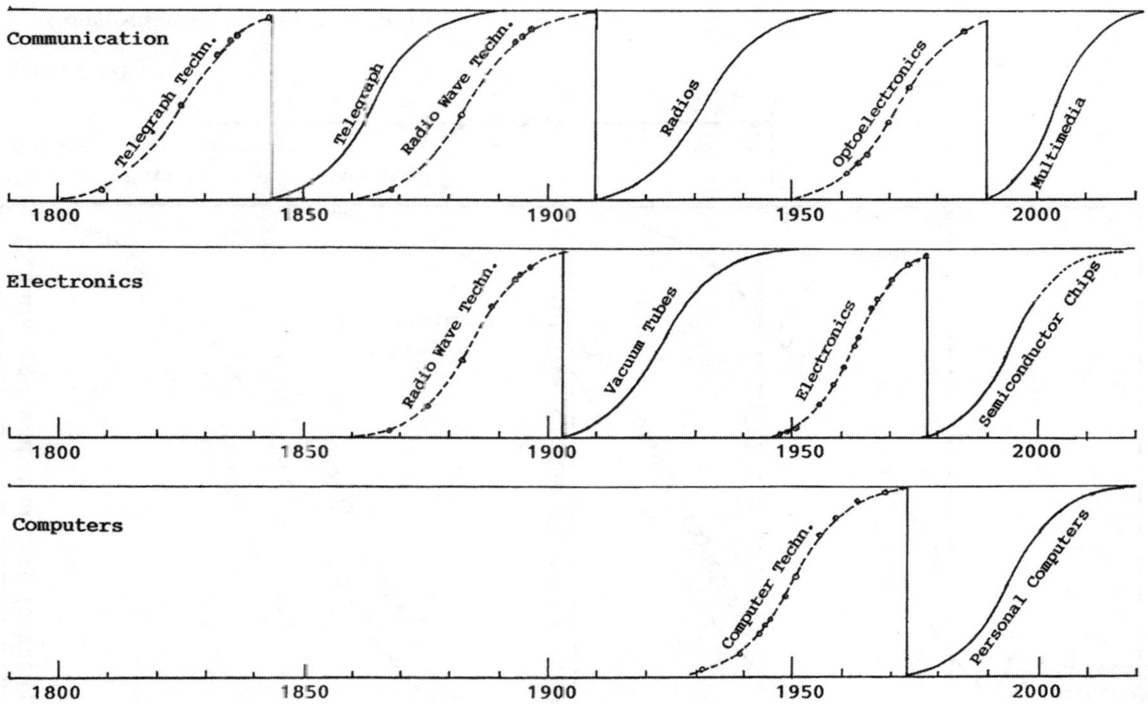

Sources: Hirooka (1998b, 1999c, 2003a, 2005b).

Figure 6.6 Cascades of technology and diffusion trajectories: information and communications technologies

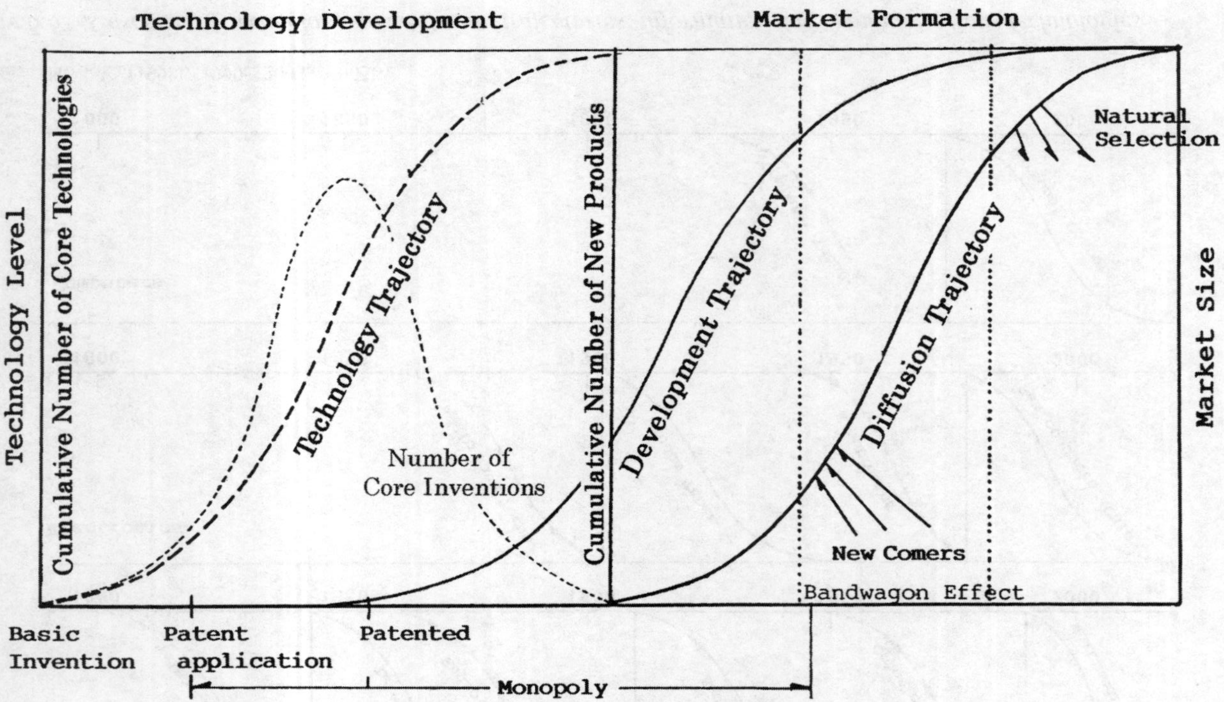

Source: Hirooka (1998b).

Figure 6.7 Composition of innovation paradigm

6.3 EVOLUTION OF INNOVATION PARADIGMS IN THE CHEMICAL INDUSTRY

Now that we have developed a method for determining the innovation trajectories, let's examine how well it works for the chemical industry.

The chemical industry came into being during the Industrial Revolution, and various paradigms related to it have since appeared. As shown in Figure 6.4, the main paradigms of the chemical industry appear together with clusters of other innovations that construct the upswing of Kondratiev cycles. As described below, it was confirmed that all of the innovation paradigms comprise three trajectories that can be easily determined.

The Dawn of the Chemical Industry: The Inorganic Chemicals

Acid/alkai industry

During the Industrial Revolution, acid/alkali factories were built to support the cotton industry by improving bleaching efficiency. The manufacture of sulfuric acid was enabled by the invention of the lead-coated chamber process by Roebuck in 1746. Demand for alkalis increased in the second half of the eighteenth century for use in making soap, making dyes for textiles and bleaching raw cloth, especially in France. Leblanc invented a process for producing soda at low cost and obtained a patent on it in 1789. Importation of this process to England triggered development of the chemical manufacturing industry. Furthermore, a more efficient 'bleaching powder' was invented in 1797.

The production of soda was initially only 320 tons per year in France; it grew to 15 000 tons in 1794 and 45 000 tons in 1825. Worldwide production increased from 150 000 tons in 1863 to 1.76 million tons in 1902, a more than tenfold increase. The diffusion trajectory of soda production is clearly located between the 1850s and 1910. The development trajectory occurred just before the diffusion trajectory, evidenced by the fact that acid/alkali technologies were drastically improved by such inventions as the Gay–Lussac tower and Glover tower for sulfuric acid production, and the Solvay process for efficient soda production in 1861. These improvements correspond well to the second Kondratiev cycle, as shown in Figure 6.8.

Electrochemicals and aluminum

Faraday discovered the principle of electromagnetic conduction in 1831 and established his law of electrolysis in 1833. This led to research on electrolysis and to the development of an electrolytic process for producing aluminum in 1886, the same year as an electrolytic diaphragm process was developed for producing soda. An electrolytic soda process using mercury was invented in 1892, and calcium carbide was synthesized by the electrolysis of lime and coal in 1892. An electric furnace was invented in 1887, and calcium carbide was manufactured using an electric furnace by Héroult in 1887. This series of inventions forms a technology trajectory with a time span of 70 years.

The elements of this technology trajectory are also composed of various aluminum technologies. The production of aluminum was 5000 tons in 1900, 27 000 tons in 1930, 2 million tons in 1945, and 16 million tons in 1980, corresponding to the diffusion trajectory. The development trajectory comprises various application processes such as the Prebaked process and the Soderberg process for aluminum smelting, and such products as duralumin, Y alloy, and so on. These trajectories are illustrated in Figure 6.9.

The Organic Chemical Industry

Synthetic dyestuffs

The development of synthetic dyestuffs was illustrated in Figure 6.1, and the products developed were listed in Table 6.1. The technology trajectory of dyestuffs is the science trajectory of organic chemistry, starting with the unification of inorganic and organic chemistry by Wöhler and its completion by Kekulé 40 years later. The development trajectory is composed of a series of new dyestuffs, starting with the invention of aniline dye (Mauve) by Perkin in 1856, as exemplified in Table 6.1. The production of synthetic dyestuffs in Germany amounted to 140 000 tons in 1913, 83 per cent of which was exported. The paradigm for synthetic dyestuffs is illustrated in Figure 6.10.

Chemotherapy – chemical drugs

At the same time as the first synthetic dyestuff was being invented, various chemicals for chemotherapy were beginning to be tested. The synthesis of 'chemical drugs' became possible due to advances in the synthesis of organic chemicals. The technology trajectory is the same science trajectory of organic chemistry as that for synthetic dyestuffs. The timeframe of the advances in organic synthesis indicates that research on chemotherapy slightly trailed that on synthetic dyestuffs. Collaborative research between academia and industry in Germany effectively advanced to form the development trajectory. The products developed included Antipyrine in 1884, Phenacetin and Aspirin in 1888, Tuberculin in 1893, and Salvarsan in 1910. The trajectories are illustrated in Figure 6.11.

Chemical Fibers and Pesticides

Chemical fibers

During the Industrial Revolution, chemicals were used to bleach textiles, as described above. John Mercer of Lancashire, England, discovered in about 1840 that caustic soda makes cotton fibers more elastic and easier to dye and also that when they are kept in tension, a more lustrous effect is obtained. Mercer also found that cotton can be dissolved in a solution of copper oxide in ammonia, a process later exploited in the manufacture of artificial silk. These technologies for finishing fibers formed a technology trajectory. The development trajectory encompasses the development of various chemical fibers. The first chemical fiber originated out of research on forming carbon filaments. Swan succeeded in preparing artificial silk from nitrocellulose by wet spinning in 1883. In the following year, Chardonnet succeeded in preparing multifilaments from nitrocellulose and developed artificial silk in 1885 that was commercialized in 1891. Bemberg silk and cuprammonium fiber were developed in 1890, and industrial production began in Germany in 1898. Viscose, the third chemical fiber, was developed by Cross and Bevan in England and industrialized in 1905. Acetate fiber of cellulose acetate was patented in 1894. Cellulose diacetate, which is soluble in acetone, was developed in 1907. Worldwide production of chemical fibers amounted to 210 000 tons in 1930 and 3.33 million tons in 1965, a 16-fold increase. The paradigm for chemical fibers is illustrated in Figure 6.12.

Pesticides

The development of pesticides was illustrated in Figures 6.1 and 6.2. The technology trajectory continues up to the beginning of the twentieth century, and the first anti-fungal product was Bordeaux mixture, a mixture of sulfur and lime, developed in the 1840s, which is effective against a powdery mildew that attacks grapevines. Thirty years later a mixture of lime and copper sulfate was developed that effectively controls the fungus that causes potato blight. In 1867, research on insecticides revealed that lead arsenate is effective against beetles. Several pesticides had already been developed by the start of the Second World War; they included thiuram (a fungicide developed in 1931), dinitrocresol (DNOC, a herbicide developed in 1932), and dichlorodiphenyltrichloroethane (DDT, an insecticide isolated in 1939 and developed in 1942), which had a significant effect. Various other pesticides have followed, resulting in the development trajectory shown in Figure 6.2. The paradigm for pesticides is illustrated in Figure 6.13.

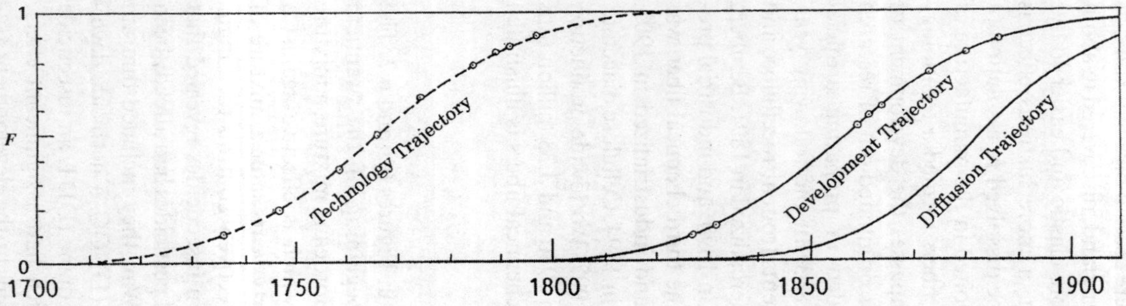

Source: Hirooka (1998e).

Figure 6.8 Acid/alkali chemical industry paradigm

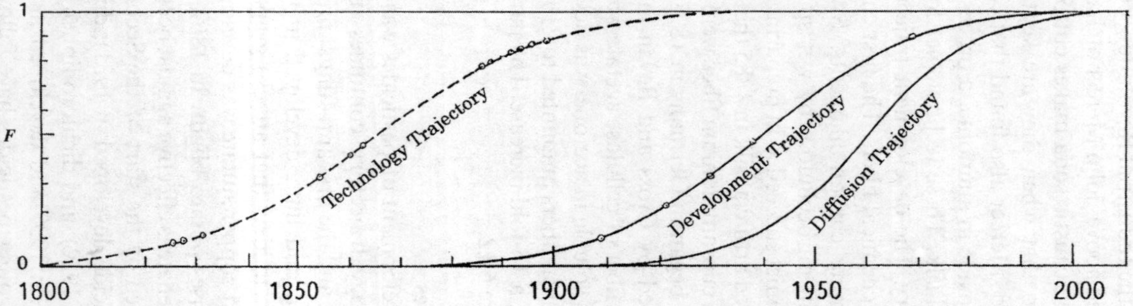

Source: Hirooka (1998e).

Figure 6.9 Electrochemical industry paradigm

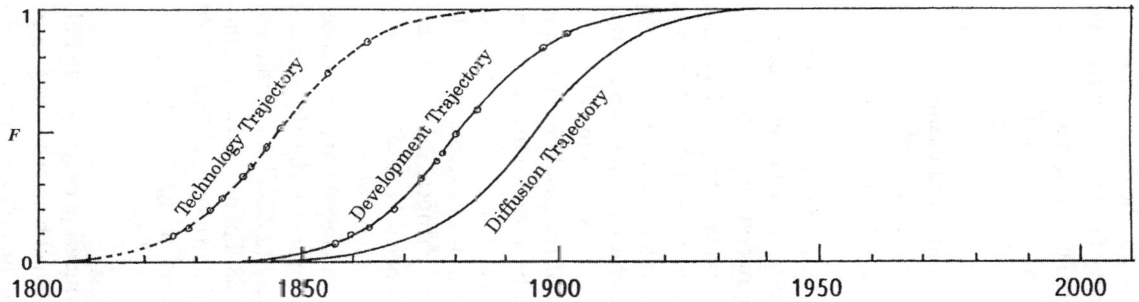

Source: Hirooka (1998e).

Figure 6.10 *Synthetic dyestuff industry paradigm*

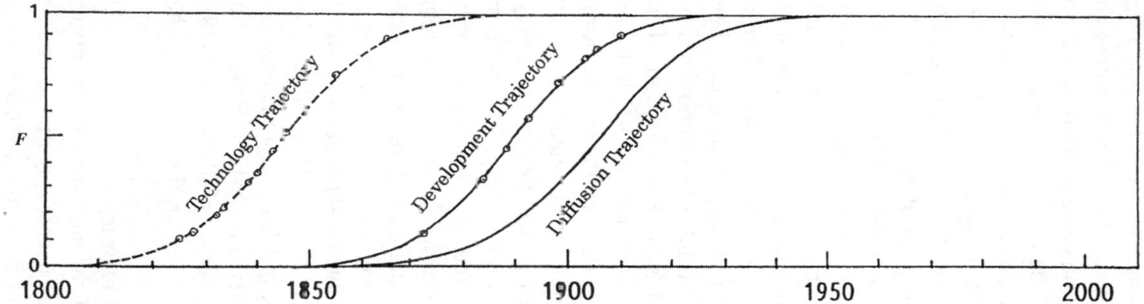

Source: Hirooka (1998e).

Figure 6.11 *Chemotherapy industry paradigm*

Innovation of Feedstocks: From Coal to Oil

Petrochemical industry

The chemical industry became a full-scale industry in the latter half of the nineteenth century. The German chemical industry in particular advanced quickly and came to dominate the world market. This success was strongly owed to the excellent activities of the universities and close collaboration between academia and industry through effective knowledge transfer. Especially important was the work done in organic chemistry in German universities, which led to major advances in dyestuffs and chemotherapy. This work is symbolized by that carried out on the uses of coal tar, and the chemical industry advanced using coal as the major feedstock.

The feedstock changed, however, to petroleum after the Second World War, which enabled the chemical industry to make rapid progress. The switch from solid coal to liquid petroleum greatly improved efficiency because batch processes could now be done as continuous operations. It also resulted in a wider variety of products. The transition from coal to petroleum is a typical example of the innovation of resources as Schumpeter referred to as the third 'neue Kombination'. The technology trajectory started with oil cracking by Burton in 1913 followed by catalytic cracking, the fluidized bed process, hydroforming and platforming. These developments initially came about due to the demand for an increased yield of the gasoline fraction in the age of motorization. Later they were driven by a wider range of applications in oil refining. The most important application of oil is the petrochemical industry which uses naphtha, a fraction of oil, as the feedstock. During the development of oil refineries, various chemical reactions were discovered, forming the technology trajectory of petrochemicals.

The development trajectory of the petrochemical industry comprises a series of chemical products derived by naphtha cracking. A naphtha cracker was first constructed to produce ethylene as the main product by BP Chemical Co., UK in 1951. Naphtha crackers in Japan were first built by Sumitomo Chemical Co. and Mitsui Petrochemical Co. in 1958. The production of ethylene in Japan was 780 000 tons in 1965 and 7.69 million tons in 1999, a tenfold increase. The paradigm for petrochemicals is shown in Figure 6.14.

Polymer industry

Some of the most important petrochemical products are polymer materials – various plastics, synthetic fibers and synthetic rubbers. The technology trajectory is again a science trajectory, and academia played an important role, in conjunction with industry, which produced important inventions. Basic research by Staudinger began in 1920 and the concept of macromolecules was established in 1930. Investigation of the polymerization mechanism through the 1960s clarified the radical and ionic polymerizations of vinyl monomers, living polymerization, stereospecific polymerization with Ziegler and Natta catalysts, and alternating sequential regulation. These advances form the technology trajectory.

This fundamental research led to the development of various polymerization products, such as polymethylmethacrylate, polyvinylchloride, polystyrene, polyethylene, and polypropylene, and of condensation products, such as nylon, polyester and polyurethane. These products, and a further variety of engineering plastics, form the development trajectory, having the time span of 35 years. The production of polymer materials in Japan increased from 1.26 million tons in 1963 to 12.63 million tons in 1993, a tenfold increase. The paradigm for polymers is illustrated in Figure 6.15.

6.4 ROLE OF INVENTION AND SCIENCE IN INNOVATION

This chapter has described the structure of the innovation paradigm and clarified the contribution of basic research to the core technologies which form the technology trajectory; a contribution that is sometimes ignored in the economics of technological change. An innovation originates from a creative invention or epoch-making scientific discovery, and a technology trajectory forms comprising the subsequent scientific work and inventions. This section focuses on the role of science and invention in the formation of the technology trajectory, first following in the footsteps of the pioneers in this field.

Discussion of Invention and Innovation by Schumpeter

Schumpeter carefully examined the role of inventions and science in innovation. Although his consideration was not always positive to these factors, his insights are informative and profound.

In *Business Cycles*, Schumpeter (1939) pointed out more than once the difference between invention and innovation. In chapter 5, he asserted that innovation is a much broader concept than technological innovation and that invention contributes only partially to innovation.

He stated that:

> Any 'doing differently' in the realm of economic life – all these are instances of what we shall refer to by the term of 'innovation'. It should be noticed at once that that concept is not synonymous with 'invention'. Whatever the latter term may mean, it has but a distant relation to ours. Moreover, it carries misleading associations . . . It is entirely immaterial whether an innovation implies scientific

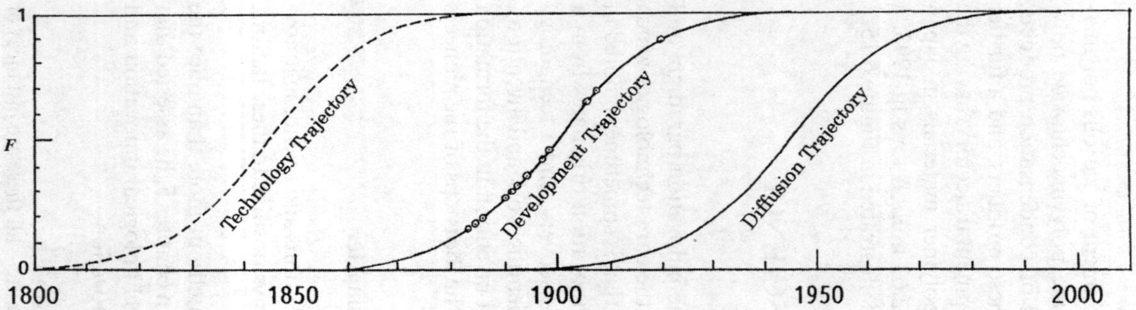

Source: Hirooka (1998e).

Figure 6.12 Chemical fiber industry paradigm

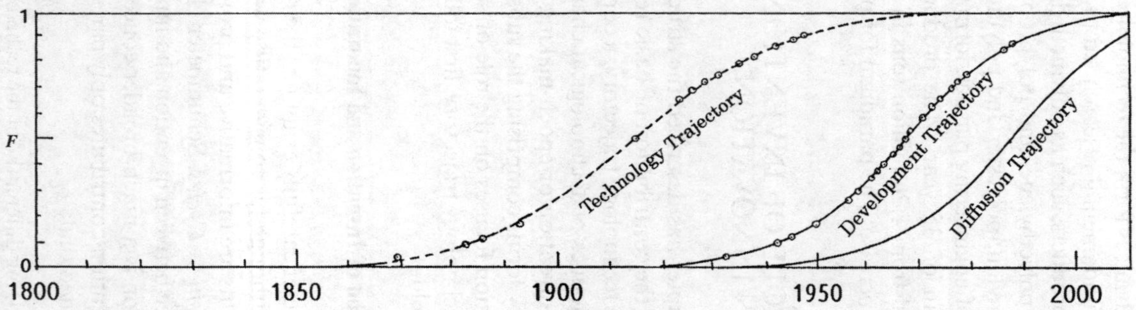

Source: Hirooka (1998e).

Figure 6.13 Pesticide industry paradigm

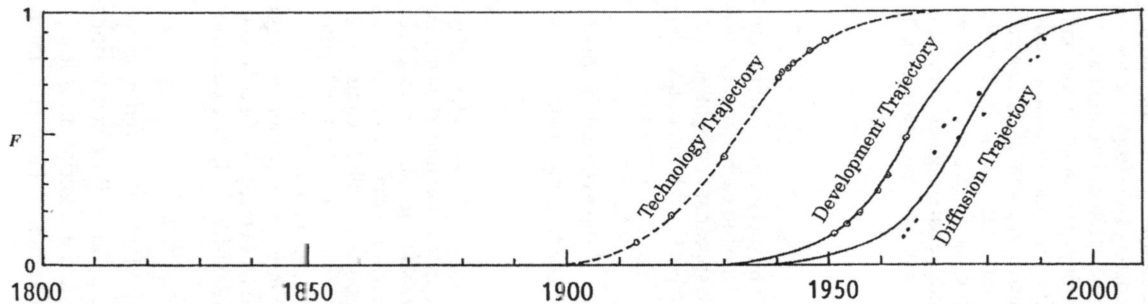

Source: Hirooka (1998e).

Figure 6.14 Petrochemical industry paradigm

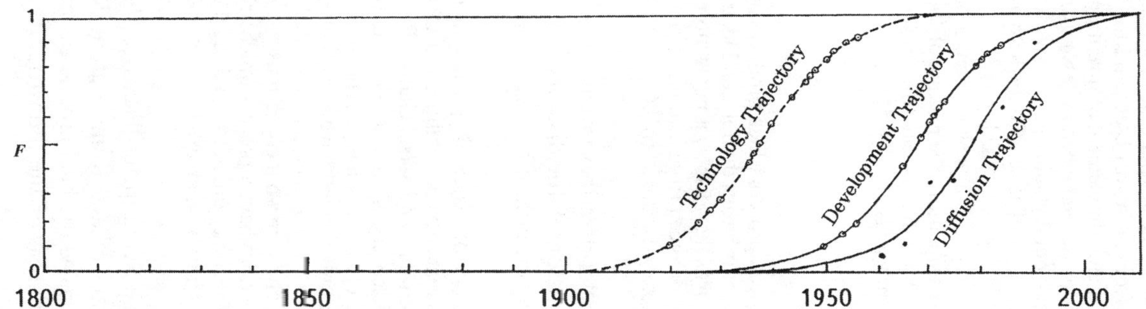

Source: Hirooka (1998e).

Figure 6.15 Polymer industry paradigm

> novelty or not . . . Innovation is possible without anything we should identify as invention and invention does not necessarily induce innovation, but produces of itself, no economically relevant effect at all. The economic phenomena which we observe in the special case in which innovation and invention coincide do not differ from those we observe in cases in which preexisting knowledge is made use of. Stressing the element of invention or defining innovation by invention would, therefore, not only mean stressing an element without importance to economic analysis, but it would also narrow down the relevant phenomenon to what really is but a part of it . . . Even where innovation consists in giving effect to a particular invention, which has either emerged automatically or has been made specifically with a view to a given business purpose and in response to a given business situation, the making of the invention and the carrying out of the corresponding innovation are, economically and sociologically, two entirely different things. (pp. 84–5)

Further:

> The social process which produces inventions and the social process which produces innovations do not stand in any invariant relation to each other and such relation as they display is much more complex than appear at first sight. As soon as it is divorced from invention, innovation is readily seen to be a distinct internal factor of change. (p. 86)

Schumpeter describes invention in his discussion of the external factors in a comparison of the discovery of America and the finding of new possibilities by invention:

> [The issue of externality] raises the questions of new countries and inventions. Both create new possibilities and are no doubt among the most important causes of economic and social change. But are they external factors in our sense? . . . The discovery of America does not thereby become an external factor, for it was not directly relevant to the course of the economic process at all. It acquired relevance only when the new possibilities were turned into commercial and industrial reality, and then the individual acts of realization and not the possibilities themselves are what concern us. . . . As soon as an invention is put into business practice, we have a process which arises from, and an element of, the economic life of its time, and not something that acts on it from without. In no case, therefore, is invention an external factor. (pp. 8–9)

Although Schumpeter discussed the irony of the fact that inventions are not always related to innovation development, he also described various inventions as the background of innovations. Such a description is found in his analysis of the Industrial Revolution in Chapter 6. He first pointed out that:

> It is necessary to guard against possible misunderstanding by making quite clear in what sense we accept the term of industrial revolution and its

> implications . . . It is not only outmoded, but also misleading, or even false on principle, if it is intended to convey either the idea that what it designates was a unique event or the idea that, unconnected with previous developments, it suddenly burst upon the world in the last two or three decades of the eighteenth century . . . It is perfectly clear that the industrial and commercial changes, which occurred in the course of our Kondratiev joined on to and grew out of developments clearly discernible before. Still less can we think of dating it by inventions, for many of the most important ones – such as Darby's, Newcomen's or Kay's – were made in the first half of the eighteenth century, while their genealogy, of course, goes back much further still, in some cases to antiquity. (pp. 253–4)

He later described this point in more detail:

> Many of the most important textile inventions – the flying shuttle, the jenny, Barker's loom, and so on – like inventions in other fields that played a role in the industrial revolution in our sense, had been made earlier, even if we date by eighteenth-century patents what really leads to much more ancient roots . . . Once more we see that invention and innovation are entirely different things, not uniquely related to each other, and that only confusion can result from trying to analyze economic processes in terms of the former. But they interact, of course, and sometimes invention is an incident in an entrepreneurial achievement. Arkwright's figure exemplifies this . . . But if invention is not the core of the matter, neither is objective opportunity . . . We readily see how every step conditions other steps – yarn and cloth, for instance, alternating in offering new demand to each other and in running up against bottlenecks, the removal of which then makes the next achievement. (pp. 271–2)

Schumpeter described the situation of the second Kondratiev and ascribed the upswing to the growth of railways and pig iron production. Interestingly, he states that steel-making technologies were developed during the second wave and substantially diffused during the third wave. This is found in his description of the series of relevant inventions: the Bessemer process, the Siemens–Martin open-hearth process, and the Thomas basic process (p. 355). Furthermore, the electric power technologies that were commercialized during the upswing of the third Kondratiev were observed by Schumpeter to have already been developed during the second wave (p. 255). The trend in electricity technologies was described in detail: induced currents by Faraday in 1831, generation of alternating current by Pixii in 1832, electromagnets by Westinghouse in 1849, shuttle-wound armatures by Siemens in 1860 and so on. The first electric industry was founded in 1895 (pp. 360–61).

These descriptions indicate that Schumpeter recognized that a cluster of core inventions is behind diffusion.

The Process from Invention to Innovation and the Contribution of Science

Schumpeter's work to distinguish innovation from invention and to provide a fairly clear perspective on invention is very impressive. Schumpeter, however, insisted that there is no clear correlation between innovation and invention. Rosenberg (1976) addressed this point by looking for a correlation between invention and innovation. He analyzed the results of the work by Enos (1962), who had first investigated this issue by examining the time lag between invention and innovation for a variety of technological innovations, mainly related to oil refining. Rosenberg found that the average time lag between innovation and invention in the petroleum refinery industry was 11 years, while it was 14 years for 35 other innovations. There were, however, large deviations among lags, which ranged from 3 to 79 years. Mansfield (1968) discussed Enos's data in his paper. Mensch (1975, 1979) also investigated the relationship between invention and innovation and found a way for describing it. In chapter 3 of *Stalemate in Technology – Innovations Overcome Depression* (1975, 1979), he expressed the time lag between invention and product commercialization as the speed of innovation. While his conclusion, 'depression trigger theory', for explaining long business cycles was criticized and not well accepted, his work is significant in that it was a quantitative evaluation of the relationship between invention and innovation.

An important challenge is clarifying the contribution of science to innovation. Rosenberg (1974, 1976, 1982, 1994, 2000) grappled with this. His first approach to the subject *Science, Invention and Economic Growth* (1974), criticized the demand pull approach of Schmookler (1966) and stressed the importance of the supply side approach. Schmookler studied investment, employment and inventive activity in the railroad, petroleum refinery, agriculture and paper-making industries in the United States from the first half of the nineteenth century to the middle of the twentieth century. He found a high degree of synchronicity between the time series for investment and patents, with the investment series tending to lead the patent series. Schmookler argued that upswings in inventive activity responded to upswings in demand; that is, demand induces the inventions that satisfy it. Rosenberg pointed out that many important categories of human wants have long gone either unsatisfied or have been very badly catered for in spite of a well-established demand. Medical progress was minimal until the great breakthroughs of Pasteur and Lister. That the explanation can be found in terms of supply side considerations seems reasonable.

Rosenberg (1982) and Mowery and Rosenberg (1989) quoted the statement by Kuznets (1966) that the distinctive feature of modern industrial societies is their success in applying systematized knowledge, that is, knowledge derived from scientific research, to the economic sphere. This has the disconcerting aspect, at least for economists, of appearing to make the central feature of modern economic growth an exogenous phenomenon. The important issue is specifying the linkages between economics and science. The dependence of technological progress on science has increased substantially in the course of industrialization. Rosenberg (1982) and Mowery and Rosenberg (1989) discussed the growing role of science in the innovation process. They cited many examples of science contributing to economic development: such scientific breakthroughs as Torricelli's vacuum experiment, Carnot's cycle, Gibbs's thermodynamics, Joule's discovery of the energy conservation law, electromagnetic theory by Maxwell, X-rays by Roentgen, identification of benzene's structure by Kekulé, and Pasteur's development of the science of bacteriology.

Mowery and Rosenberg (1989) discussed the importance of the appropriability of basic science research. The key elements were proposed by Nelson (1959) and Arrow (1962b), who considered the economics of knowledge production through investment in industrial research. Neoclassical theory's emphasis on the appropriability of the results of research must be complemented with an analysis of the conditions supporting the utilization of these results. In the absence of an analysis of the utilization of the fruits of R&D, economic theory is hard pressed to provide answers to a broad array of increasingly urgent policy questions.

Endogeneity of Science and Technology

Endogeneity of technology by Rosenberg

Rosenberg (1982) developed his discussion by asking the question, 'How exogenous is science?' He described how daily economic life has become closely linked with science. The industrialization process inevitably transforms science into a more and more endogenous activity by increasing its dependence upon technology. A promising model for understanding scientific advances is one that combines the logic of scientific progress with a consideration of the costs and rewards that flow from daily life and are linked to science through technology.

In his Graz Schumpeter Lecture, Rosenberg (2000) pointed out the importance of the endogeneity of technology. Both science and technology have been rendered a great deal more endogenous in the course of the twentieth century; both play roles of increasing importance in economic life. The growing endogeneity of scientific research goes beyond the function of engineering disciplines in strengthening the incentives to perform scientific research. The development of new technology of considerable commercial significance provides a powerful stimulus to systematic scientific research.

This proposition is surprising only if one is already committed to a rigid, linear view of the innovation process, one in which causality is always expected to run from prior scientific research to 'downstream' product design and engineering development. Rosenberg described how a series of innovations in steel-making (the Bessemer converter, Siemens' open-hearth furnace, and the Gilchrist–Thomas lining) were achieved by using scientific knowledge of chemistry held within the industry. Another example of the endogeneity of science is the invention of the transistor, which occurred in an industrial laboratory using scientific knowledge of quantum physics. The knowledge obtained quickly spread from industry to academia. Rosenberg also addressed the endogeneity of science to university activities in the United States. He focused on the activities of departments of engineering and applied sciences in universities. In particular, he explained the activities related to chemical engineering in the universities and chemical industry. The chemical industry has routinely utilized knowledge of chemical engineering for their business, and university professors have advanced their science based on their experience as consultants to industry. Rosenberg referred to this endogenous activity of chemical engineering as a general purpose technology, GPT.

Nelson and science

Nelson (2000) analyzed the contribution of science to invention and innovation in Part III, 'Science and Technical Advance', of his book *The Sources of Economic Growth*. He first described how the rise of fundamental sciences during the last decade of the nineteenth century was accompanied by several developments that changed the nature of industrial technological innovation and pointed out three important trends: the rise of industrial laboratories, the increased importance of university training programs and the growth in the number of applied sciences and engineering disciplines. In particular, chemical and electrical engineering developed as fields of research as well as training. Research was undertaken in academia and industry. Both Rosenberg and Nelson addressed the increasing importance of the role of universities in the fields of engineering and applied sciences and the increasing endogeneity of science. Nelson also discussed the invention of the transistor at Bell Laboratories, an industrial research facility.

Another important subject considered by Nelson (2000) was the collaboration between academia and industry and the role of the government.

> Over the last decade, debate over the role of American universities in fostering technical advance has intensified. On the one side are those who argue that universities can and should play a larger and more direct role in assisting industry. Such enterprises as Stanford's Center for Integrated Systems, and hundreds of

> centers like it around the country, show a cluster of firms and a university attempting to make their connections more intimate and more effective. The percentage of academic research funded by industry was estimated to be about 6.9 per cent in 1990, up considerably from 3.9 per cent a decade earlier (National Science Foundation, 1991). A recent study estimates that 19 per cent of university research is now carried out in programs that involve linkages with industry in a fundamental way . . . Much of the discrepancy between 6.9 per cent and 19 per cent in the figures reported above is accounted for by government support of these programs. On the other hand, many academics and others see these developments as a threat to the integrity of academic research. They despair that greater involvement with industry and commerce will corrupt academic research and teaching, divert attention away from fundamental research, and potentially destroy the openness of communication among university scientists that is such an essential component of academic research . . . (p. 189).

In his conclusion to Part III, he writes:

> We began this essay by remarking on the significant increase in the fraction of academic research funded by industry over the past two decades, and the rapid growth in the number and size of university-industry research centers. Many in universities clearly see all this as just the beginning, and anticipate a significant further increase of industry funding of academic research . . . Many academics clearly have a firm belief in what has been called the 'linear model' of technological advance, seeing unfettered research by academics as providing the basis for technological innovations in industry, which the academics actually do. The new government programs buy into some of this, but increasingly are insisting upon significant industry involvement in the processes by which research funds get allocated, and therefore influence over the composition and nature of academic research, as well as strong links to ensure 'technology transfer'.
>
> We believe that expectations held by some about what university research can contribute directly to industrial innovation, are quite unrealistic . . . We believe that American universities can help restore competitiveness in those technologies that research illuminates. However, it is important to sort out when universities are capable of helping and where, while there may be problems, university research does not seem to be an appropriate answer . . . While much attention recently has been on the weakness of American industry, we think it a mistake to see universities as a likely source of solution here. Less attention has been given to the erosion of industrial research . . . (pp. 224–5).
>
> Industry has also undertaken some quite fundamental research, and in some fields a good deal. Corporate research laboratories such as Bell Labs, IBM Yorktown, DuPont Central Lab, and others have performed at or sometimes even above the level of top universities. But, many companies have been cutting back on their research. Universities will remain the dominant site of such research . . . Sustained strong public support of university research in fields such as electrical engineering, computer science, and material science will continue to benefit mostly the high-tech industries . . . A shift in emphasis of university research toward more extensive connections with the needs of civilian industry can benefit industry and the universities if it is done in the right way. That is to respect the division of labor between universities and industry that has grown up

> with the development of the engineering disciplines and applied sciences, rather than one that attempts to draw universities deeply into a world in which decisions need to be made with respect to commercial criteria. There is no reason to believe that universities will function well in such an environment, and good reason to believe that such an environment will do damage to the legitimate functions of universities. On the other hand, binding university research closer to industry, while respecting the condition that research be 'basic' in the sense of aiming for understanding rather than short-run practical payoff, can be no enduring benefit of both (pp. 227–9).

Science being endogenous and routinization

As discussed above, Rosenberg and Nelson well described the recent pervasive penetration of science into industry as a tool for dealing with technologies and sometimes playing an important role in innovations. This penetration was expressed by Rosenberg as the endogeneity of science and technology and by Nelson as the institutionalization of science. The development of science forms a science trajectory in the same way as technological innovation, which is described in Chapter 11. In the maturity period, the science is absorbed into society as a common tool for technologies and even various routine activities.

An example of the contribution of academia is chemical engineering, as pointed out by Rosenberg and Nelson. Chemical engineering was established on the basis of chemical physics, thermodynamics and physical chemistry. The science trajectory of chemical engineering has been already established and is now in a maturity period. Thus, the knowledge of chemical engineering has been routinely used to analyze chemical processes, to construct chemical plants and to operate chemical apparatuses. These sciences of chemical engineering and electrical engineering have been utilized by industry as tools. This is the reason why these sciences have been more endogenized and institutionalized.

An important consideration is whether a particular scientific matter is an element on the science trajectory or not. A scientific creation should be on a science trajectory. If not, it is a routine tool rather than an element of the innovation trajectory. Once a science has been established, it becomes simply one of an established discipline in a routine situation, but is not a creative scientific activity.

The Scientific Technology Trajectory

Consequently, science has penetrated into industry, which can carry out basic research in its laboratories, because of the enhanced level of research even in industry. Let's discuss in more detail basic research in industries as described by Nelson.

While the transistor, a core technology of the electronics paradigm, was developed in an industrial laboratory, it was the result of a series of scientific studies. That is, as will be shown in Figure 7.4, the invention of the transistor was based on the science precursor trajectory of semiconductors, which had a time span of 25 years and consisted of various scientific findings, such as the discoveries of the semiconductor, the rectification phenomenon, the rectification mechanism, the electroconductive mechanism, the tunnel effect in rectification, the Schottky barrier concept and the p-n junction phenomenon. The transistor was invented on the basis of the semiconductor science trajectory. The place where the invention occurred does not matter.

The discovery of nylon 6,6 in a condensation polymerization reaction by Carothers at DuPont in 1928 is an element on the polymer science trajectory, which started with the recognition of macromolecules by Staudinger in 1920. While nylon fibers were first commercialized in 1935, it took over 13 years before sufficient commercial production was achieved (DuPont, 1952, p. 107). Another element on the polymer science trajectory is the invention of polyethylene in the research laboratory at ICI Co., UK in 1932. While they soon discovered that polyethylene is an excellent insulator for radar, its potential as a general purpose plastic was not so evident because of the severe polymerization conditions for preparing it (extremely high pressure and high temperature). It wasn't until 23 years later, in 1955, that ICI launched substantial mass production of polyethylene in Wilton, England. This demonstrates that even industry cannot explicitly forecast the future market for innovation products and that it takes time for full-scale commercial production to begin.

These examples show that even industry can carry out basic research that contributes to the technology trajectory. They also indicate that it is difficult to foresee the future possibilities of markets based on a technology trajectory. This means that the course of a technology trajectory has an implicit nature.

On the other hand, science created in academia has become more and more utilized in industry and is apt to become easily industrialized. We have seen many cases in which the technology trajectory is a science trajectory that emerged in academia. In other words it is actually a pure science trajectory created at universities. As described above, the diffusion trajectory starts when the technology trajectory is almost finished. This means that industry often has no chance to contribute to the formation of the technology trajectory, indicating the exogenous nature of the technology trajectory. While science has penetrated the industrial sphere, becoming endogenous, the scientific discoveries that are elements of the technology trajectory are substantially exogenous in terms of the innovation paradigm. This means that the emergence of science is serendipitous and cannot be triggered by market demand. Therefore, to promote innovation, we should become more involved with academic science at an earlier stage.

Governmental Policy for Promoting Scientific Activities

The United States Government strongly recognizes the increasing importance of science to the economic society, especially for emerging industries, which will become pioneers and strengthen international competitiveness. At the end of the Second World War, Vannevar Bush (1945), the Director of the Office of Scientific Research and Development in the United States, submitted a report entitled 'Science: The Endless Frontier' to the President. In it he stressed the importance of science for stimulating national activities and strengthening international competitiveness. He concluded that a National Research Foundation should be created in order to promote national interest in scientific research and scientific education.

He addressed the importance of basic research. Basic research is performed without thought of practical ends. It results in general knowledge and an understanding of nature and its laws. This general knowledge provides the means for answering a larger number of important practical problems, though it may not give a complete specific answer to any one of them. The function of applied research is to provide such complete answers. The scientist doing basic research may not be at all interested in the practical applications of his work, yet the further progress of industrial development would eventually stagnate if basic scientific research were long neglected (Bush, 1945, p. 13).

Bush specifically described the matter of research and industry in chapter V of the Appendix:

> One important fact is clear . . . the process of transition from pure research to its practical applications does not work equally effectively in all industries . . . The National Research Foundation should be charged with the responsibility of studying the process of technological development in industry and of experimenting with methods of aid to industrial research. The following suggestions are tentative and submitted with the thought that they might be of assistance to the Board in meeting this important challenge.
>
> A. Assistance to Technological Clinics for Small Business Enterprise
> B. Grants to Nonprofit Industrial Institutes for Fundamental Research
> C. Encouragement for New Scientific Enterprises
> D. Strengthening the Patent System

The United States established the National Science Foundation in 1950 to promote progress in science; to advance the national health, prosperity, and welfare; and to secure the national defense.

In 1959, President Carter sent a message on innovation to Congress, appealing for a strengthening in the international competitiveness of industries. The Stevenson and Wydler Technology Innovation Act of 1980 was designed to stimulate technology transfer from academia to industry and to encourage high-tech small businesses.

A series of such actions – certainty based on a linear model of innovation – became the basis for the splendid achievements of the United States in science-based frontier industries.

Distinction between Technology Development and Diffusion Periods

The innovation paradigm is composed of three interrelated trajectories, as described above. The nonlinear nature of the innovation process was disclosed by Rostow (1960), van Duijn (1983, 1984), Marchetti (1979) and Perez (2002), all of whom described the innovation process using an S-shaped curve reaching a maturity period.

Perez (2002) in particular described the innovation paradigm, starting from the initial 'big-bang' origin of innovation technology to the maturity of the market, as a comprehensive system that she formulated using a single S-shaped curve. She divided the innovation paradigm into two periods: 'installation' and 'deployment'. Each lasts between two and three decades, and each consists of two phases. The first period, from the big-bang origin to the high growth of innovation diffusion, consists of two phases, 'irruption' and 'frenzy', which result in a bubble economy. There is a transitional 'turning point' which corresponds to the collapse of the bubble. This is an adjustment stage preceding the coming deployment period. The early phase of the deployment period is 'synergy', in which the innovation paradigm flowers; this is followed by the 'maturity' phase. This is Perez's scenario for the innovation paradigm. In this scenario, the Great Depression occurs at the end of the frenzy phase. This scenario matches ours, described in Chapter 5, in which an infra-trajectory extends for more than 50 years across two Kondratiev cycles, as shown in Figure 5.10. The synergy phase corresponds to the mass-consumption stage of Rostow (1960).

Perez (2002) identified big-bang origins in 1771, 1829, 1875, 1908 and 1971. These are the inventions of Arkwright's spinning machine, the 'Rocket' steam locomotive, the Bessemer furnace, the Model-T automobile, and the microprocessor. According to our analysis, these inventions are just the origins of development trajectories brought about by previous technology trajectories, so that Perez's surges start from development trajectories and do not include technology trajectories (which she refers to as 'gestation periods' but does not discuss at all). Before this long trajectory of Perez's, there are exogenous technology trajectories.

The origin of innovation technology is believed to have no causal relationship to certain conditions of the economy; rather the emergence is serendipitous in an exogenous condition. When a chaotic soup of scientific knowledge reaches some critical level, a serendipitous idea pops up that leads to an invention or an epoch-making discovery. The scientific technology trajectory becomes visible in the economy after the accumulation and concentration of related knowledge leads to a development trajectory.

Exogeneity of Emergence of Innovations

As discussed above in this section, industrial innovations follow a series of relevant inventions increasing scientific intensity. On the other hand, an innovation always originates from a creative invention or an epoch-making discovery. That is, the origin of innovation is exogenous and serendipitous. Let us clarify the endogenization of science in innovations and exogeneity of emergence of innovation.

Enos (1962), Kuznets (1966), Mansfield (1968) and Mensch (1979) investigated the correlation between inventions and innovations and determined the time lag between invention and commercialization. This is a direct recognition of the causal relation between innovation and invention. Rosenberg (1974, 1976, 1982, 1994, 2000) discussed the correlation between science and innovation, and suggested that the supply side considerations will frequently play a critical role in determining the breakthrough of innovation.

The recognition by the US government of the importance of science and knowledge transfer from academia to industry since Vannevar Bush's proposal has led to the implementation of various measures. Nelson (2000) described how this trend of academia–industry collaboration has been accelerated and how it is closely related to international competitiveness. This means that the future of industry strongly depends on the fruits of academia and that the supply side will become more important.

Using our model of innovation having three trajectories, we can interpret the above as follows.

Enos, Kuznets, Mansfield and Mensch dealt with the development of inventions to innovations, which is the transition from a development trajectory to a diffusion one. They did not describe a technology trajectory or the origin of the emergence of innovation. While Perez's big bang is also the beginning of the development trajectory, it is not the beginning of the technology trajéctory.

Schumpeter (1939) recognized a series of relevant inventions, which correspond to the existence of development trajectories, in the case of the steelmaking and electricity innovations. He pointed out that a single invention does not initiate innovation; instead a series of inventions or discoveries are the precursors to innovation. More important is that he implicitly recognized that core technologies were behind a series of inventions of the spinning machine. He thus recognized the technology trajectory as well, although he was not aware of it.

Rosenberg (2000) specifically pointed out that there had been scientific developments in the areas of chemistry and chemical analysis prior to the series of inventions forming the steel-making innovation. This is unambiguous recognition of the science trajectory behind the development trajectory, which included inventions by Bessemer, Siemens and Thomas. This indicates that the innovation in steel manufacturing was achieved thanks to the science trajectory of chemical analysis. This is the same relationship found for synthetic dyestuffs, as being described in Chapter 7, in which the trajectory is the science of organic chemistry.

Rosenberg also criticized Schmookler's demand side argument and stressed the importance of the supply side of science and technology. This suggests a seed-oriented approach for a technology trajectory. The United States government recognized the importance of activities in academia and knowledge transfer from academia to industries. This is the recognition that cutting-edge innovations frequently originate in academia and that the technology trajectory develops more effectively in academia, particularly when academia collaborates with industry. This supply side concept is connected with the recognition of the exogeneity of innovation emergence.

In short, innovations have depended on academic science. An important aspect is the trigger of innovations, which is becoming more and more science-intensive. Even creative basic research in industry could be exogenous because of serendipity. This trend does not contradict the endogenous trend of scientific intensity in industrial innovation because of the issue of the development process of innovations. That is, innovations emerge exogenously and develop with the increasingly endogenous trend of science.

6.5 SUMMARY

In this chapter, the technology development period in the innovation paradigm was intensively analyzed and an attempt was made to describe it with a concrete equation. The comprehensive structure of the innovation paradigm was clarified.

I. Innovation trajectories

1. The innovation paradigm consists of three trajectories: technology, development and diffusion in that order.

2. Each trajectory has a nonlinear nature that can be expressed using a logistic equation, which means it matures within a definite time span.
3. The technology trajectory is a cluster of core technologies starting from a basic invention or an epoch-making discovery. The development trajectory is a cluster of new products developed using core technologies.
4. The logistic nature is shown by the fact that the distribution of the number of relevant innovation patents or number of new products in the course of innovation forms a logistic curve. All developed products are distributed throughout the logistic curve of number of innovation products or patents.
5. Due to the nonlinear nature of the trajectories, the result of (4) indicates that the logistic curve can be determined by evaluating the cluster of inventions (innovation elements) arranged in time sequence. That is, the logistic curve can be defined by the determination of the time span between the first invention and the last one.

II. Analysis of each trajectory

6. The time span of technology trajectories has grown shorter year by year since the Industrial Revolution, from about 60 years to 25–30 years.
7. Technology trajectories emerge to make a cluster in each epoch, and each cluster corresponds to the upswing of a Kondratiev cycle. This cluster formation of innovations seems to be the origin of Kondratiev cycles.
8. The diffusion trajectory starts when the technology trajectory is nearly completed, in a cascade fashion.
9. A series of innovation paradigms in the chemical industry were identified and each paradigm was found to consist of three trajectories: technology, development and diffusion, in that order.

III. Role of science and the origin of innovations

10. The role of science and invention in innovation was discussed. Innovations have increased scientific intensity and science has been endogenized deeply into industry.
11. The recognition of the technology trajectory is the most important aspect of this study. Although the technology trajectory has not yet been sufficiently recognized, the emergence of innovations depends deeply on academic activities being exogenous, and science plays a crucial role in the technology trajectory.
12. Even when basic research is carried out in industry, it is hard to recognize its significance as the origin of subsequent innovation. That is, the technology trajectory is implicit at the beginning, even in industry, and has an exogenous nature.

7. Comprehensive description of the innovation paradigm and the role of the development trajectory

The previous chapter determined that an innovation paradigm consists of three trajectories: technology, development and diffusion. This chapter describes comprehensive paradigms comprising these trajectories and discusses the role of each trajectory in detail. The development trajectory is the junction of knowledge transfer from the technology trajectory to the diffusion trajectory and plays a key role in the creation of new products; venture firms emerge explosively during the course of the development trajectory. The behavior of firms is described over the course of the innovation paradigm and Schumpeter's 'bandwagon effect' is illustrated through the number of firms that emerged.

7.1 IDENTIFICATION OF THE INNOVATION PARADIGM

In the previous chapter, the three trajectories were shown to exist in a series of chemical industry paradigms. The development period of innovation technology consists of two trajectories. After core technologies are completed, new products start to be developed. This is also a time of application research and market development. Three innovation paradigms in particular are now described in more detail, and the mutual relationships between the three trajectories are discussed using as examples the synthetic dyestuffs paradigm in the latter half of the nineteenth century and the present day biotechnology and electronics paradigms.

Synthetic Dyestuff Paradigm

While England was the dominant player during the Industrial Revolution, innovation technologies were gradually transferred to Germany and the United States in the second Kondratiev cycle, that is, in the latter half of the nineteenth century. Germany quickly developed and industrialized during this period. The chemical industry did particularly well in Germany and acquired a strong international competitiveness. Success was especially remarkable in the synthetic dyestuffs industry.

The science of chemistry sprang from the outstanding achievements of Lavoisier in France in 1777 and was advanced by a series of further discoveries. Organic chemistry progressed mainly in Germany following Wöhler's discovery that organic chemistry is the same as inorganic chemistry when he synthesized the urea of organic substances from inorganic compounds. Leading German scholars, such as Liebig, Bunsen, Hoffmann, Würtz, Kekulé and Baeyer, then proceeded to establish the science of organic chemistry in the first half of the nineteenth century. Brilliant achievements in organic chemistry formed the technology trajectory, followed by the development of synthetic dyestuffs. That is, when it became possible to analyze the structure of organic substances due to the development of organic chemistry, synthetic dyestuffs began to be developed. Thus, the technology trajectory of synthetic dyestuffs was actually a science trajectory.

Since the earliest times, all dyestuffs had been produced from natural materials (plants, animals and minerals), for example, alizarin from madder root, indigo dye (indigotin) from the indigo plant, and purple dye (purpura) from murex (shell fish). As these natural dyestuffs were very expensive, synthetic, less expensive dyestuffs had been sought for. The first synthetic dyestuff, mauve (aniline dye), was invented by Perkin in 1856 at the London College of Chemistry. The next, fuchsine, was developed in France in 1859. The diazo reaction was discovered by Griess and the first azo dye (Bismark Brown) was invented by Martius. The brilliant invention of synthetic alizarin was carried out in the Berlin Institute of Technology, and the patent was granted to Caro at Badische Anilin und Soda Fabrik (BASF). Eosin and methylene blue were developed in 1873 and 1877, respectively. The most desired dyestuff, indigo, was first synthesized in 1880 by Professor Baeyer at the Berlin Institute of Technology and was commercialized by Hoechst in 1897 and further by BASF in 1901. The production of synthetic indigo in Germany increased from 337 tons in 1898 to 4320 tons in 1906, while the production of natural indigo in India fell drastically from 167 tons in 1898 to 70 tons in 1906 and 22 tons in 1912. This is a typical example of what has been called the 'gale of creative destruction'.

This development history of new products formed the development trajectory of synthetic dyestuffs, which had a time span of 40 years. The diffusion trajectory extended about 15 years behind the development trajectory. The production of tar dyestuffs in Germany was valued at 72 million marks in 1896, 72 per cent of the worldwide value; it reached 120 million marks in 1900, 82 per cent of the worldwide value. The comprehensive synthetic dyestuff paradigm is shown in Figure 7.1.

Biotechnology Paradigm

Biotechnology can be defined two ways. It can mean production technology using organisms in the wider sense, including fermentation and all production systems aided by living resources. It can also mean technologies aided by molecular biology, the so-called 'new biotechnology'. Biotechnology here refers to the latter.

The trajectory of biotechnology depends on scientific research in academia. It started with the breakthrough discovery by Avery in 1944 that DNA (deoxyribonucleic acid) is the substance of genes, the key to the scheme of heredity. Research developments forming the elements of the technology trajectory include the discovery of the double helix structure of DNA by Watson and Crick in 1953, of the central dogma (the mechanism of the formation of proteins from DNA) by Crick in 1958, and of recombinant DNA by Cohen and Boyer in 1973. The development of recombinant DNA technology led to the synthesis of human insulin in 1978. The development trajectory began on the basis of this success. This trajectory is the locus of knowledge transfer from academia to industry and also the locus of the development of new products for commercialization. After the success of the synthesis of human insulin, various biotechnology products have been developed, such as human growth hormone, α interferon, erythropoietin, interleukin-2 derivatives, A-type hepatitis vaccine, and insulin growth factor I. The three trajectories of the biotechnology paradigm are illustrated in Figure 7.2.

Electronics Paradigm

The technology trajectory of electronics was shown in Table 6.2 and Figure 6.3 in the previous chapter. Most of the elements were advances made by industry, starting with those at Bell Laboratories by Shockley et al. and continuing through those made at Texas Instruments, Fairchild and IBM. The development trajectory is described by the trend in the degree of chip integration. After the development of submicron lithography by IBM, the development trajectory took off and included the development of 16K DRAM in 1975, 64K in 1978, 256K in 1980, 1M in 1984 and 256M in 1994, following the predictions of Moore's law, that is, the degree of integration steps up about four times every 3 to 4 years. Along this development trajectory, microprocessors, workstations and various types of software were developed together with various application technologies. The diffusion trajectory is described by the consumption of integrated circuits. These trajectories each have a time span of 25 years, as shown in Figure 7.3.

The bigger contribution by industry to the elements of the technology trajectory of electronics compared to those for synthetic dyestuffs and biotechnology can be explained by the precursor paradigm of transistors before the electronics trajectories of integrated circuits; that is, the transistor paradigm, or more precisely, the research paradigm of rectifying property of semiconductors, formed by a science trajectory. This science trajectory of transistors started in 1915 with the postulation for the mechanism of the point-contact-type rectifying property of germanium by Benedicks, and was advanced by various discoveries and theories as elements; the rectifying property of germanium by Merritt in 1925, the Wilson model of semiconductors by quantum mechanics in 1931, the concept of the Schottky barrier forming the background of semiconductor physics, postulation of diffusion rectification theory by Mott and Schottky in 1939, and the discovery of the p-n junction by Scaff and Theurer in 1941. This science trajectory had a time span of 25 years, its development trajectory was a series of semiconductor rectifier developments between 1940 and 1965, and the diffusion trajectory started with the production of transistors in 1954. The development trajectory switched to the paradigm of electronics. In that sense, the trajectories for transistors were precursor trajectories for electronics. These relationships are shown in Figure 7.4. That is, the transistor and electronics paradigms can be considered to be a twin paradigm. Shockley was invited to Bell Laboratories by Kelly in 1935 and devoted himself to developing a completely new amplifier to replace vacuum tubes.

7.2 ROLE OF ACADEMIA IN THE INNOVATION PROCESS

It is clear that the technology trajectory is often composed of the results of basic research, carried out mostly at universities, that is, scientific research. The technology trajectory of synthetic dyestuffs is the outcome of research in the field of organic chemistry, which is pure science, at universities. Biotechnology was established through basic research in the field of genetic science at universities, starting with the discovery of genes by Avery and ending with the breakthrough discovery of recombinant DNA by Cohen and Boyer. The technology trajectory of polymer technology is also a kind of scientific trajectory in which research was mostly carried out at universities, starting with the establishment of the concept of macromolecules by Staudinger. The technology trajectory of electronics was formulated mostly through research by industry, but the precursor trajectory was mostly composed of basic science. Furthermore, many universities were involved in the establishment of electronics technologies, especially Stanford University. More importantly, the development trajectory encompasses technology transfer from academia to industry.

Now, let's look at the collaboration between academia and industry for some innovation paradigms.

In the case of synthetic dyestuffs, the first invention of a dyestuff occurred at the Royal College of Chemistry in London. In the first half of the nineteenth century, advances in organic chemistry were made mainly in Germany. Professor Justus von Liebig of Giessen University was invited to London University and based on his advice, the Royal College of Chemistry was founded in 1845. Professor W.H. Hoffmann from Germany was appointed the first director and his assistant, W.H. Perkin, discovered the first synthetic dye, mauve (aniline dye obtained from coal tar), in 1856. This was an epoch-making event – Queen Victoria wore a dress dyed with mauve at the opening of the International Exhibition of 1862, and penny postage stamps were printed with mauve.

The next important event was the invention of synthetic alizarin (natural alizarin had been prepared from madder grown by farmers, mainly in France). This brilliant red dye was first synthesized in the laboratory of Professor Baeyer at the Berlin Institute of Technology by Graebe and Liebermann. This achievement was based on the discovery by Kekulé that benzene has a hexagonal ring structure. Following this finding, Graebe and Liebermann were able to work out the structure of alizarin by identifying the structure of naphthalene and to develop a process for preparing synthetic alizarin. They submitted patent applications to the Prussian and British patent offices and also entered into a patent license agreement with BASF (Badische Anilin und Soda Fabrik). Caro at BASF improved this process for commercialization, and submitted a patent application on 25 July 1869 after some trouble with the attorney in England. Meanwhile, in England, Perkin also prepared alizarin and submitted a patent application on 26 July 1869, one day after that by Caro. An agreement was subsequently reached between Perkin and BASF giving Perkin the right to manufacture alizarin in England under license from BASF. The madder-growing industry fell into rapid decline and soon ceased to exist. The production of madder in France fell from 25 000 tons in 1862 to 500 tons in 1878.

Professor Adolf Baeyer of the Berlin Institute of Technology played a key role in the development of dyestuffs, and the research staffs at BASF and Hoechst, both chemical companies in Germany, collaborated with him. The development of eosin was the fruit of collaboration between Caro at BASF with Professor Baeyer's laboratory. Success with synthetic alizarin stimulated chemists to turn their attention to indigotin. Professor Baeyer identified the structure of indigo and succeeded in synthesizing indigo in his laboratory in 1880. This process was licensed to BASF and Hoechst. Professor Baeyer was awarded the Nobel Prize in chemistry in 1905 for his outstanding contributions to the development of organic dyestuffs and

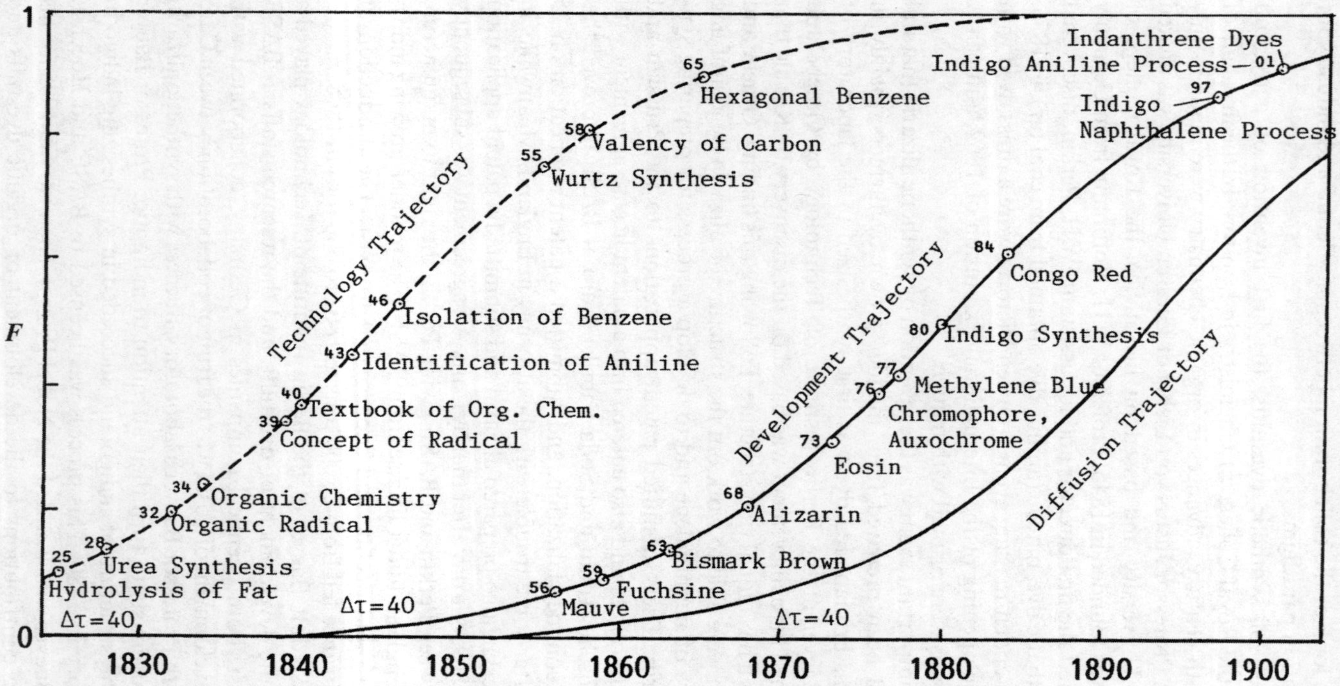

Sources: Hirooka (1998b, 1999c); data from Aftalion (1991), McNeil (1990) and Kaku (1986).

Figure 7.1 Innovation paradigm of synthetic dyestuffs

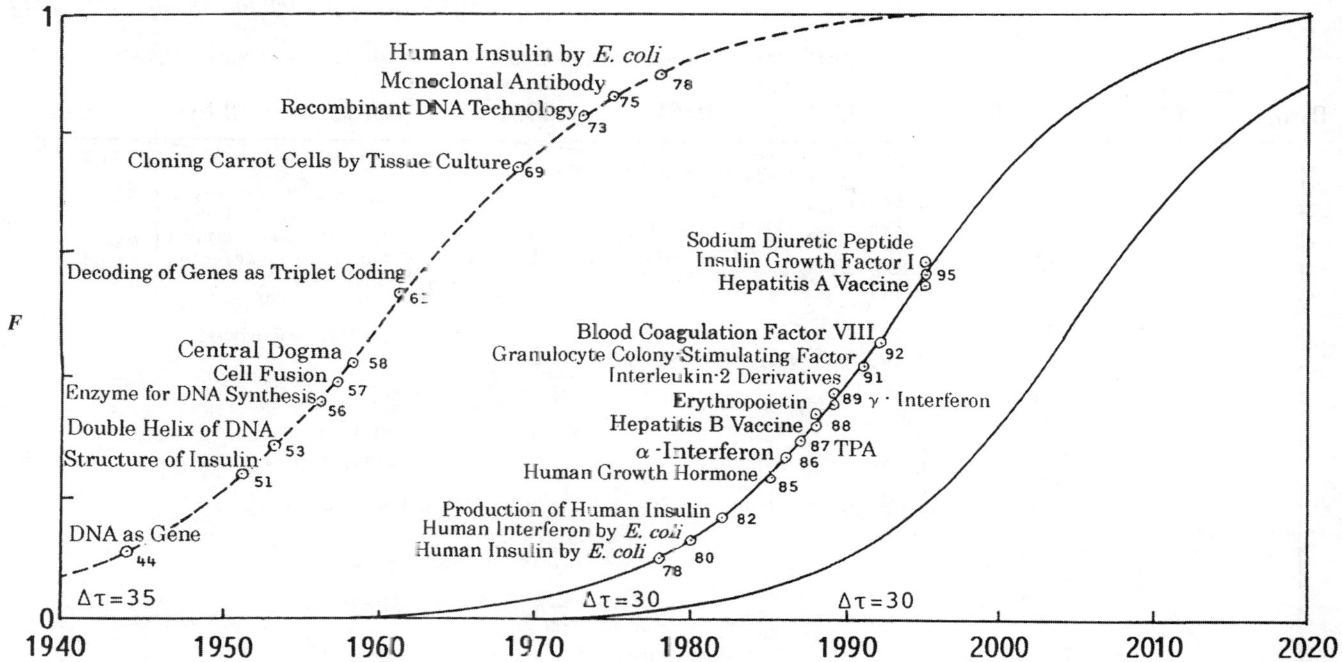

Sources: Hirooka (1998b, 1999c); data from Walker Gingold (1985), and JBA (1999).

Figure 7.2 Innovation paradigm of biotechnology

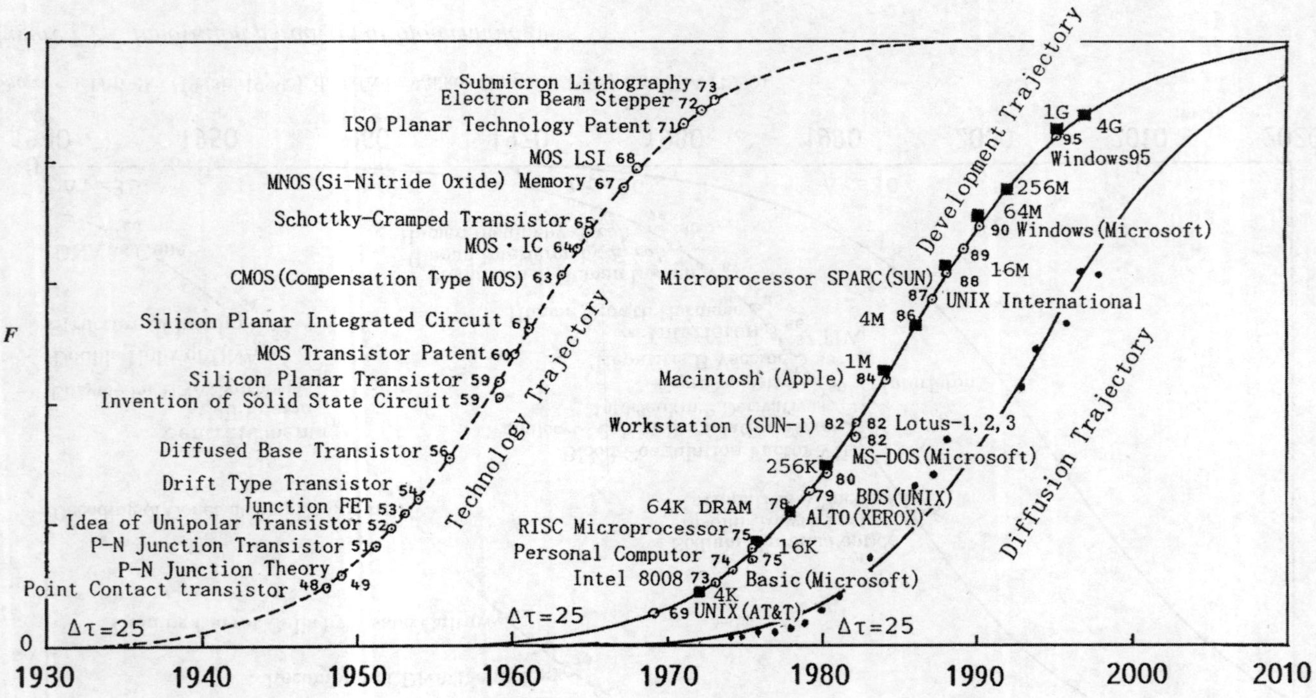

Sources: Hirooka (1998b, 1999c); data from Kisaka (2001), Shimura (1992) and MITI (2001).

Figure 7.3 Innovation paradigm of electronics

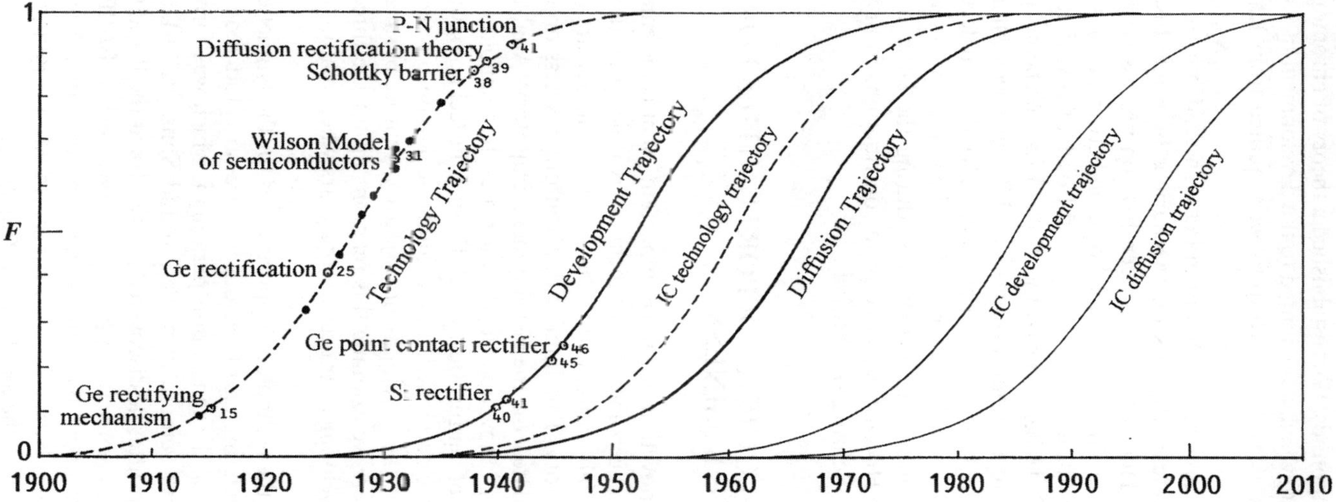

Source: Data from Kisaka (1984).

Figure 7.4 Transistor paradigm: precursor trajectories for electronics

hydroaromatic compounds. Witt established a theory of chromophore and auxochrome in 1876, which contributed to the development of various azo dyes. (Descriptions of synthetic dyestuffs were quoted from Kaku, 1986, and McNeil, 1990.)

There are many cases of collaboration between academia and industry in Germany. The research results for saccharides by Emile Fischer were commercialized by Hoechst. The nitrogen fixation process for producing ammonia came about through the collaboration between Professor Haber of the Karlsruhe Institute of Technology and Bosch of BASF; it was commercialized in 1913. Haber received the Nobel Prize in chemistry in 1918 for his contribution to ammonia synthesis, and Bosch received it in 1931 for his contribution to high-pressure chemistry and its industrial development. In the field of medicine, the development of chemotherapy owed much to contributions from academia in Germany during the latter half of the nineteenth century. Hoechst benefited from contributions from academia, especially by Koch, Behring and Ehrlich.

7.3 TECHNOLOGICAL OPPORTUNITIES FOR VENTURE BUSINESS

In the course of innovation, the most important event is the emergence of venture business. As seen above, most technology trajectories developed in academia, while collaboration with industry became important during the development trajectory. This implies that venture business opportunities should be accumulated on the development trajectory. Analysis of the logistic dynamics of innovation clearly illustrates the technological opportunities and timing of venture businesses. During the technology trajectory, core technologies are developed, and technological opportunities open up. At the beginning of the development trajectory, venture businesses begin to be launched. Here we examine this relationship for the three cases by plotting on the development trajectory the years when venture businesses were established.

Figure 7.5 plots the technology opportunities for the synthetic dyestuffs industry – the launch timings of various ventures are plotted on the development trajectory of the innovation paradigm, which was formulated in Figure 7.1. The first venture firm, Perkin and Sons, was established by Perkin just one year after he discovered mauve dye at the Royal College of Chemistry. Leading chemical companies of today, such as BASF, Hoechst and Bayer, were set up at that time as small venture firms. Note that the launching of venture businesses was concentrated in a period of 10–15 years in the first half of the development trajectory.

BASF (Badische Anilin und Soda Fabrik AG) originated from Chemische Fabrik Dyckerhoff Clem AG, which was founded by Engelhorn, Clem, Sonntag and Dyckerhoff in 1861 for producing tar dyes, and became BASF in 1865. Friedrich Bayer learned the process for producing synthetic tar dyes and set up Bayer AG with his friend Weskott to produce anilin blau from fuchsine in 1863. Eugen Lucius purchased land and constructed a factory in Hoechst to produce dyestuffs and founded Meister Lucius AG with his brother-in-law Meister in 1863. They changed the name to Farbwerke Hoechst in the same year. As dyestuffs were very expensive, many get-rich-quick schemers crowded in, and the number of venture businesses exceeded 25 within the 1860s; those maintaining close contact with academia were the ones that survived.

In the case of the biotechnology paradigm, the technological opportunities were also concentrated within a period of 10–15 years during the first half of the development trajectory, which extended from 1980 to 2005, as shown in Figure 7.6. Most of the active firms in biotechnology, such as Cetus, Genentech, Biogen and Amgen, were founded during that period.

The electronics paradigm is shown in Figure 7.3, and the development trajectory comprises the locus of advancements in DRAM subsequent to the completion of the core technologies. Venture businesses were launched from the beginning of the development trajectory, and the major firms of the contemporary electronics industry, such as Intel, Microsoft, Apple Computer and Sun Microsystems, developed rapidly during the first half of the development trajectory of electronics, as shown in Figure 7.7.

These results clearly indicate that opportunities for venture businesses, as reflected in the innovation paradigm, tend to be concentrated within a limited period in the first half of the development trajectory. This common behavior of venture businesses coincides with what Schumpeter called the 'bandwagon effect'. Utterback (1994) also recognized a sharp increase in the number of firms at the beginning of innovation diffusion.

7.4 FIRM BEHAVIOR IN THE INNOVATION PARADIGM AND LOCK-IN

The innovation paradigm has a common narrative: an entrepreneur obtains a basic patent and enjoys the exclusive right to produce the innovation product. Once an attractive new market develops, the business enters a high-growth period, and many firms rush to enter the market; this is Schumpeter's bandwagon effect, stirring up a hornet's nest of activity. Once the market matures, natural selection of firms takes over, and the innovation paradigm comes to an end.

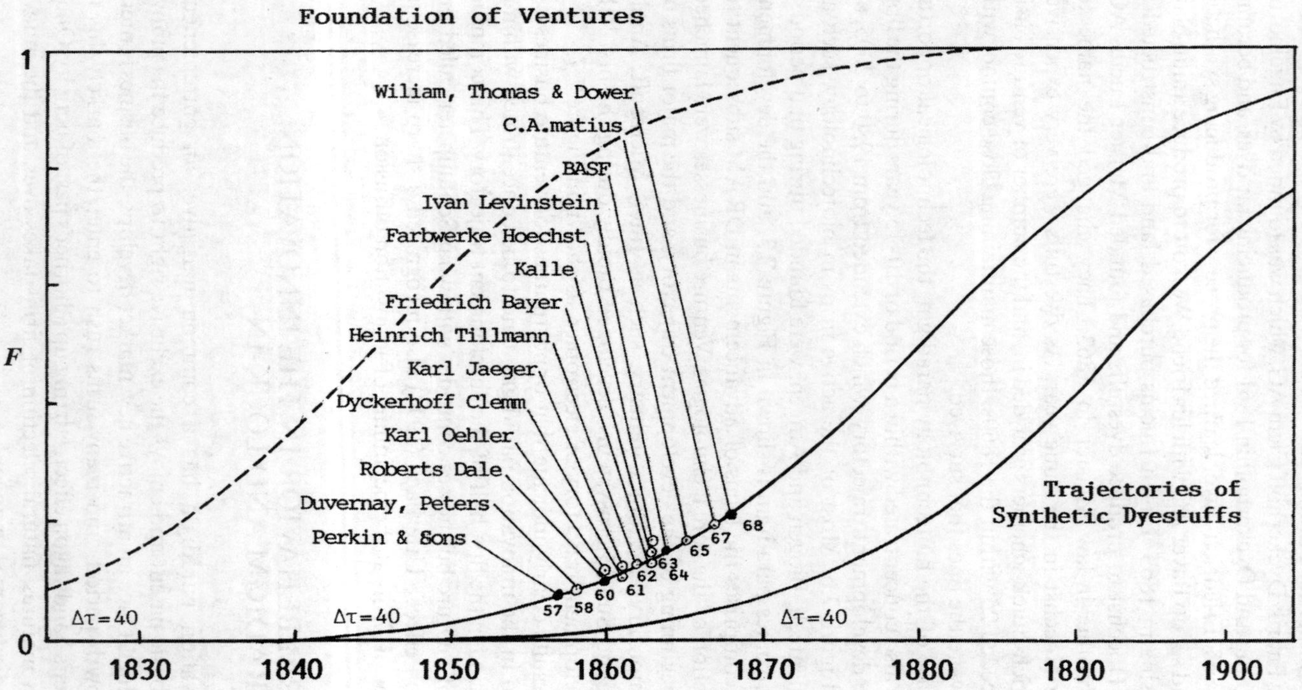

Source: Hirooka (1998b, 2000b).

Figure 7.5 Timing of venture business in dyestuff paradigm

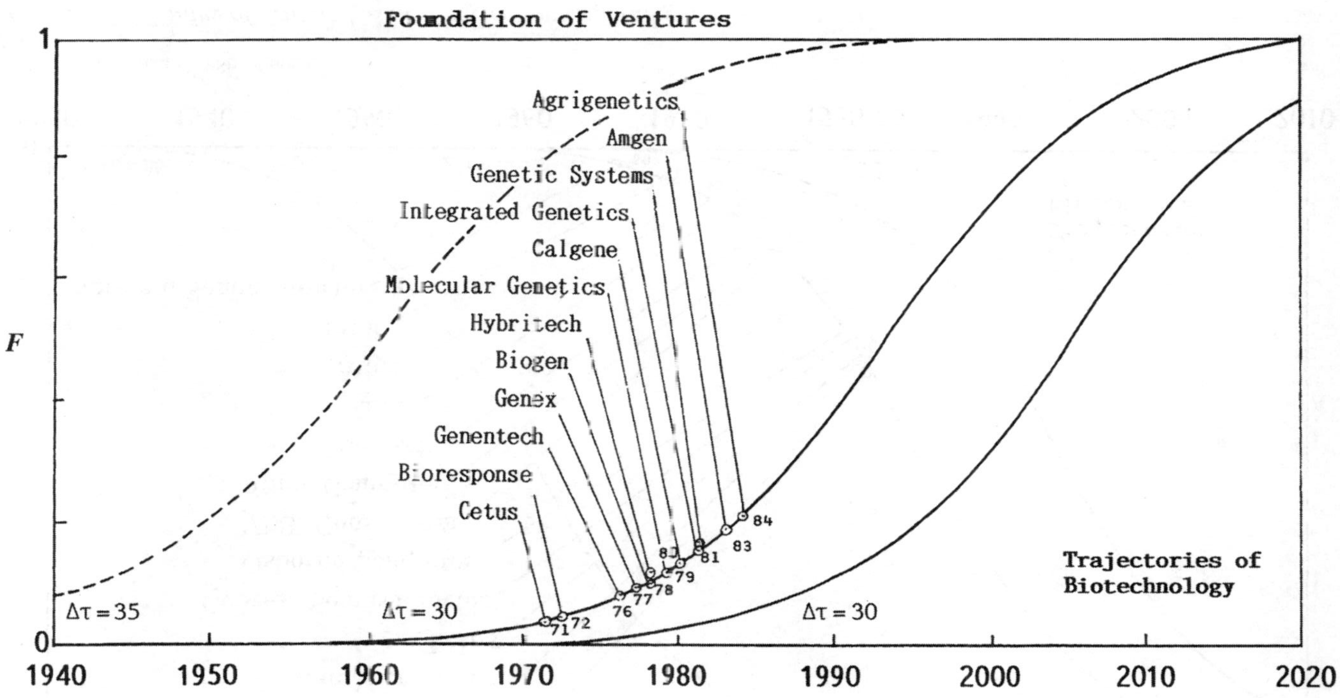

Source: Hirooka (1998b, 2000b).

Figure 7.6 Timing of venture business in biotechnology paradigm

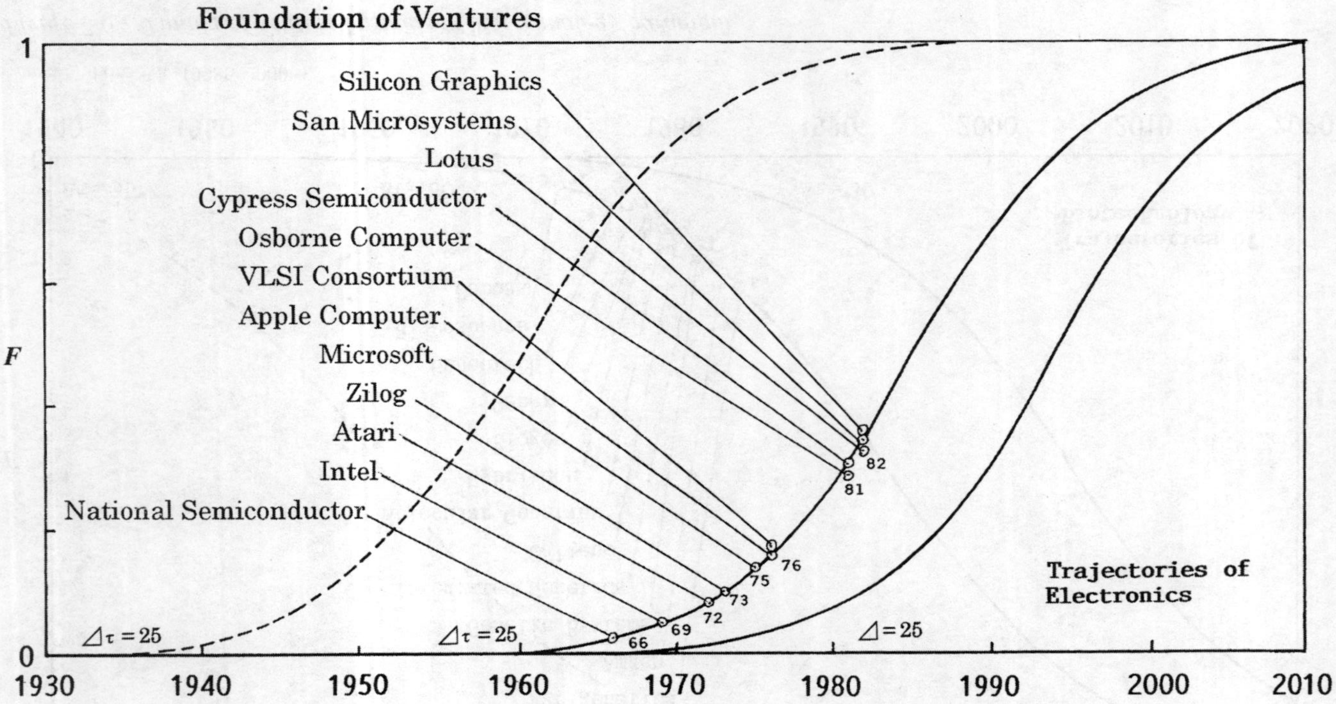

Source: Hirooka (1998b, 2000b).

Figure 7.7 Timing of venture business in electronics paradigm

Interesting studies by Utterback and Abernathy (1975), Utterback and Suárez (1993) and Utterback (1994) revealed that, during the life of an innovation, versatile products appear during the first period of innovation and then converge toward a specific design with which a firm can obtain a competitive edge. They called this design the 'dominant design'. The analysis by Utterback (1994) showed that the number of firms sharply increases as the innovation progresses, showing a bandwagon effect. Once a dominant design emerges, the number of firms begins to decrease due to natural selection. Utterback and Abernathy described many examples of such behavior. Utterback (1994) suggested the concept of innovation dynamics as an innovation paradigm consisting of product innovation followed by process innovation, which has three phases: fluid, transitional and specific. This differs from our scheme of trajectories.

We examined the trend of the number of firms by applying our innovation paradigm comprising three trajectories. We look at three cases of innovation in particular: typewriters, automobiles and synthetic dyestuffs. The first two cases were analyzed using Utterback and Suárez (1993), and the third was analyzed using statistical data by Murmann and Homburg (2001).

The Typewriter Paradigm and Firm Behavior

Utterback (1994, p. xx) states that:

> At first an innovation such as the typewriter may be almost entirely a combination of design elements tried out in earlier uses or prototypes. As the market grows, greater emphasis will usually be given to development of components tailored especially for the product itself. Ultimately these may be synthesized into a model that includes most features and meets most user requirements, a model we have termed a 'dominant design'. . . . A dominant design usually takes the form of a new product synthesized from individual technological innovations introduced independently in prior product variants. A dominant design has the effect of enforcing or encouraging standardization so that production or other complementary economies can be sought. Then effective competition begins to take place on the basis of cost and scale as well as of product performance.

According to Utterback (1994), the keyboard arrangement on the typewriter became fixed in the current QWERTY order in 1874, and the Underwood No. 5 model became the dominant design in 1899. As there is not enough information for the elements of a technology trajectory to be traced to the patenting of the writing engine by Henry Miller in 1714, the starting point of the development trajectory of the typewriter is taken to be around 1870, with the development of the Sholes typewriter; therefore, the development trajectory is located in the 30 years between 1870 and

1899. The diffusion trajectory was described using the trend in the number of typists. The innovation paradigm of the typewriter, together with the number of relevant firms shown by the axis on the right, is illustrated in Figure 7.8. When Sholes produced his first keyboard typewriter in 1874, there was only one firm, Remington. The number of firms began to increase: Yost Caligraph was launched in 1878, there were 30 firms by 1895, and a peak of 36 was reached in 1907. The entry of firms became brisk in the 1880s, and the rush continued until 1895, for 15 years. These trends coincide well with the timing of technology opportunities for venture business during the first 10 to 15 years of the development trajectory ascribed by us. Based on the positions of our trajectories, the timing of the appearance of the dominant design, the Underwood No. 5, in 1899 is located at the 82 per cent point of the development trajectory and at the 25 per cent point of the diffusion trajectory. Following the peak in 1907, the number of firms abruptly decreased, from 36 to 6, as natural selection proceeded. These analyses explicitly indicate that the trend in the number of firms reflects the progress of the innovation paradigm. Thus, the description of innovation using trajectories becomes more informative.

The Automobile Paradigm and Firm Behavior

The innovation paradigm of automobiles started with the development of the internal combustion engine. During the development of steam engines, automobiles equipped with steam engines were also tested. While this combination did not succeed, the experience did establish some design concepts. The development of automobiles equipped with internal combustion engines began with the development of the gasoline internal combustion engine by Lenoir in 1860, which followed the design of the free-piston engine by Barsanti in 1854, the Otto four-stroke engine in 1876, and the gasoline engine automobile by Daimler and Benz in 1885. Thus, the technology trajectory lay in the 30 years from 1855 to 1885. The development trajectory was in the 30 years between 1885 and 1915: the first half comprised a series of improvements in the gasoline engine, and the latter half comprised the development of various parts and equipment for automobiles. Automobile production was started in 1882 by Benz, and carried out by only three companies – Benz, Daimler and Peugeot – in the 1880s. The number of producers then grew, reaching, for example, 75 in France by 1899. Thus, the 15-year period from 1885 to 1900 was the take-off period for venture enterprises in Europe. This period coincides well with the first half of the development trajectory, which we recognize as the period of concentrated technology opportunity for venture business in the innovation paradigm, as discussed in the previous section.

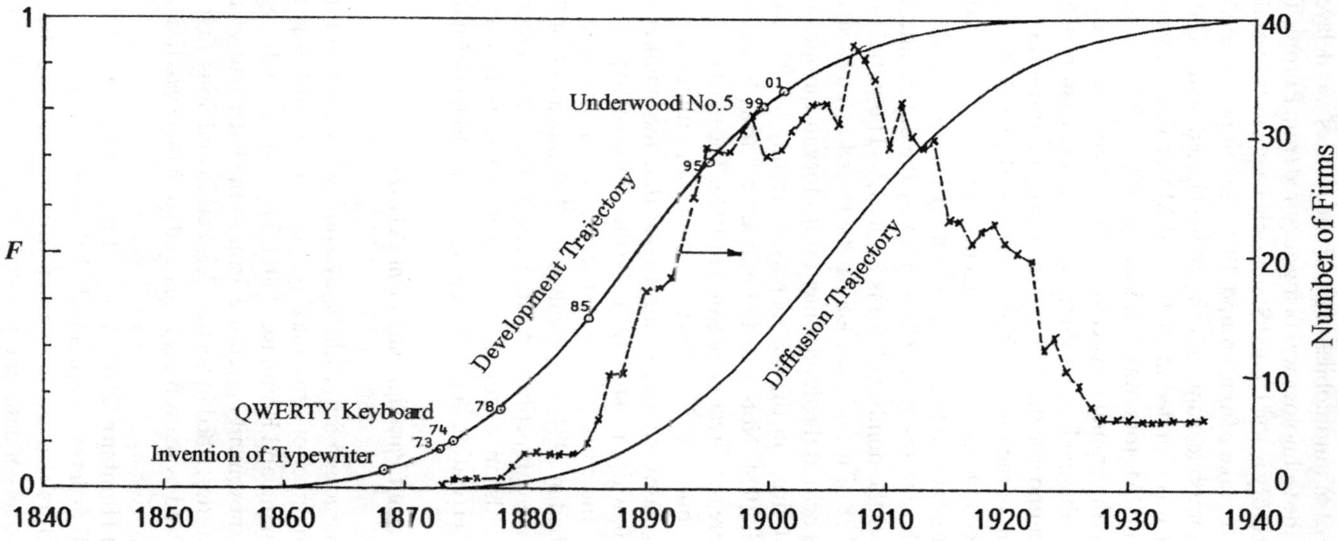

Sources: Hirooka (2003b); number of firms from Utterback and Suárez (1993).

Figure 7.8 Innovation paradigm of typewriters and number of firms

The development of automobiles in the United States was delayed about ten years because the technology was being acquired from Europe. Thus, the development trajectory in the United States only began in 1895 and various ventures were undertaken from around 1900. Schumpeter described the foundation of automobile manufacturing in the United States: Olds Motor Works of Detroit was founded in 1900, Ford Motor was incorporated in 1903, and General Motors was founded in 1908 (Schumpeter, 1939, pp. 414–15). As to the annual registered number of new cars, the complete data series can be obtained, for example, from the US Historical Statistics (1989). These data underlie the diffusion trajectory of automobiles.

The innovation paradigm of automobiles is depicted in Figure 7.9 together with the number of firms, taken from Utterback and Suárez (1993). The development trajectory is shown for Europe and the development of venture businesses in the US is given by the dotted line. The axis on the right shows the number of firms in the US. The initial increasing trend in the number of firms in the US can be traced by an S-curve which is about ten years behind the development trajectory in Europe.

The number of firms in the US increased steadily for 25 years, from around 1900 until 1924. Note that the number of firms began a sudden decrease well before the Great Depression in 1929. The market is estimated to have matured around 1925, so this trend coincides with the saturation of the market. It is, however, interesting to see that the number of firms reached 44 in 1910, which is estimated to be the period of 10 per cent to the saturated market, and corresponds to 60 per cent of the maximum number of firms, 73, reached in 1924. The situation in 1910 indicates that there was significant excessive competition. At its peak in 1924, the integration of firms began towards an oligopolistic market. These trends are clearly grasped by the interrelation between trajectories as shown in Figure 7.9.

The Synthetic Dyestuff Paradigm and Firm Behavior

The innovation paradigm for synthetic dyestuffs was shown in Figure 7.1. The trend in the number of firms making synthetic dyestuffs was analyzed in detail by Murmann and Homburg (2001). These are brought together in Figure 7.10. The movement of venture firms in the latter half of the nineteenth century was described in Section 7.1. Perkin and Sons, first founded in 1857, was followed by many more firms, and soon there was heavily excessive competition.

Murmann and Homburg (2001) analyzed the evolutionary dynamics of synthetic dyestuffs firms by comparing the behavior of firms by country and identified the factors that led to German firms dominating the industry. They collected two sets of data: one included the number of subsidiaries and

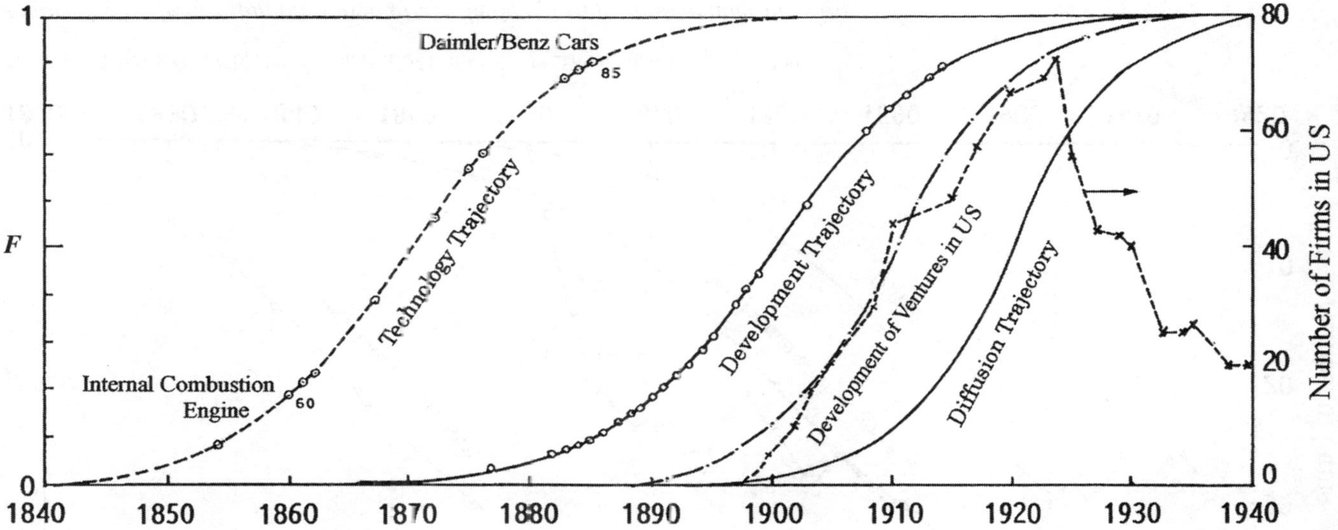

Source: Hirooka (2003b); Number of firms from Utterback and Suárez (1993).

Figure 7.9 Innovation paradigm of automobiles and number of firms

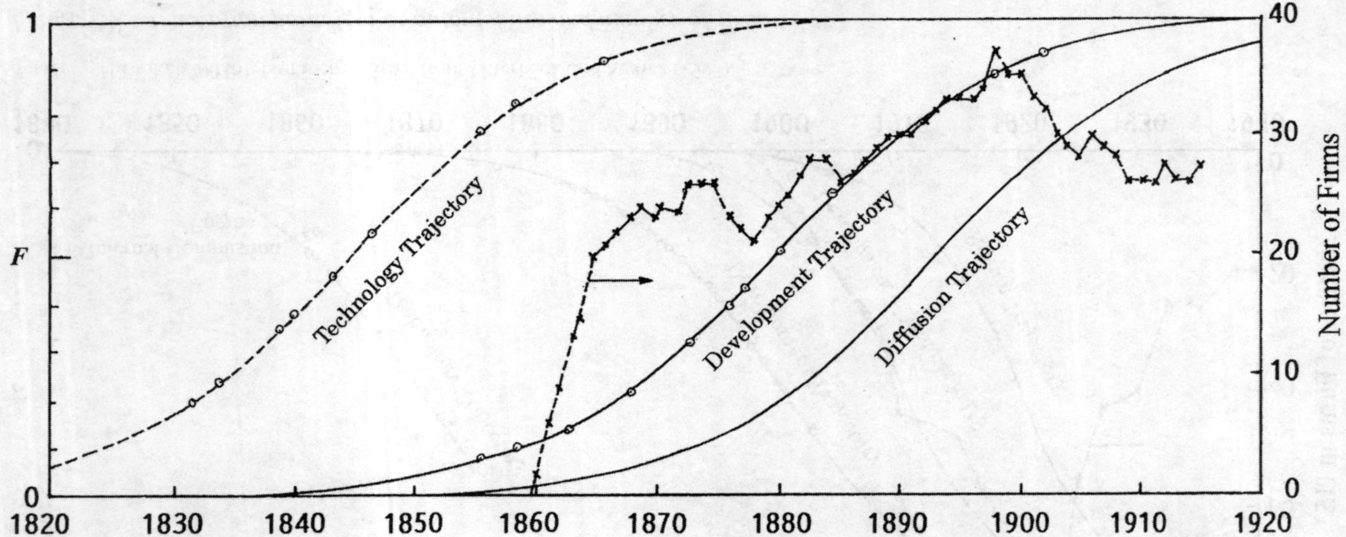

Sources: Hirooka (1998b, 2003b); number of firms from Murmann and Homburg (2001).

Figure 7.10 Innovation paradigm of synthetic dyestuffs and number of firms

the other did not. We used the latter for Germany. As shown in Figure 7.10, during the first take-off period, the number of firms abruptly increased within the first 12 years to reach 26, almost 70 per cent of the 37 at the maximum in 1898. However, the market size was only 10 per cent of the ultimate maximum. This corresponds to the bandwagon effect of Schumpeter. This trend continued, and the number of firms reached the maximum in 1898, when the market was 62 per cent of the ultimate maximum ($F = 0.62$).

After that, the number of firms decreased gradually, in contrast to the rapid decreases in firms producing typewriters and automobiles. The decrease was faster for typewriters and automobiles because natural selection was allowed to function. For dyestuffs, however, firms in Germany formed cartels to maintain prices, then amalgamated to form big 'Konzerns', and eventually formed the giant IG Farben. The dyestuffs industry developed smoothly from 1870 to 1900; the growth rate then slowed down as the market matured. While BASF, Hoechst and Bayer managed to maintain their dominant position during the first half of the trajectory, they began to organize business alliances during the second half to prevent drop-out. In 1903, they started to talk about forming an alliance, and two groups were formed by 1907; one was an alliance of Hoechst, Cassella and Kalle, and the other one was an alliance of BASF, Bayer and Agfa. After the First World War, these groups were united into IG Farben AG in 1925 to create a dominant empire.

The trend of venture business during the development trajectory was described in the first part of this chapter. The trend of the number of firms shown in Figure 7.10 can be translated as the actual number of firms emerging, which corresponds well to the tendency for venture businesses to emerge during the first half of the development trajectory. That is, the venture firms are clustered in the 15 years between 1855 and 1870.

Dominant Design and Lock-in Phenomenon

Utterback's 'dominant design' describes the way in which, during the progress of innovation, various kinds of products are developed and converge to a specific design. David (1985, 1992b) discussed the selection processes for the letter arrangement on the typewriter that became fixed in the QWERTY arrangement and for the VHS system for video tape recording that won out against Sony's Betamax system. Cusumano et al. (1992) asserted that VHS became the dominant design even though its technology was not superior to that of Betamax. He ascribed its success in gaining the dominant share to a better marketing strategy. Cowan (1990) analyzed the process of a nuclear power station being selected for a specific reactor using uranium and proposed a concept he called 'lock-in'. These analyses indicate that the dominant design is selected and standardized not only according to the quality of the product but also on various other factors such as market share and governmental regulation.

When innovation products begin to diffuse into the market, various kinds of products are developed in a competitive process of trial and error. Each firm can use any kind of design, and products are successively refined and improved. Many firms rush into the market, resulting in a bandwagon effect. Any one firm is unable to dominate the market, and too many firms are apt to enter it. Customers have little means of distinguishing the quality of products and cannot request any proper qualification. According to Utterback, producers and customers evolve by learning together. Through this trial-and-error process, the products merge into a dominant design. This does not always result in the best quality product, and various strategies to expand the firm's market share can lead to success.

In the case of video tape recording, Victor Japan expanded its market share by licensing its technologies to others, while Sony, which supposedly had a superior product, insisted on retaining exclusive rights. Victor's strategy resulted in its product becoming the de facto standard, leading to the dominant design.

Once a dominant design is settled upon, the market changes rapidly to encompass it. A dominant design becomes sovereign in the market. The market then becomes established on that design, making it difficult for another design to take over. Because much energy is required to change the dominant design, the situation is inescapable. This is the mechanism of 'lock-in'.

According to the study by Utterback (1994), there were two dominant designs in the development of the typewriter. One was the QWERTY keyboard arrangement in 1874, and the other was the Underwood No. 5 in 1899. The QWERTY keyboard arrangement was fixed at the starting point of the development trajectory, and the Underwood No. 5 appeared at the 82 per cent point of the development trajectory. As the latter corresponds to the 32 per cent point of the diffusion trajectory, the Underwood company had much time to enjoy their dominant position. The dominant design has a big impact during the innovation process. After the dominant design is fixed, the number of firms quickly decreases due to natural selection.

While only four automobiles were produced in the US in 1895, by 1905 there were 20 companies producing automobiles in the US. Henry Ford designed the Model-T in 1909 and implemented a conveyer belt assembly line in 1913 to produce it. This brought the price down to less than \$600, and, as a result, it became the dominant design in the US. Ford was producing 535 000 cars a year by 1916, and more than half the world's total production in 1920. This production process has become the prototype since used by most car manufacturers. The Model-T became the dominant design at the 86 per cent point on the development trajectory, the same relationship as that for the typewriter, the Underwood No. 5. The number of car producers in the United States began a sudden decrease in 1924, ten years after the Model-T became the dominant design. This was still five years before the Great Depression began in 1929.

In the case of automobiles, the dominant design was actually the manufacturing process, rather than the product as was the case of the Underwood No. 5. Many manufacturers imitated the assembly line system and mass production systems became widespread. Some enterprises, however, failed to achieve profitability due to the heavy investment required and went bankrupt.

In the case of synthetic dyestuffs, there was no dominant design as such because each dye is a creative product, protected by a patent. However, among the dyestuff products, indigo, which is a major product with a large share in the market could be called a dominant design. Synthetic indigo was first commercialized by Hoechst using a naphthalene process in 1897, and, four years later, in 1901, BASF commercialized the aniline process. This happened at about the 90 per cent point on the development trajectory of synthetic dyestuffs.

These considerations are summarized in Table 7.1. It shows that there is a common feature among the three cases of innovation: dominant design and lock-in. That is, the dominant design appears at around the 80 to

Table 7.1 Development of firms and innovation trajectories

	Typewriters	Automobiles	Synthetic dyes
Foundation of the first firm			
F-value on development trajectory	0.1	0.1	0.1
Entry rush of firms			
No. of firms/Max no. of firms (%) at $F = 0.1$ of diffusion trajectory	47	61	70
Interval of entry rush (years) until $F = 0.1$ of diffusion trajectory	11	10	12
Peak of firm numbers			
F-value on diffusion trajectory	0.60	0.70	0.62
Dominant design			
F-value on development trajectory	0.82	0.86	0.88
F-value on diffusion trajectory	0.32	0.20	0.60

Sources: Hirooka (2003b); data based on Utterback and Suárez (1993) and Murmann and Homburg (2001).

90 per cent point on the development trajectory, which corresponds to around the 20 to 60 per cent point on the diffusion trajectory. Therefore, a firm acquiring a dominant design can enjoy a fairly lengthy monopolistic advantage. The number of firms entering begins to increase at about the 10 per cent point on the development trajectory and then increases explosively for about 10 to 12 years until reaching about the 10 per cent point on the diffusion trajectory. This is Schumpeter's so-called bandwagon effect. At this point, the number of firms is about 50 to 70 per cent of the largest number of firms during the innovation paradigm while the sales of the product are only about 10 per cent of the eventual maximum sales. That means there is serious excessive competition among the firms. Towards the end of this rush period, a dominant design appears, and about 10 years later the number of firms begins to decrease. This narrative seems to be common for the observed innovation paradigms. This means that trajectory analysis can provide useful information about the behavior of firms during the course of innovation.

Recognition of Trajectories and Path Dependence

Schmookler (1966, 1972) described the trend in technologies using changes in the number of patents and academic papers. This kind of analysis of innovation using a series of patents and academic papers has been developed as a bibliometric analysis and frequently carried out by many economists, for example, K. Pavitt. The statistical data gathered by Achilladelis, as shown in Figure 6.1, were obtained by such a survey. Andersen (2001) conducted this kind of trial and systematically analyzed the trend in patents in the US, and found that it can be described by a logistic equation. These studies are closely related to the recognition of the existence of trajectories.

As already pointed out, many economists have analyzed the process of innovation and have proposed various concepts regarding trajectories or clusters of innovations. These include the 'natural trajectory' of Nelson and Winter, the 'technological paradigm' and 'technological trajectory' of Dosi, the 'T-cluster' of Rosenberg and Frischtak, and the 'new technological system' of Freeman. Utterback (1994) pointed out that in the early stage of innovation, products are very versatile, and a large number of firms emerge. Later, however, the products converge toward a 'dominant design' as the result of 'lock-in', and the number of firms is greatly reduced due to a kind of natural selection. These observations indicate that there is a definite cluster of technologies and suggest a kind of path dependence. The concepts of 'path dependence' and 'lock-in' were proposed by David (1985, 1988) and Arthur (1985, 1987, 1988, 1989). The analysis here suggests that the concepts of path dependence and lock-in can be more clearly interpreted in terms of trajectories.

7.5 SUMMARY

Comprehensive studies were carried out on the innovation paradigm, which comprises three trajectories, and the role of each trajectory was analyzed.

1. Innovation paradigms were analyzed for three cases, synthetic dyestuffs, biotechnology and electronics, and three trajectories were precisely identified.
2. The technology trajectory comprises fruits of basic research, which is often carried out in academia. The knowledge obtained is then transferred to industry during the development trajectory through academia–industry collaboration.
3. The development trajectory plays the most important role in the innovation paradigm. Launches of venture businesses are concentrated during the first 10 to 15 years of the first half of the development trajectory.
4. The number of firms drastically changes during the course of innovation. Three cases, typewriters, automobiles and synthetic dyestuffs, were studied. The number of firms abruptly increased during the initial stage – the so-called 'bandwagon effect'. Once a 'dominant design' arose, the number of firms quickly decreased.
5. The common features of the trend in the number of firms is summarized as follows:
   i. The number of firms reaches 50–70 per cent of the maximum at the 10 per cent point on the diffusion trajectory.
   ii. The dominant design appears at about the 80–90 per cent point on the development trajectory, this is still at less than the 60 per cent point on the diffusion trajectory.
   iii. The number of firms reaches a maximum at about the 60–70 per cent point on the diffusion trajectory.

8. National systems of innovation and industrial competitiveness

In previous chapters, we described how the innovation paradigm consists of three trajectories and how the development trajectory situated in the middle plays particularly important roles. Both knowledge transfer from academia to industry and the establishment of venture businesses occur during the course of the development trajectory. In this chapter we discuss the role of firms and governments during the course of innovation. The studies are concerned with the electronics paradigm at the time of the electronics industry take-off and focus on the behavior of firms and the actions taken by government to support the take-off. Such collaboration between firms and government has often been observed in Japan and other Asian countries and can be described as a national system of innovation. Analysis of innovation dynamism on the logistic map of the innovation paradigm, focusing on the development trajectory, produces interesting results.

8.1 BACKGROUND OF NATIONAL SYSTEMS OF INNOVATION

Lundvall (1985) developed the concept of 'national systems of innovation', taking into account the importance of institutional factors which originated from Veblen. This concept is important when considering the emergence of technological gaps between countries and the rate of reducing or eliminating the gap. Lundvall (1992) summarized the concepts of 'national systems of innovation', emphasizing that institutions have a strong impact on technological changes, but that at the same time they are normally quite rigid and do not change easily. For a country to strengthen its international competitiveness, it must deal with technological change by learning, adapting, and changing its institutional framework.

Perez's (1983) 'techno-economic paradigm' is based on Schumpeter's concept of technological revolutions and creative destruction, and is brought about by combinations of radical innovations in products, processes and organizations. These interrelated clusters of innovations occur infrequently (about every half century), but when they do occur, they necessitate changes in the institutional and social frameworks. The overwhelming importance of such technological transformations is that if the adequate institutional adaptation and structural change can be established, a potential for competitiveness on a fully prepared infrastructure is acquired. Freeman (1984b) developed his notion of a 'new technology system' using the work of Perez (1983).

In this kind of institutional change, government plays an important role in helping industries adjust to the new paradigm. The first explicit use of the phrase 'national systems of innovation' was by Freeman (1987), describing how innovations in Japan came about through the support provided by MITI. A few years later, Nelson (1993) described US innovation systems in *National Innovation Systems – A Comparative Analysis*, which focused on the combined public and private character of technology and the role of private firms, government and universities in the development of new technology in various countries.

Niosi and Bellon (1996) synthesized the definition of the 'national systems of innovation' by summarizing the concepts of several authors:

> A national system of innovation is the system of interacting private and public firms (either large or small), universities and government agencies, aiming at the production of science and technology within national borders. Interaction among those units may be technical, commercial, legal, social or financial, inasmuch as the goal of the interaction is the development, protection, financing or regulation of new science and technology.

In the age of the Industrial Revolution, inventions were made by individuals, as exemplified by Watt, Arkwright, and Stephenson. In the middle of the nineteenth century, German chemical companies were the first to have their own research laboratories to carry out their own research and development. Since that time, innovations have endogenously emerged from within firms, and firms have become entities of innovation. After the Second World War, Japan worked hard to catch up with the West and enjoyed a period of high growth during the 1960s. In the course of this catch-up, the Japanese government played an important role in supporting industry. The Ministry of International Trade and Industry (MITI) in particular strived to create a national system of innovation. As Freeman (1987) observed, the behavior of the Japanese government was recognized as a new approach to a system of innovation.

This chapter describes such national systems of innovation for electronics in the cases of Japan and other Asian countries, using the innovation paradigm comprising the three trajectories. This new approach has revealed several interesting features of innovations. National systems of innovation have been adopted around the world and their implementation in the United States and Europe is also described.

In this study, it is useful to express the various actions taken by government and firms by inscribing them on the trajectories of the electronics paradigm. The resulting figures clearly show their situation and timing.

8.2 CHARACTERISTICS OF THE INNOVATION PARADIGM OF ELECTRONICS

As described in the previous chapter, the innovation paradigm of electronics has a precursor paradigm of transistors. The electronics paradigm is for integrated circuits, and the precursor paradigm of transistors predates it by about 30 years, starting around 1915. Transistors were thus commercialized within the technology trajectory of electronics. Furthermore, since transistors are alternatives to vacuum tubes, the transistor paradigm is also related to the vacuum tube paradigm. That is, these paradigms encompass a family of products having a rectifying function and have evolutionary relationships. In a broader sense, electronics covers two paradigms of vacuum tubes and transistors, including integrated circuits. Because the transistors paradigm was absorbed into the integrated circuits paradigm, we express the integrated circuits paradigm as simply the electronics paradigm together with the precursor transistors paradigm.

Precursor Transistor Paradigm

Almost all the core technologies of electronics were developed in the United States, starting with the invention of the transistor by Shockley et al. in 1948 and ending with the development of submicron lithography by IBM in 1973, forming the technology trajectory of electronics. The core technologies of transistors, however, had already developed as technologies of semiconductors with a rectification function. They began when Benedicks disclosed the rectifying mechanism of germanium in 1915 and continued with a series of inventions that formed a precursor paradigm, as shown in Figure 7.4 in Chapter 7. The development trajectory of transistors started with the development of germanium rectifiers as crystalline wave detectors. Brattain and Bardeen invented the point contact transistor in 1948, and Shockley submitted a patent application for the p-n junction transistor in 1948, which represents the first practical application of the transistor. The development trajectory of transistors was absorbed by the technology trajectory of electronics, which was based on the development of integrated circuits by Kilby (1959) and Noyce (1961). Actual commercialization of transistors started in the United States in 1952 and in Japan in 1954. This narrative clarifies why transistors were commercialized during the course of the electronics technology trajectory. The actual paradigm of transistors was described and compared with that of electronics, that is, that of integrated circuits, in Figure 7.4.

Behavior of Firms in Electronics Paradigm

The transistor paradigm was absorbed by the electronics paradigm after integrated circuits were invented in around 1960. Thus, the behavior of electronics firms can be described on the electronics trajectories, as summarized in Figure 8.1, which compares events on the technology trajectory between US and Japanese firms.

Companies in the US started producing transistors in 1952. They were initially produced for military use. Their first civilian application was to hearing aids; Raytheon sold about one million for such applications in 1953. Shockley set up a research laboratory, the Shockley Research Institute, in 1956 and recruited many researchers, including Robert Noyce and Gordon Moore. When the institute was dissolved the following year, Noyce and Moore established another company, Fairchild Semiconductors, in 1957. Noyce invented monolithic integrated circuits in 1961, which became the basis for the development of subsequent integrated circuit technologies. IBM replaced vacuum tubes with transistors in computers and introduced the transistorized IBM 1401 in 1960. IBM launched the seminal IBM 360 computer in 1964; it was equipped with static translinear (STL) circuits, which became the prototype for integrated circuits in the 1970s. In 1970, IBM finally developed the IC equipped IBM 370 series.

Companies in Japan quickly followed the US in introducing transistors. When Masaru Ibuka of Tokyo Communication Engineering Company (later to change its name to Sony Corporation) visited the United States, he learned about transistors. His company quickly obtained the license for transistors from Western Electric Company in 1952 and began producing transistors in Japan in 1954, followed by the production of transistor radio receivers in 1955. Through the success of transistor radios, the Sony name spread worldwide. At around the same time, Leona Esaki, a researcher at Sony, discovered the tunneling effect of semiconductors and invented the Esaki diode; he was awarded the Nobel Prize in Physics in 1973. The production of transistors was carried out not only by Sony but also by Fujitsu, NEC, Toshiba, Hitachi and Kobe Industries, a total of six companies, in 1954 and additionally by Matsushita Electric, Mitsubishi Electric and Oki Electric in 1957 and by Sanyo Electric in 1959. Just one year after Noyce invented the silicon monolithic planar integrated circuit, NEC (Nippon

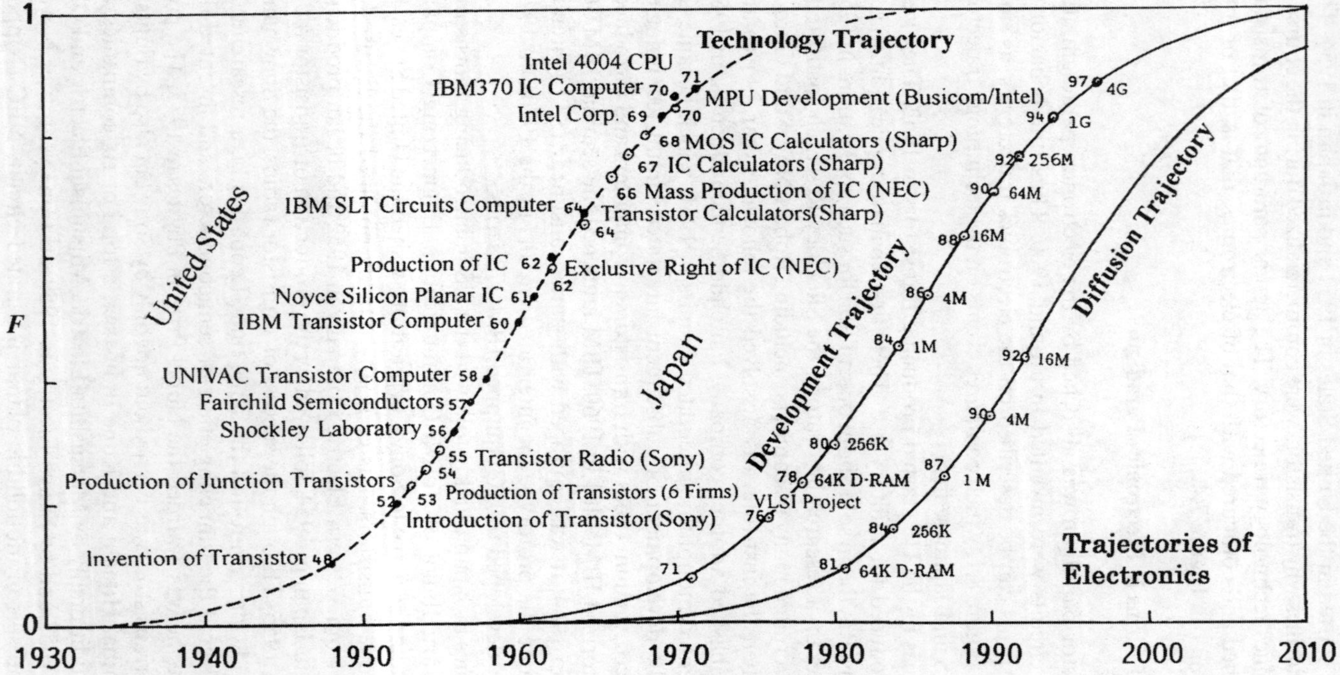

Source: Hirooka (1999e)

Figure 8.1 *Behavior of firms in the US and Japan in electronics innovation*

Electric Company) introduced this IC technology from Fairchild in 1962 and began producing integrated circuits in 1966.

Japan took the lead in applying transistors to calculators. While the United States concentrated its efforts on the development of computers using transistors, Japanese companies concentrated on developing electronic calculators. Sharp Corp. began to sell transistorized calculators in 1964, and Canon and Ooi Electric Co. followed suit. Japanese electronic calculators spread around the world. After they were equipped with integrated circuits in 1967, many companies rushed into this business. Hitachi developed a calculator that could handle the four fundamental rules of arithmetic (addition, subtraction, multiplication and division), and Sharp introduced a calculator equipped with six chips MOS-IC. By 1971, all calculators were equipped with LSI chips, and Japanese calculators dominated the market, with a sales share of 80 per cent worldwide. Busicom Corp., a venture business in Japan, developed a calculator small enough to fit in the palm of the hand; it triggered the popularization of hand-held calculators.

As shown in Figure 8.1, these developments related to transistors and integrated circuits can be described along the electronics technology trajectory. This is because the electronics technology trajectory closely corresponds to the diffusion trajectory in the transistor paradigm.

The Trigger to the Development Trajectory

Busicom Corporation, which developed electronic calculators as described above, was a small venture business. As most calculator producers in Japan were large corporations, Busicom had to compete using its own technologies. It thus set up a system engineering group. At that time, a major problem standing in the way of LSI chips being widely adopted was how to generalize the operation of LSI devices. The engineers at Busicom concluded that LSI operation would best be carried out by separating the functions of ROM, RAM, CPU and I/O into different components. They contracted with Intel Corp., which had been established by Noyce and Moore in 1969, to develop a new microprogramming system.

Busicom dispatched Dr Masatoshi Shima to Intel to coordinate its development. During development there, his colleague, M. Ted Hoff, suggested that the program should be designed by the simple operation of four-bit calculation as the central processing unit of common device. Shima and Federico Fazin used this idea to develop a microprogramming device that became the first microprocessor, the Intel 4004, which was introduced in 1971.

Although Busicom had exclusive sales rights, they gave up on submitting the relevant patent applications for the microprocessor as it was perceived to be lacking in originality because the stored programming system and LSI were already known. Busicom did develop and sell the first calculator equipped with a microcomputer, the 141 PF. When they went bankrupt after the oil crisis in 1973, the exclusive sales rights of microprocessors were transferred to Intel. Shima was recruited by Noyce and joined Intel, where he developed the Intel 8008 and 8080 microprocessors. He later moved to Zilog, where he developed the Z-80 microprocessor. This development of microprocessors became the origin of the development trajectory of electronics and triggered the widespread diffusion of personal computers. Microprocessors, most recently the Pentium, became Intel's core product, producing huge earnings and profits. While the development of microprocessors made it possible to develop personal computers, their application has expanded way beyond computers. Now, almost all electrical and electronic products are equipped with microprocessors, including electrical appliances, games, robots, automatic cameras, printers, car-electronics, POS terminals and mobile phones. The development trajectory of electronics comprises the history of these developments.

8.3 NATIONAL SYSTEMS OF ELECTRONICS INNOVATION

The development history of electronics in Japan is that of unusually early development of products even on the technology trajectory. There was a national system of innovation strongly supported by the government. This section discusses this system as it related to the electronics industry in Japan and corresponding actions by the United States and Europe.

Fostering New Industries by the Japanese Government

Japanese industry grew quickly after the Second World War, especially during the 1960s, aiming at catching up with the West. Such a splendid leap by Japanese industries was made possible by support from the Japanese government. Post-war rehabilitation from burnt-out ruins was achieved by prioritizing the production systems; steel and coal production were selectively promoted by giving them the highest priority for the small amount oil allotted to Japan. This leadership by the government effectively promoted Japanese rehabilitation and infrastructure reconstruction. After the Korean War in 1950–53, Japanese industries entered a period of high growth and achieved a GDP growth rate of more than 10 per cent during the 1960s. During this period, a protective trade policy was implemented by applying a high tariff rate on imported goods to protect Japanese industries, especially computers and automobiles. In the face of strong opposition to such protectionism, the Japanese government began liberalizing the rules on foreign investment and reducing the tariff rates in the latter half of the 1960s. Trade was completely liberalized by 1976.

Most industries – other than computers and electronics – however, were not directly protected or given assistance by the government. One of the important governmental policies was to identify new industries resulting from innovations and to support their take-off. We thus analyze the government policies adopted and measures taken to promote the development of the computer and electronics industries. To defend Japanese industry against the great surge of American computers from IBM, the Ministry of International Trade and Industry (MITI) took timely countermeasures. The barrier that was raised gave Japanese computer and electronics companies a chance to survive in Japan by avoiding monopolization by IBM, which differed from events in other countries. Recognizing the importance of computers and electronics as promising industries, MITI took various countermeasures in addition to imposing high tariffs on imported computers. The establishment of a rental company specifically to offset IBM's advantage in rented equipment was particularly important. The Japan Electronic Computer Company (JECC) was established in 1961 and was jointly owned and run by the seven leading computer firms. It received financing from the Japan Development Bank as well as commercial banks, and about half the domestic computers purchased in the 1960s were financed through JECC.

MITI moved for various acts to be set up for promoting the electronics industry. The Provisional Act for the Promotion of Electronics Industries (PAPEI) was set up in 1957 to stimulate the development of IC technologies, to stimulate domestic demand for electronics, and to secure the future of electronics parts producers. It led to the creation of the Association for Promoting Electronics Industries in 1958 to foster the business. The Act for Research Associations was passed to promote collaboration between firms. It led to the creation of the Research Association for Computer Technologies and the launching of the Large Scale Research Project for Advanced Electronic Computer Technology in 1961. This was at the time when Noyce invented monolithic integrated circuits. The PAPEI Act was transferred to the Provisional Act for the Promotion of Specified Machinery and Electronics Industries (PAPSMEI) in 1971 and then became the Provisional Act for the Promotion of Machinery and Information Industries (PAPMII) in 1978. The government, MITI, only funded money to stimulate the take-off of newly rising industries by aiding research on technology developments. The research funds were mostly subsidies, which were to be refunded to the government if the research succeeded. The total amount of subsidies for R&D remained fairly small – only US$11 million were provided between 1957 and 1972. The real significance of these measures was that MITI proclaimed a policy to foster the domestic computer and electronics industries.

The most important thing is that the actions by the Japanese government and firms related to electronics technologies were taken early in the course of the technology trajectory of the electronics paradigm. As shown in Figure 8.2, they were taken at about the 40 per cent point of the trajectory. The actions taken by MITI are marked on both the technology and development trajectories to emphasize the timing.

Success of the VLSI Project on the Development Trajectory

This early action by the Japanese government and firms enhanced the activities of the Japanese electronics industry and created the basis for the success of the VLSI project, carried out between 1976 and 1979, in which five leading companies in the Japanese electronics industry collaborated, with the support of the government through MITI, to generate the technological knowledge needed for producing large-scale integrated circuits. This project was driven by the fear of IBM developing an advanced computer equipped with 64K DRAM that might defeat the Japanese firms. MITI launched the VLSI project in which it provided ¥35 billion to industry over four years. This project resulted in 64K DRAM being developed in Japan in 1981, before the United States. The timing of these actions can be clearly described on the development trajectory of electronics, as shown in Figure 8.2. Note that the timing of the VLSI project corresponds to the period of highly concentrated technological opportunities for venture businesses discussed in Section 7.3. Logistic analysis is thus quite effective in interpreting this kind of situation. Building on this success, the Japanese electronics industry took the lead in developing and marketing semiconductors worldwide for more than a decade.

The Role of MITI in the National System of Innovation

The role of MITI was precisely analyzed and evaluated by Freeman (1987). According to his analysis, the splendid advance of Japanese industries can be ascribed to a well-functioning national system of innovation at the national, industrial (*keiretsu*), and company levels. Freeman (1987, pp. 77–8 and pp. 87–9) described this as follows.

It is not always easy for companies (or countries) to adapt to a change of paradigm and there is a great deal of inertia in social institutions, as pointed out by Perez (1983). When a new techno-economic paradigm, however, becomes so predominant that it is an established technological regime, then

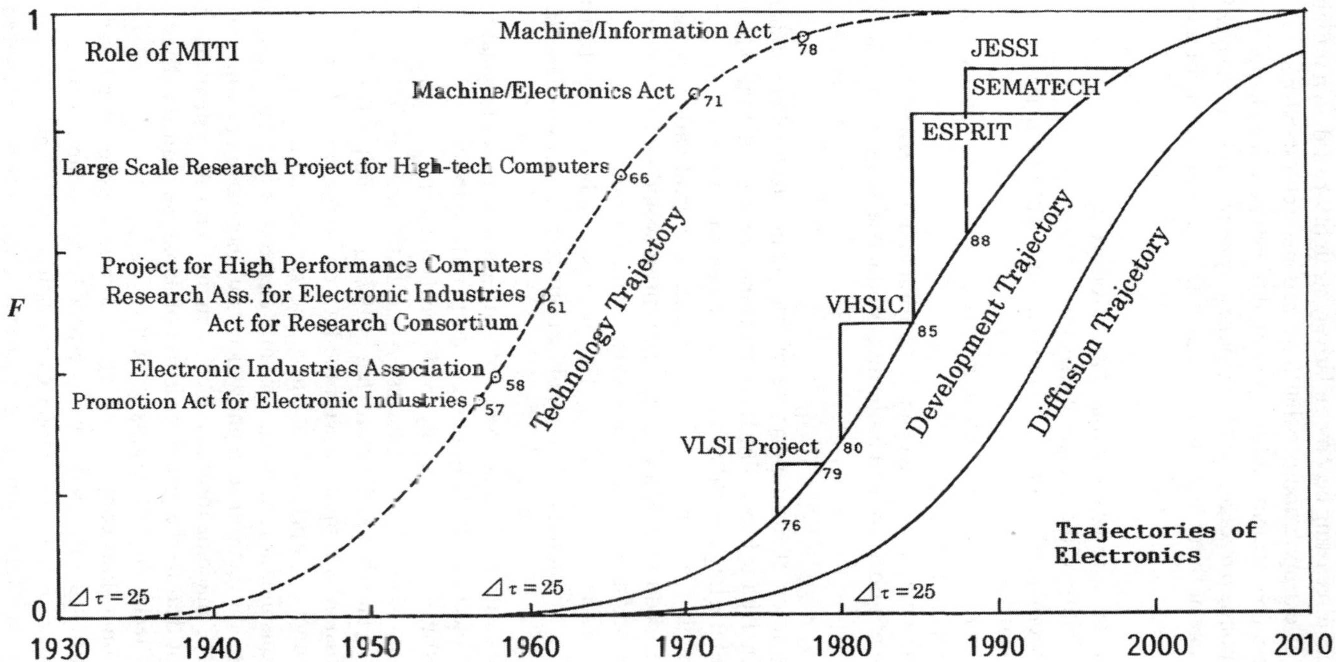

Sources: Hirooka (1999e, 2001a); data from Freeman (1987), HTSRG (1990).

Figure 8.2 Role of government and collaboration projects in electronics

technological forecasting becomes relatively straightforward, as it is mainly a question of extrapolation of what have become well-established trends. The Japanese government had the potential to estimate the change in the techno-economic paradigm, and the Japanese system seems well adapted to take advantage of the enormous potential of this new paradigm, for several reasons:

1. The systems approach to process and product design.
2. The flexibility of the industrial structure.
3. The capacity to identify crucial areas of future technological advance at national and enterprise level.
4. The capacity to mobilize very large resources in technology and capital in pursuit of strategic priorities.
5. The horizontal flow of information within and between firms.
6. The use of information in innovative production systems, such as 'just in time'.

The recognition of the importance of information technology occurred very early in Japan. Especially, the importance of the national Japanese system of forecasting lay in the diffusion and generalization of these expectations through a large number of companies in a great variety of industries. This helped to create a climate where firms would make investments in new products and processes associated with the new technology on a much larger scale than elsewhere. This was especially important in the diffusion of information and communications technologies to new industrial sectors.

As pointed out by Kodama (1985, 1986), the Japanese economy has been particularly successful in adapting the new paradigm to the needs of some other industrial sectors, especially mechanical engineering and machinery. Kodama pointed out that the expression 'mechatronics' was first coined in Japan in 1975 and that even before that in 1971 an explicit policy designed to induce 'fusion' was initiated with the Law on Temporary Measures for the Development of Specific Machinery and Electronics Industries which spoke of 'consolidation of machinery and electronics into one', and 'systematization' of them. The highly successful Japanese development of robotics, CNC machine tools and flexible manufacturing systems has also owed a great deal to the group structure of Japanese industry. These are examples of technological fusions of electronics.

Another important contribution of the Japanese government was the long-run infrastructural policy with respect to telecommunications. The Nippon Telegraph and Telephone Corporation (NTT) has pursued its program for an information network system (INS), which is envisaged as ultimately providing a connecting web for information services to private and industrial users throughout the country, with a digital broadband infrastructure. This project was examined by prototype test line in Mitaka in the early 1980s. This was an intentional challenge to a techno-economic paradigm change through construction of a new infrastructure; however, its actual achievement was delayed.

Freeman's analysis noted that the 'visions' of the future produced by MITI, STA, NIRA and numerous other government, and private sources do not pretend to be accurate predictions, nor do they commit companies to inflexible plans. They chart the broad direction of advance for the economy and for technology and give companies sufficient confidence to make their own long-term investments in research, development, software, equipment and training.

Collaboration Consortiums in the United States and Europe

The United States was shocked by the astonishing advances made by Japanese industry after the high-growth development period that resulted in the US/Japan trade friction, as described later. Japan's defeat of America's 'final fortress', the electronics industry, was especially irritating to the US government. At that time, collaborative research between companies in the same industry was prohibited by antitrust laws in the US. After President Carter's 'presidential message on innovation' in 1979, the United States eased the laws against firms conducting collaborative research and development. In response to Japan's promotion of collaboration, represented by the VLSI project, the US Department of Defense launched the five-year VHSIC project in 1980 to encourage leading electronics companies to develop very high-speed, sophisticated, integrated circuits. A collaborative research organization, the Semiconductor Manufacturing Technology Initiative (SEMATECH), was launched in 1988. It was a consortium of the Department of Defense and 14 semiconductor device manufacturers: IBM, Hewlett Packard, Intel, LSI Logic, National Semiconductor, AMD, DEC, TI, AT&T, Motorola, NCR, Rockwell International, Microtechnologies and Harris. The aim was to regain enough competitiveness to enable American firms to take the lead in developing 4M DRAM by 1993 through the development of manufacturing technologies for advanced semiconductor devices and through technology transfer between member firms. These activities are shown in Figure 8.2 on the development trajectory of the electronics paradigm.

The governments of the European Community paid close attention to these activities and in response set up a European collaboration project: the European Strategic Program for Research and Development in Information

Technology (ESPRIT) in 1984 as a ten-year project. The aim was to develop advanced integrated circuits in order to catch up with Japan and the US, and a total of 1.5 billion ECUs, including 750 million ECUs provided by the EC governments, was spent in the first five years, with 9300 personnel involved in total. Collaboration among industries, universities and governments was aimed at developing technologies that would be at the same level as the world's leading technologies within ten years. Research and development were focused on advanced microelectronics, software technologies, advanced information processing technologies, office systems and CAM, covering five fields.

Furthermore, in 1988, at about the same time that SEMATECH was launched in the US, the Joint European Submicron Silicon (JESSI) project was launched aimed at developing 64M DRAM by 1996 through collaboration between Siemens, Philips and SGS Thompson, the core members. The project provided 3.8 billion ECUs (US$4 billion), involved 21 400 participants, and was supported by a fund to which the EC governments contributed 25 per cent, individual countries contributed 25 per cent, and involved firms contributed 50 per cent.

The JESSI project did not have much success, and IBM proposed to the EC governments that the core members of JESSI collaborate with SEMATECH. This proposal was accepted, and the EC and US companies began collaborating in 1990. These activities are also shown in Figure 8.2 on the electronics development trajectory, and the timing of these projects can be compared at a glance. National systems of innovation have now become prevalent throughout the world.

National System of Innovation in the United States

The rise of Japanese and German industries led to a decline in the competitiveness of US industry due to the hollowing-out phenomenon. President Carter's 'innovation message' to the Congress in 1979 presented several proposals for improving the competitiveness of US industries: fostering innovative SMEs (small and medium enterprises), granting patents to private firms for inventions derived from research done under government contract, and easing antitrust laws to enable collaborative research between firms within the same industry. These proposals took form in the Stevenson–Wydler Technology Innovation Act and Bayh–Dole Act of 1980 and other acts within a few years.

In 1985, during the Reagan administration, the President's Commission on Industrial Competitiveness produced the so-called Young Report, which proposed various actions for improving global competitiveness. One proposed action was promoting technology transfer from academia to industry.

A plan was approved in 1987 for the National Science Foundation (NSF) to set up engineering centers in various fields, and 25 such centers were established by 1991. Another proposed action focused on a long-term strategy for future industrial competitiveness. It called for promoting basic science in the research fields important to future industries as well as promoting innovative and interdisciplinary activities. Thus, President Reagan decided in 1987 to set up science centers to promote systems in which basic research in academia would lead to breakthroughs and serve as catalysts for next-generation industries. While the engineering centers were designed to strengthen existing industrial technological competitiveness, the science centers were designed to gain the initiative in next-generation industries. Eleven science centers were set up in universities in 1989, and 14 more were added in 1991.

Another action by the US government was the creation of the Federal Technology Panel belonging to the Office of Science and Technology Policy to the President. It prepared a list of important technologies in 1991. According to this list, the Department of Defense summarized a 'Critical Technology Plan' by which technological competitiveness between the US, the EU and Japan was evaluated and actions to strengthen technology development activities were proposed. In 1989, MIT (the Massachusetts Institute of Technology) organized a survey committee to study the industrial behavior of the US compared to that of Japan and to indicate the direction to be taken by the US. These actions were taken to develop concrete strategies for enhancing US industrial competitiveness.

Thus, the United States began strongly to promote a national system of innovation.

8.4 NATIONAL STRATEGIES OF ASIAN AREA FOR ELECTRONICS INDUSTRY

As described in Section 3.5 of Chapter 3, the international development of innovation can be described on a logistic S-curve of the diffusion. 'Follower countries' can produce innovation products after a time lag and then export them to the 'inventor country' during the last part of the diffusion trajectory, as shown in Figure 3.7. This correlation can be interpreted using Vernon's life-cycle theory. Japan, for instance, followed the Western countries with a fairly long time lag and caught up by 1970. That is, innovation products are produced in follower countries at a later timing than in the original country, as shown in Figure 3.9. The time lag between inventor and follower countries, however, has become shorter with recent innovations. Especially in the case of electronics, innovations have been developed almost simultaneously worldwide. This trend has also been observed in other cases.

As discussed above, the Japanese electronics industry developed successfully in the 1980s with the aid of a national system of innovation. This so-called 'flying geese model' has been followed by other Asian countries repeating the Japanese success. It is noteworthy that electronics developments in every country and region in Asia have followed the same development trajectory. This means that developments in electronics spread worldwide at the same time and were synchronized. The Korean and Taiwanese governments actively supported the electronics industry in the first half of the development trajectory and made them self-supporting in the latter half. The success of the electronics industries in both countries is well known.. Singapore, Malaysia and India developed their electronics industries in their own way but following the same development trajectory.

Actions by Firms and Government for Electronics in Japan

The actions taken by the Japanese government and firms in the development of the electronics industry were discussed above. Figure 8.3 illustrates the correlation between the actions taken by government and those taken by electronics firms in Japan. It shows that private firms acted first and that the government quickly followed by providing appropriate support schemes. This is the counterpart of methods followed in other countries, as discussed below. Figure 8.3 shows that both Japanese industry and government took actions unbelievably early during the take-off stage of the electronics industry and that their actions were closely cooperative and timely.

The Electronics Industry in Korea and Actions by Government

The development of the electronics business in Korea has been described in various papers: Tanimitsu (1999), the Nomura Research Institute (1996), Ching and Hayashi (1995), Kaneko et al. (1999), Fukagawa (1989) and Shimura (1992). What follows is a comprehensive summary of these papers.

As the Korean electronics industry started at the root of the development trajectory of the electronics paradigm in which the trajectories are the same as those in Figure 7.3, the trend tracks the development trajectory, as shown in Figure 8.4. The development of the electronics industry in Korea was about ten years behind that in Japan, and, unlike in Japan, the industry developed along the development trajectory instead of the technology trajectory.

The Korean government decided to promote the electronics industry in 1966 and first allowed foreign companies to build production factories. Motorola and Fairchild of the US built factories to produce transistors in 1966. The government organized the Association of Electronics Industries

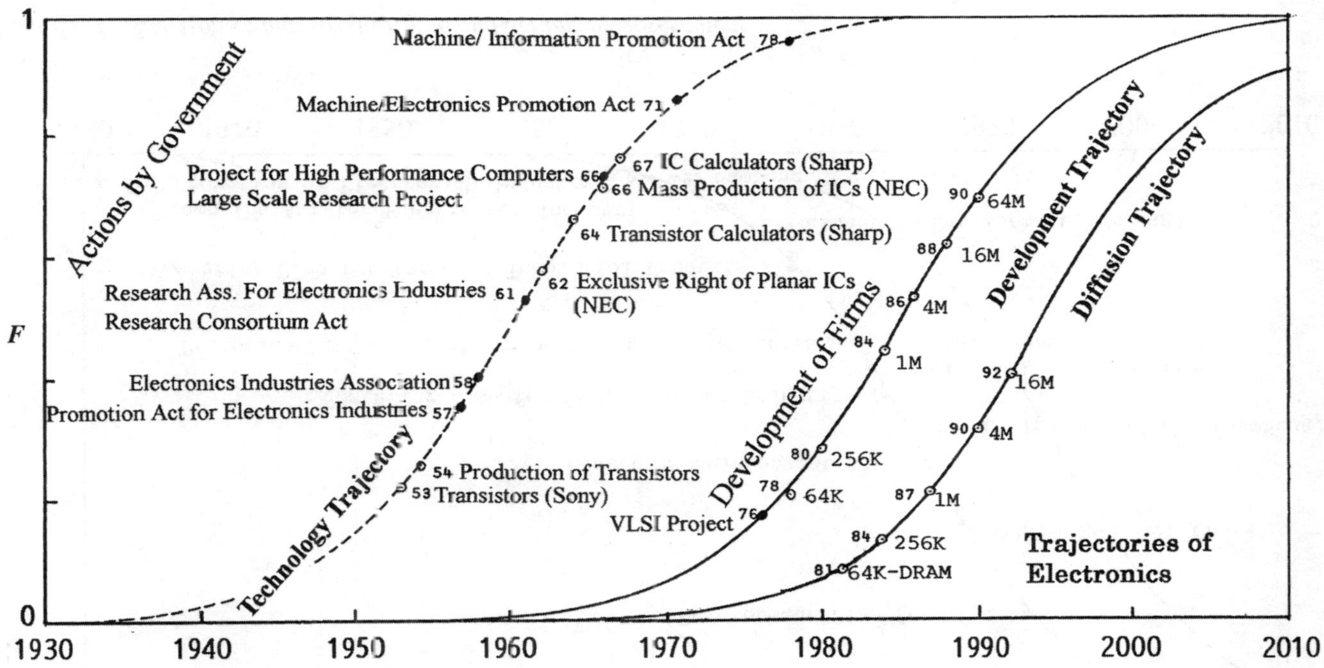

Source: Hirooka and Nitta (2000).

Figure 8.3 *Electronics industry in Japan and actions by government*

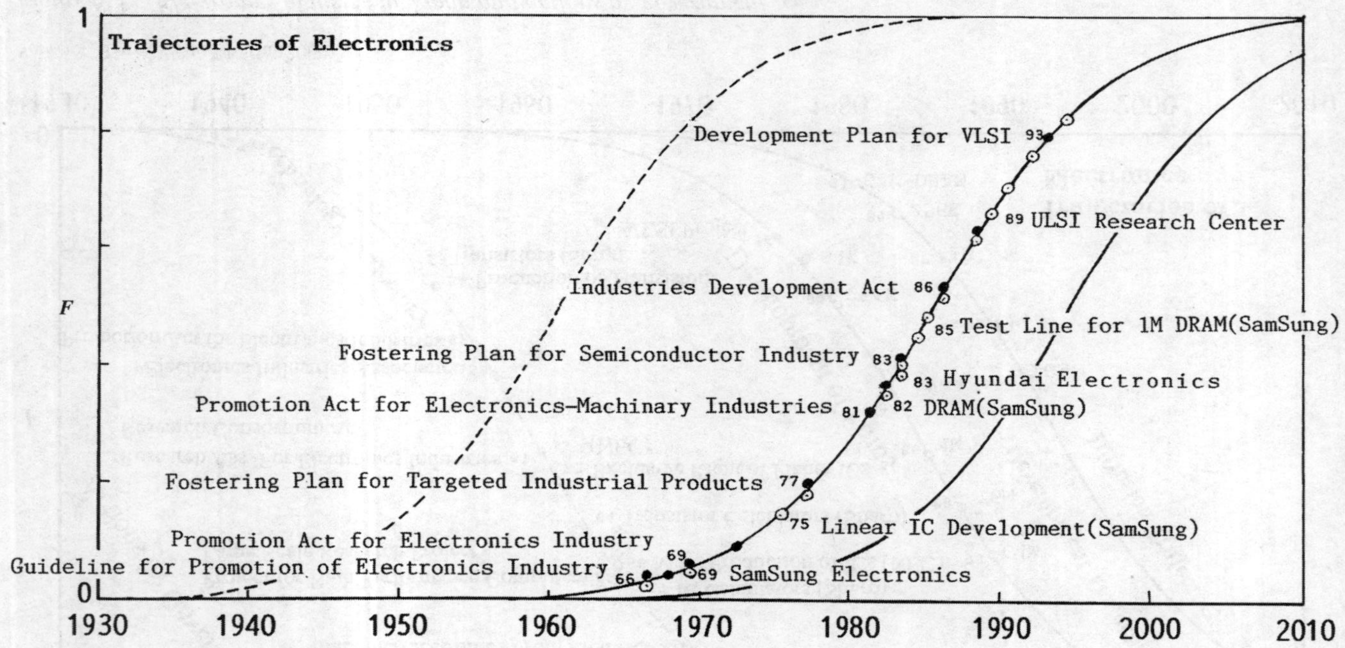

Source: Hirooka and Nitta (2000).

Figure 8.4 *Electronics industry in Korea and actions by government*

in 1967 and passed the Act for Promotion of Electronics Industries in 1969. Japan had already set up the Provisional Act for the Promotion of Electronics Industries in 1957. The Korean government passed the Promotion Act for Technology Development in 1972 and nominated 77 areas, including semiconductors and computers, as important promotion areas in 1977. The Act for Promotion of Electronics and Machinery Industries was passed in 1981, and a fund for the Promotion of Electric and Electronic Industries was set up in 1982. The Industry Development Act was passed in 1986. The Korean government thereby took actions to support the development of the electronics industry for 20 years, from 1966 to 1986. It also developed the Fostering Plan for Electronics Industries in 1983 and launched the VLSI Research Association in 1986 to establish and develop the basis of the industry. The Semiconductor Manufacturing Equipment Association was launched in 1988 to promote the domestic production of semiconductor manufacturing equipment. The most important action taken by the Korean government was establishing a four-year, collaborative research project in 1993 for the development of ultra-large-scale integrated circuits of 256M DRAM, involving semiconductor manufacturers, a governmental research institute (the Korean Electronics and Communications Institute) and universities.

The trend of Korean electronics firms is as follows.

Samsung Electronics Co. was launched in 1969 as the first electronics firm and developed into a core electronics company in Korea. It became a symbol of success, growing as much as 60 per cent a year for more than ten years. Samsung began to produce CMOS semiconductors using the technologies introduced from the American ICI Co. and pattern masks supplied by them in 1974. Samsung and Gold Star Co. started to produce TV sets in 1977. Samsung began developing 64K DRAM in 1982 and finished the development by obtaining technical support from Microntechnologies and Sharp Corp. in 1983. It developed 256K DRAM in cooperation with Microntechnologies in 1984. It constructed its own experimental lines for 1M DRAM in 1984 and finished the development in 1986. Samsung finally caught up with Japan when both countries started producing 1M DRAM in the same year, in 1992. It went on to produce 4M DRAM in 1988, 16M in 1990, and 64M in 1992. Though these were OEM production for NEC, Japan, Samsung entered into a strategic alliance with NEC in 1994 to develop 256M DRAM. This means they had caught up with Japan.

Economic development in Korea was carried out mainly through the *zaibatsu* (*chaebols*), large business groups of various types of companies, under strong guidance from the government. The development of the electronics industry was exposed to the struggle between these groups, with each one setting up its own electronics company, which resulted in excess competition. Gold Star Electron Co. was set up in 1983, less than a year after Samsung was created. It started by introducing technologies from AT&T. Hyundai Electronics was also incorporated in 1983. These three soon began battling Japanese and US companies under conditions of complex international alliances.

An important action in this battle was the creation of a local Samsung subsidiary in Silicon Valley in 1983. It became an important window for the acquisition of VLSI technologies, a gateway for the export of Samsung products, and a channel for obtaining information about advanced technologies and market developments. The test line for Samsung's 1M DRAM was constructed at this subsidiary by their engineers, creating a comprehensive organization for development and production. Samsung also recruited young engineers with experience of working in Silicon Valley, and domestic organizations in Korea were enhanced by these excellent talents. The company then established a research laboratory for developing ULSI circuits in their own factory in Korea in 1989, establishing their own development organization.

The narrative followed by Korea in developing its electronics industry shows that the industry developed along the same development trajectory as in the US and Japan. There was close correlation between the actions of government and industry all along the trajectory. Electronics companies in Korea began to stand on their own during the latter half of the development trajectory and acquired strong international competitiveness thanks to the strong governmental support.

The Electronics Industry in Taiwan and Actions by Government

Taiwan likewise efficiently developed its electronics industry. The history of the electronics industry in Taiwan has been described by Shimura (1992), Aoyama (1999), Ko and Tabata (1996), Kato (1997, 1998) and the Nomura Research Institute (1996). What follows is a comprehensive summary of these papers. The logistic analysis is shown in Figure 8.5.

The development of the electronics industry in Taiwan was directly supported by a strong government policy. The Taiwanese economy in general progressed rapidly due to the export-oriented policies of the 1960s and 1970s. After the bitter experience of the oil shock, the government recognized the need to reduce its dependence on labor-oriented industries and to begin developing a technology-oriented electronics industry. It thus inaugurated the Industrial Technology Research Institute (ITRI) in 1973 and set up the Electronics Research & Service Organization (ERSO) as the research laboratory in 1974. ERSO came to play the central role in the development of the electronics industry. The first plan for an electronics industry as a national project was started, and the Semiconductor Research Center was

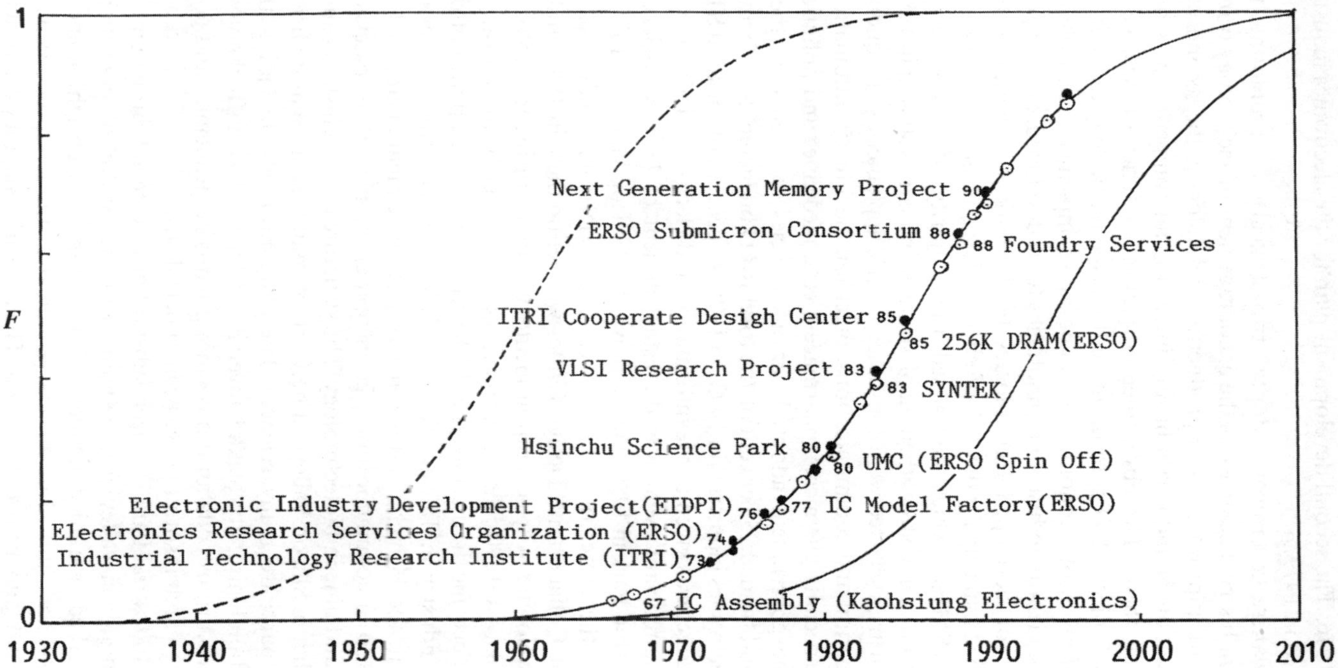

Source: Hirooka and Nitta (2000).

Figure 8.5 Electronics industry in Taiwan and actions by government

opened in 1976. The second development plan for the electronics industry was launched in 1979.

The Taiwanese government adopted two strategies for promoting the electronics industry. The first was the encouragement of technology transfer to support improvements in production technologies; the second was the identification of industries having international competitiveness and strengthening them. The government organized a series of Science & Technology Projects (STPs) to promote domestic technology transfer. In 1979, the Ministry of Economic Affairs began to implement new technology projects, identify industrial technologies to be developed, and support the transfer of developed technologies to domestic companies. These actions helped make up for the lack of effective R&D at small and medium enterprises (SMEs), which comprised the majority of enterprises in Taiwan. Selection and strengthening of strategic industries started in 1982. The ten industries selected were consumer electric appliances, semiconductors, information and communications, precision machinery, automation, aerospace, specialty chemicals, pharmaceuticals, medicines and pollution control. The government adopted various policies to strengthen these industries and funded 25 per cent of the total R&D budget in the information and electronics industries in 1993. In 1995, 41 per cent of the total STP program budget was given to the information and electronics industries.

Another important event was the establishment of Hsinchu Science-based Industrial Park in 1980. This park was the 'Silicon Valley' of Taiwan and became the center of the information and electronics industries in Taiwan. The Collaborative Design Center was set up there in 1985, and a plan for developing next-generation memory manufacturing technologies was put in place in 1990. These aggressive measures taken by the Taiwanese government became the driving force of the remarkable development of the electronics industry in Taiwan. By 1996, 180 industrial firms had set up operations in the Hsinchu Industrial Park, 56 in semiconductors, 41 in computers, 30 in communications, 26 in optoelectronics, 18 in precision machinery and nine in biotechnology. The structure of this cluster is quite similar to that in Silicon Valley – a network of mainly small venture firms. Inside the park are many science-related organizations, including the Transportation University, Seika University and the Industrial Technology Agency, and various collaborations among industry, academia, and governmental research institutes have been undertaken.

Hsinchu Industrial Park has been substantially managed through interchange with Silicon Valley. A large number of the researchers in Hsinchu have also worked or studied in Silicon Valley. Many of the bright, talented students who had studied in the US remained in the US after receiving their PhDs and launched venture firms. The networks that they established have played important roles and contributed to the activation of Hsinchu Park and the electronics industry in Taiwan. There are now many venture capitalists and angel networks in Silicon Valley supporting the electronics industry in Taiwan.

The historical development of the Taiwanese electronics industry was comprehensively described by Aoyama (1999). It started in the 1960s with the construction, using foreign capital, of factories to produce transistors and IC packages. These were Kaohsiung Electronics (MICROCHIP) by General Electric, Taiwan Philips by Philips, Taiwan TI by Texas Instruments and RCA. Kaohsiung Electronics started the assembly of transistors in 1966, and the assembling of integrated circuits and production of IC packages the following year.

Professor Tho of the National University of Transportation launched the first electronics venture firm in 1971. The Electronics Research & Services Organization (ERSO) was organized within the Industrial Technology Research Institute (ITRI) in 1974, and 37 staff members were dispatched to RCA in the United States in 1976 to introduce CMOS-IC technologies, including IC design and manufacturing and wafer fabrication. Taiwan then entered the age of technology introduction. Semiconductor development was carried out mainly within ERSO. The Semiconductor Research Center was set up in 1977 where a model factory was established and succeeded in trial production of CMOS-IC. The following year the production of photomasks started in ERSO. Following these successes, 13 staff members left ERSO to establish a venture company, the United Microelectronics Company (UMC), in 1980 and started producing IC in 1982.

Many manufacturing companies of semiconductor chips were set up in 1987, and Taiwan entered the era of full-scale construction in the electronics industry. In addition to UMC, various other companies such as Hualon and the Taiwan Semiconductor Manufacturing Company (TSMC) were established. The 1980s was a period of VLSI development, and Taiwan entered the arena of severe international competition. ERSO changed its strategy from introducing technology by paying expensive lump-sums to developing their own technologies through collaboration with development firms, such as the Chinese firms in Silicon Valley. This resulted in its development of 256K DRAM in 1985.

Small enterprises in Taiwan could not make the big investments to construct manufacturing plants, like the large Western companies could, so they constructed one large-scale manufacturing plant for all companies to use in common. This is a foundry plant, that is, a service plant which produces electronics products for customers on commission but does not produce on its own account. Dutch Philips was asked to organize this undertaking, and a company, TSMC, was founded in 1987. It operated the first foundry plant in the world. Many staff members moved to TSMC from ERSO, and foundry operations started under an agreement between ERSO and TSMC in 1988.

The setting up of this foundry led to the appearance of many companies that designed products but had no factory. Its establishment triggered the emergence of many venture companies because design companies could produce ASIC (ICs for specific applications) products without the anxiety of production. At around the same time, the Industrial Technology Research Agency set up a venture capital firm, the Industrial Technology Investment Company. This led to an increase in the number of semiconductor companies, from only two in 1987 to ten by 1994. Taiwan established mass production technologies for 16M DRAM by the end of 1994, less than one year behind the more advanced countries. At that time, the supply of ICs to the US from Japan decreased due to trade friction. This enhanced Taiwan's business opportunity of capturing more of the DRAM market in the US. A five-year project for developing submicron lithography technologies was launched in 1990. ERSO proceeded successfully on this project by organizing a consortium of enterprises and universities. This led to the establishment of an advanced integrated circuits company, Vanguard International Semiconductor Corporation (VISC), another spin-off from ERSO.

This initial competition in semiconductor chips has developed recently into competition in custom-made ASIC products. Taiwan's strategy of having many small, flexible firms, like the organizational arrangement in Silicon Valley, has given it a big advantage in this area, enabling it to attain a competitive edge worldwide. Taiwan's electronics industry has grown rapidly and now produces about one-third of the personal computers sold worldwide. As of 1997, Taiwan had worldwide market shares of 70 per cent for scanners and personal computer cases, 60 per cent for keyboards, motherboards, mice, monitors and modems, and 30 to 40 per cent for notebook computers.

In short, the electronics industry in Taiwan started at the beginning of the development trajectory of the electronics paradigm, developed along the trajectory, and became self-supporting by the middle of the trajectory, resulting in its strong position internationally. Figure 8.5 clearly shows the timing and interrelationships between the actions of firms and the government in this development.

The Electronics Industry in ASEAN and Actions by Governments

As described above, the electronics industries in Japan, Korea and Taiwan were built up, with support from their governments from an initial position of importers of foreign technologies. Many other countries in Asia have also been involved in the electronics industry. Here we examine the situation and governmental actions in two ASEAN countries, Singapore and Malaysia. They organized their electronics industries differently, by inviting foreign enterprises to set up sites in their countries.

Singapore

The electronics industry in Singapore started when American companies set up production sites as their supply bases in Asia. National Semiconductors and Texas Instruments began operations in Singapore in 1968, Harris Semiconductors in 1969 and Hewlett Packard in 1970. In the 1970s, Japanese manufacturers of electrical appliances began to set up production plants in Singapore: NEC Semiconductors in 1976 and Matsushita Electronics in 1978.

The Singaporean government began actively to promote the development of an advanced information society, recognizing that Singapore should become the center of total business in Asia. They invited multinational enterprises to establish international hub functions in the Asian economy. The National Committee of Computers was organized in 1980, and the National Computer Bureau (NCB) was set up in 1981. The government announced its 'National Information Technologies Plan' in 1986 and an advanced information plan, 'IT2000', in 1992. The aim was to construct a comprehensive national information infrastructure covering households, enterprises, schools and government organizations, making Singapore an 'intelligent island'. The Computer Bureau announced the 'Singapore One Plan' in 1996 for connecting all of Singapore by optical fiber, thereby enhancing administration, education, amusement, electronic commerce and tele-shopping, under the catchphrase 'one network for everyone'.

This aggressive governmental policy and promotion resulted in the creation of an international hub mechanism. Various active centers were built, such as two Regional Control Centers set up by Unisys Asia Pacific and Datacraft Asia and one by Deutsche Telecom, and the Technical Center by Apple Computer and Silicon Graphics. The Singapore Digital Media Consortium was set up to promote the development of technology. Various electronics information enterprises constructed production facilities in the 1980s. A number of companies invested in Singapore to promote their business: Motorola in 1983, Silicon Systems and AMD (Advanced Microdevices) in 1984, Lucent Technologies in 1985, AT&T and Fujitsu in 1986, and Linear Technology and Hitachi Asia in 1989. The entry by foreign enterprises began in the 1960s and was reactivated in the 1980s, well along the development trajectory of the electronics paradigm. As shown in Figure 8.6, Singapore shared the same trajectory as Japan, Korea and Taiwan.

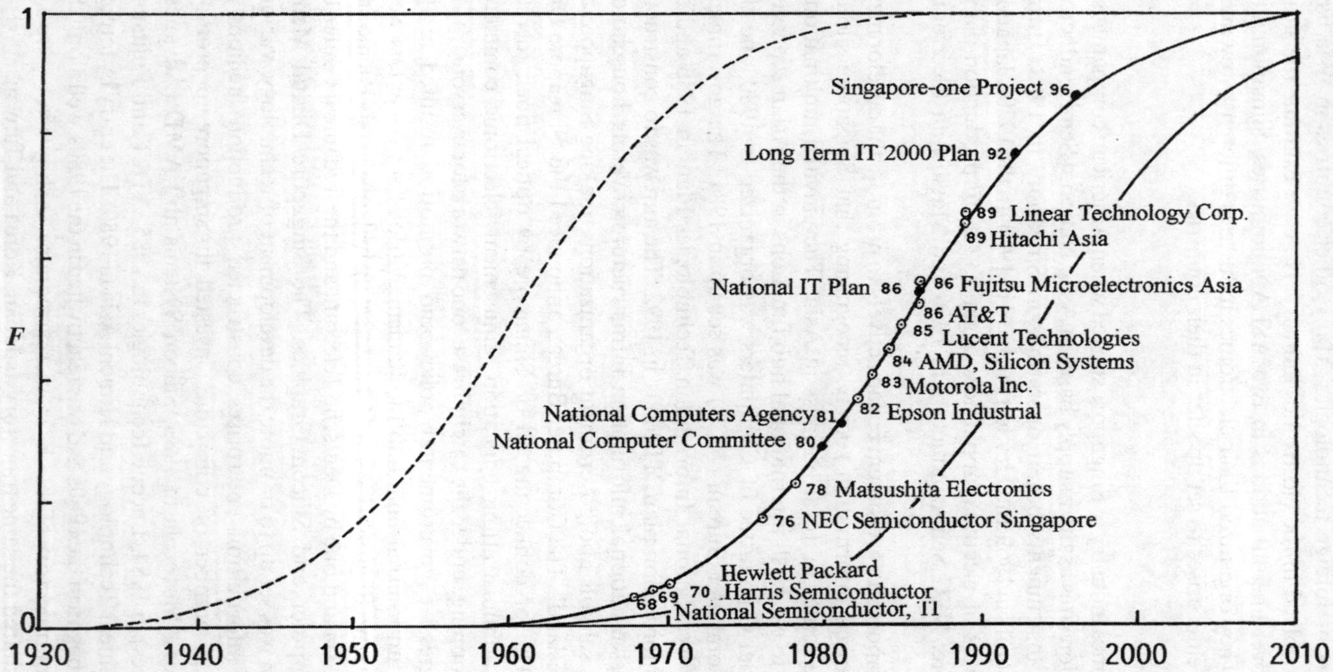

Source: Hirooka and Nitta (2000).

Figure 8.6 Electronics industry in Singapore and actions by government

Malaysia

The development of the electronics industry in Malaysia has been well studied (Yoshimura, 1998; Kato, 1997; Okamoto, 1998). Since gaining independence in 1957, Malaysia has implemented various economic policies to promote industrialization. To promote new industries, the Pioneer Industries Act was enacted in 1965; to encourage investment in export-oriented industries, the Investment Incentive Act was enacted in 1968, and the Free Trade Special Ward Act was enacted in 1971. Among the new economic policies, the Promotion Act for Electronics Industry was set up to provide large incentives to foreign electronics enterprises. The Free Trade Special Ward Act in particular stimulated investment by foreign firms.

The electronics industry made a big advance after the development of microprocessors and the advances of integrated circuits in the 1960s. While semiconductors are high-tech products, the assembly processes are simple and labor-intensive, and the electronics industry in the US sought to locate this work in developing countries with low labor costs. The advance of multinational electronics enterprises into the Asian area started with Fairchild's foray into Hong Kong in 1961, followed by substantial advances into Taiwan and Korea. National Semiconductors opened factories in Singapore in 1968 and in Malaysia's free trade ward of Penang in 1971. They were followed by six firms in 1972, seven in 1973, and four in 1974, a total of 18 companies in Malaysia. Intel came in 1974, and Motorola in 1979. Japanese electronics firms also established operations: Hitachi Ltd. in Penang in 1972, and Toshiba in 1973 and NEC in 1974 in Selangor.

Malaysia has several ethnicities (races), and the antipathy between the Malay and Chinese communities has escalated. The economic dominance of the Chinese led to the implementation of a new economic policy in 1971 – the 'Bumiputera policy', which proclaimed that the Malay were to be given priority in areas of education and employment. The Industrial Coordination Act of 1975 required manufacturing firms to obtain an operating license from the government and to comply with the Bumiputera policy.

The Malaysian economy was stagnant between 1982 and 1986. The government announced its Medium and Long-Term Industrial Master Plans, IMP, the latter a ten-year plan, in 1986 for revitalizing the economy and energizing the private sector. It also deregulated foreign investment in 1986 and 1988, allowing 100 per cent foreign ownership, to stimulate involvement by foreign enterprises. The National Development Policy was adopted to encourage participation by foreign enterprises through deregulation and favorable treatment of foreign investment. These measures brought a horde of American enterprises into Malaysia – Hewlett Packard, Motorola and Intel were the first to come, in 1987. It is noteworthy that the factories opened by these companies handled all the steps in the manufacture of semiconductor chips, unlike the factories opened in the first wave. Japanese electronics firms also rushed into Malaysia – nine in 1988 and seven in 1989. The largest investment by foreign firms from 1970 to 1990 was by Japanese manufacturers, especially after the Plaza Accord in 1985, which greatly accelerated the advance of Japanese enterprises. The number of Japanese-operated factories in Malaysia increased from 276 before 1985 to 841 in 1990. By 1995, the number of Japanese-related enterprises had increased to 1070, of which 546 were manufacturing factories and 244 were electrical and electronics ones. Related to these activities, by 1995 38 Japanese firms of electric appliances and 29 electronics companies operated in Malaysia of which 39 companies arrived after 1986.

As shown in Figure 8.7, the development of the electronics industry in Malaysia followed the same trends as in the other countries described above and their activities were distributed along the development trajectory of the electronics paradigm.

In short, these two ASEAN countries developed their electronics industries in their own way, based on their recognition of the importance of electronics innovation, and their governments played crucial roles.

The Electronics Industry in India and Actions by Government

The development of the Indian information and electronics industries has been described by Taniura (1990), Itoh (1999), Kojima (1995), Mitsubishi Research Institute (1997) and Yamazaki (1997).

The Indian economy was stagnant for a long time after the Second World War. In 1967, a committee for technology introduction recommended strict regulations on the introduction of technology, and the next year a committee on foreign investment was set up and regulations became still stricter. Moreover, the Foreign Exchange Regulation Act of 1973 reduced the ratio of investment by foreign capital to less than 40 per cent. A technology evaluation committee was set up in 1976 to advise on dealings with foreign enterprises from the technological side. Under these circumstances, recognition grew that these regulations were impeding economic development. Movements for deregulation thus began to take off. Industrial policy deregulation began in the latter half of the 1970s. Foreign investment was deregulated in 1981, and foreign investment of up to 100 per cent was permitted for export-oriented enterprises. The technology policy announced in 1983 emphasized the importance of effective absorption of foreign technologies. The Industrial Approval Agency was unified and functioned as the window for approving the introduction of foreign investment and technologies.

Modernization under Prime Minister Gandhi was via the computer policy announced in 1984. This policy recognized the importance of enhanced

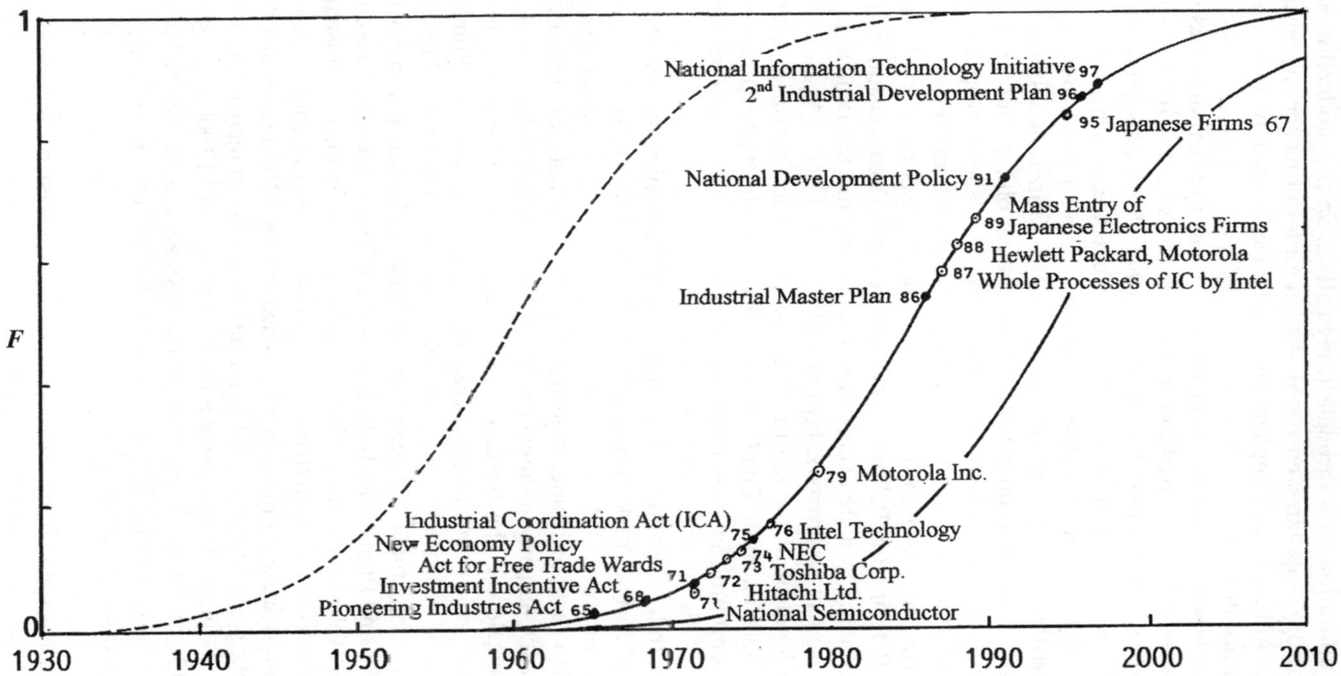

Source: Hirooka and Nitta (2000).

Figure 8.7 Electronics industry in Malaysia and actions by government

computer utilization, and through it, regulations on the production and importation of computers were eased. A plan for fostering comprehensive information industries was implemented, and a Department of Electronics was established in 1986.

Attention was then focused on the software industry as a result of which seven software technology parks were constructed. An association of software enterprises was established in 1988 to help formulate government policy. The National Information Science Center, established in 1986, played an important role in developing the Indian software industry. It coordinated the development of management information systems, the introduction of information systems to government offices, and the development of communications networks. The Advanced Science and Technology Center was established in Bangalore in 1989 and software technology parks were constructed in Pune and Bangalore in 1990. The Economic Reformation Policy of 1991 initiated India's conversion from a planned economy to a free competition economy, and the regulation of foreign investment was remarkably reduced.

The development of the information and electronics industries started with the establishment of Tata Consulting Services, which played a central role. Tata Consulting Services, which is part of the Tata conglomerate established in 1907 as an iron and steel company, is now the largest software company in India. Another such company is WIPRO Systems, established in 1981. Intel and Microelectronics advanced in 1978, and IBM and Unisys set up joint ventures with Tata. The government's information fostering plan of 1986 led to Texas Instruments opening the TI Semiconductor Design and Development Company in Bangalore in 1987, which triggered the development of Bangalore. Following this wave of activity, various electronics businesses, mostly American, set up operations in the Bangalore area. Starting with Motorola in 1989, various enterprises set up software development operations, among them, Siemens, Hewlett Packard, British Telecom, City Corp, AT&T and Fujitsu. With the flourishing of Bangalore, the Tata conglomerate developed a huge industrial park, with an excellent infrastructure, in Bangalore, with help from the Singaporean government.

The software industry in India was valued at only $210 million in 1991, but, just four years later, it reached $1.1 billion, a six-fold increase. In 1994 there were about 1000 software companies with around 12 000 employees. By 1996, 131 of *Fortune*'s top 500 companies had commissioned Indian firms to develop software. The success of the Indian software industry can be ascribed to several factors:

1. India has an abundance of talented well-educated people, many of whom had worked in Silicon Valley.

2. The official language is English.
3. Wages are relatively low.
4. There is a 12-hour time difference between the US and India, so work can be done in India while workers in the US are sleeping.

As shown in Figure 8.8, the electronics and information industries in India developed along the same electronics paradigm trajectory as in Japan, Korea, Taiwan, Singapore and Malaysia. The reason for the low level of activity during the first half of the electronics development trajectory can be explained by the many Indians working in Silicon Valley and other overseas areas during that period. This can be considered an incubation period for the Indian industry. One of the important things is that the government recognized the importance of electronics and information and took proactive measures, focusing on software technologies.

The analysis in this section indicates that the actions taken to develop the electronics industry in various countries followed the same trajectories of the electronics paradigm. This contradicts the generally held belief that there have been rather long time lags between the activities taken in advanced countries and those taken in developing ones, as shown in Figure 3.9, for example. This gap can be interpreted to mean that innovations have become more globalized so that the time lag has been shortened to virtually nothing. The world economy and industrial activities have become synchronized and globalized. Moreover, the actions taken by governments have played an important role, and national systems of innovation have become a prerequisite for the taking off of modern industries.

On the other hand, the above data suggest that not only a series of patents and products but also a series of such bibliographic events as actions taken by governments and industrial firms can be used as elements related to innovation for identifying the trajectories.

8.5 THE FALL OF THE JAPANESE ECONOMY

The international competitiveness of a country is generally evaluated according to the country's ability to generate and utilize innovations. A useful tool for explaining the competitiveness of latecomer countries during the mature stage of innovation is Vernon's life-cycle theory. Here we discuss another factor governing international competitiveness – the exchange rate of a country's currency. We also discuss the recent collapse of Japan's economy from a new angle.

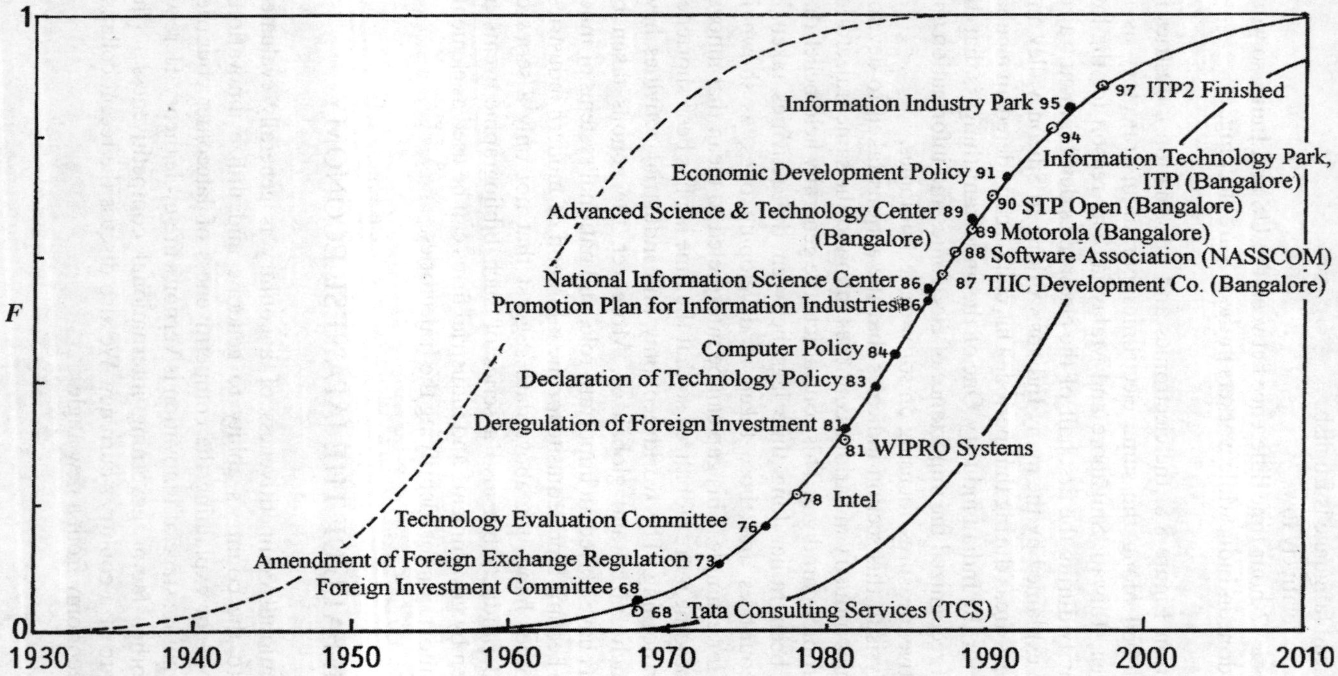

Source: Hirooka and Nitta (2000).

Figure 8.8 Electronics industry in India and actions by government

Another Factor Governing International Competitiveness

Various innovations have produced modern civilized society, creating long business cycles. The first two cycles were driven by technological advances made in the UK, while the last two cycles were mainly dominated by the US. During the upswing of a Kondratiev cycle, a cluster of innovations diffuse, leading to developments in the economy and the relevant innovations deeply related to the competitiveness of the country. As shown in Chapter 4, during the latter half of an upswing, the economy becomes overheated due to speculation after a long period of prosperity, a bubble forms, and the economy crashes at the peak of the cycle, resulting in a long depression. This phenomenon has been repeated in every Kondratiev cycle. During the upswing, industries are prosperous, and technological leadership plays a crucial role in international businesses. Subsequently, international competitiveness enters a different phase. In particular, the disparity between countries is affected by another factor – the exchange rate. This trend gained strength following World War II as the globalization of world trade expanded.

Relative Competitiveness of the United States and Japan

The United States enjoyed prosperity during the upswing of the fourth Kondratiev cycle as a victor of the Second World War, its economy accounting for about one-third of the world's GNP. Japan suffered from the heavy damage of the war, but after a period of reconstruction, recovered to enjoy a period of long economic growth during the 1960s, making the transition from a developing country to a developed country by introducing Western technologies.

After the Second World War, the United States was a great power and the dollar was strong and overvalued in comparison with other currencies. The yen/dollar exchange rate was fixed at 360 ¥$ while the actual purchasing power parity was about 300 ¥$. As a consequence, the United States found overseas production to be cheaper than domestic production. Many multinational companies thus established production abroad, while domestic manufacturing began to decline under a low rate of operation, leading to a hollowing out of the US economy. Domestic steel production plants, for example, operated using old, lower efficiency prewar facilities. The international competitiveness of US industries thus weakened. On the other hand, a huge amount of dollars were issued, and these so-called 'eurodollars' flooded Europe to pay for purchases of European real estate and companies. This put downward pressure on the value of the dollar. The excess supply of dollars also undermined the gold standard and the value of the dollar declined until the purchasing power of dollars became 245 ¥$. The US abandoned the gold standard in 1971, the so-called 'Nixon shock' and the money market shifted to a floating exchange system.

In contrast, Japan rebuilt its economy by constructing new factories with imported advanced technologies, starting from the bottom due to the devastation of the war. By the end of the 1960s, Japanese industries fulfilled domestic demand and began to export some products to the US and Europe, thanks to the advantage of having a lower valued currency. At that time, the upswing of the fourth Kondratiev cycle was still under way, and world trade was increasing along the upswing. This 'following wind' favored Japanese export-oriented industries. These factors benefited the Japanese economy. After the shift to a floating exchange rate system, the value of the yen remained at the level of purchasing power parity, and Japanese exports remained competitive.

Scenario of US–Japan Trade Friction

As described above, Japanese industries began exporting products in the latter half of the 1960s, taking advantage of the 'following wind' of a cheaper yen. Exports started with cotton clothes and steel products, followed by electrical appliances, represented by TV sets, which quickly accelerated. Trade friction, especially with the US, became a serious problem. In 1968, the US claimed that Japan was dumping electrical appliances, especially TV sets, on the US market and imposed punitive tariffs on Japanese TV sets in 1971. In 1977, Japan placed a limit on the number of TV sets that Japanese companies could export to the US. Japan also placed a limit on the export of steel to the US in 1969, and Japan and the US concluded a textile agreement in 1971. Although the oil crisis of 1973 sent the world economy into decline, the export of Japanese compact automobiles thrived due to their energy-saving performance and high reliability. The high level of automobile exports to the US raised even more friction, so that Japan was forced to limit exports of automobiles to the US to 2.3 million cars per annum, starting in 1981. In the same year, Japan beat the US in the race to develop 64K DRAM IC chips. Subsequently, Japanese electronics products and machine tools achieved dominance in the market, and export controls were imposed in 1986.

The effect of trade friction between Japan and the US is illustrated on the S-curve in Figure 8.9, along with the amount of world trade. World trade expanded steadily for more than 20 years, from the latter half of the 1960s to the early 1990s; the total value of international trade grew 30 times in just the decade of the 1980s, reaching $50 billion a day. This trend is interpreted to mean that the world economy was on the upswing of the

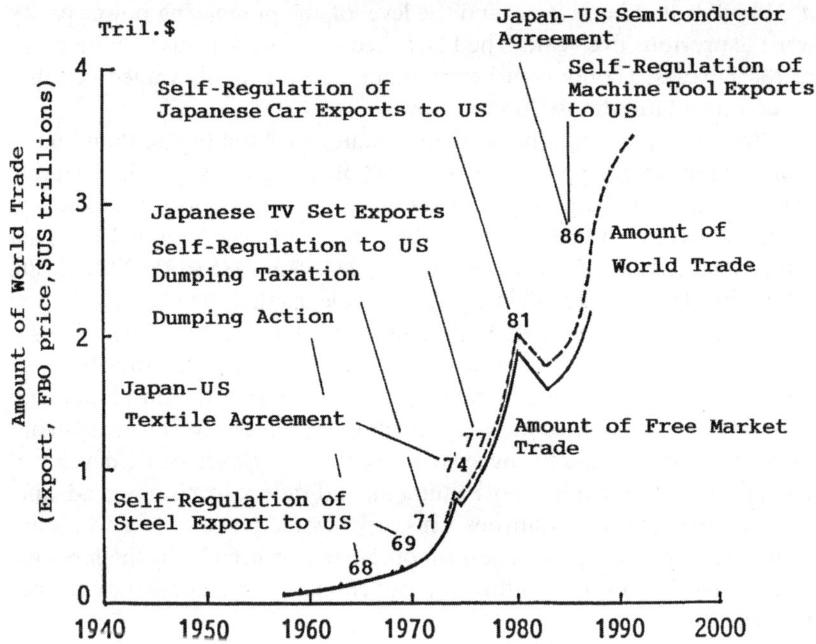

Source: Hirooka (1994a, 1999e).

Figure 8.9 Expansion of world trade and US/Japan trade friction

fourth Kondratiev business cycle, and world trade of various innovation products was accelerating along the upswing. Japanese industries readily rode this fortuitous uptide of international trade. The exchange rate remained at the level of purchasing power parity, and there were no substantial currency barriers. This gives a rough outline of the response of Japanese industry to the trade friction between Japan and the US.

The Collapse of Japanese Industry due to the Appreciation of the Yen

The increasing surplus in Japanese trade led to growing resistance in Western countries. The United States, on the other hand, faced growing deficits in both international trade and internal revenue, which led to a financial crunch in May 1985. The financial ministers of the five major countries (the US, the UK, Germany, France and Japan) met at the Plaza Hotel in New York in September 1985 and agreed to adjust the exchange rate by raising the value of the yen against the dollar. Before the accord, the exchange rate was 234 ¥$, slightly lower than the purchasing power parity of 218 ¥$. It had been at around the level of the purchasing power parity over the previous five years. The Plaza Accord was a serious turning point for the Japanese economy, ushering in an era of a highly valued yen that has continued for almost 20 years now.

As the rate of inflation in the United States has been higher than that in Japan, the purchasing power parity (PPP) of the Japanese yen has steadily increased against the dollar. The actual exchange rate for yen, however, increased at a much higher rate, beyond the increasing level of PPP. That is, the exchange rate of yen increased from 234 ¥$ in 1985 to 169 ¥/$ in 1986, 145 ¥/$ in 1987 and 128 ¥/$ in 1988, which means that the value of the yen doubled in only three years. Export-oriented industries in Japan faced a serious situation and quickly lost their international competitiveness. At that time, computers were being introduced into production plants, and manufacturing industries were implementing advanced operating systems, FMS (flexible manufacturing system) technologies were being developed to cope with the difficult export situation, and rationalization was advancing faster than in other countries. This effort would prove helpful to regaining international competitiveness to some extent, especially by the Japanese car industry. However, this effort was not enough to overcome the currency problems.

To cope with the highly valued yen, the Japanese government implemented policies to stimulate the domestic economy and increased the money supply. As the domestic market in Japan had almost reached the mature period of the upswing of the fourth Kondratiev cycle, the money mainly went to purchase company stock and real estate in Japan and to purchase real estate in the United States, which saw a flood of Japanese money. This behavior was similar to that of the United States in the 1960s when it issued large amounts of dollars that were used to buy European companies and real estate. Japan inevitably developed a bubble economy, as described in Chapter 4, and money raised by equity financing was readily invested in excessive domestic plants and building facilities, leading to serious bad debt problems after the bubble economy collapsed. As the bubble economy continued to expand, the government decided to take vigorous action to suppress it. The government raised the interest rate sharply and shrank the money supply. Stock prices collapsed, and the bubble economy crashed in the early 1990s.

On the other hand, the higher valued yen resulted in the same situation as had the higher valued dollars in the last part of the 1960s. Japan's export-oriented enterprises went abroad to construct production plants through direct investment. Electrical and electronics firms mainly went to East Asia, and automobile companies mainly went to the United States.

The Highly Valued Yen even during Depression

After the crash in 1990, the Japanese economy deteriorated and a deep depression took hold. The trade surplus, however, did not decrease – it nominally increased due to the rising value of the yen. After the crash, the value of the yen continued to rise, reaching its highest value of 79 ¥$ in April 1995.

Why was the value of the yen so high during a continuing depression? According to the analysis by Werner (2001), the decisive factor in determining the exchange rate is not the difference in interest rates between two countries but the difference in the real liquidity of the money supply. If the money supply increases and the liquidity of credit creating increases, the exchange rate decreases; if the liquidity of credit shrinks, the exchange rate increases. After the crash, the Bank of Japan enforced a tight money policy, which greatly reduced monetary liquidity. In contrast, the United States faced a serious recession after the crash in 1987, and the growth rate of the economy was negative, −1.2 per cent, in 1991. To fight the recession, the Federal Reserve Board greatly increased the supply of dollars and successfully avoided dropping into the recession by dealing with bad debt. This put the US on the road to recovery. The Japanese government lowered the interest rate and issued a huge amount of government bonds to overcome the depression. These actions, however, did not increase yen liquidity, and the exchange rate continued to increase. The money supply shrank drastically from August 1994, with the tightest supply coming in March 1995. The exchange rate reached its highest value of 79 ¥$ in April.

Another factor determining the exchange rate is closely related to the movements of hedge funds in the US. Along with the growth in international trade and the globalization of the world economy, currency speculation has become rampant given the floating exchange rate system. The variety of derivatives and the number of hedge funds have increased exponentially since 1993, with both having a large effect on exchange rates. Barings Securities failed due to losses from derivatives trading, and Daiwa Bank in Japan lost one billion dollars by dealing derivatives in 1995. Thailand's economy crashed in 1997 due to a concentrated attack by hedge funds selling the baht.

The amount of the settlement of international trade was only 22 per cent of the total exchange dealings in 1995 in which the breakdown is 12 per cent for international trading and 10 per cent for direct investment. In 2000, the rate of real trading accounts was only 3 per cent of total exchange dealings. Most of the money dealing was handled for speculative purposes. This unusual situation certainly poses a serious problem for the world economy.

The exchange rate for the yen has remained at the highest level for almost 20 years since the Plaza Accord in 1985.

The Hollowing-out of Japanese Industry and the Rise of East Asia

After the Plaza Accord, Japanese export-oriented industries turned their attention abroad to avoid the effect of the highly valued yen. The electrical and electronic industries moved their bases to the East Asian region, particularly Thailand and Malaysia in ASEAN. These countries had developed rapidly and offered incentives to Japanese companies. They developed infrastructures and offered tax incentives to entice Japanese companies. This mutual interest accelerated the movement overseas of Japanese firms and the development of the host-nation economies. The crash of the bubble economy in Thailand was triggered by the hedge fund assault in 1997 and quickly spread to the rest of East Asia.

The continuing shift of Japanese manufacturing overseas since 1985 has drastically decreased domestic production of various goods, as shown in Figure 8.10. This has seriously hollowed out the Japanese economy, resulting in drastic structural changes in the economy. More than 25 per cent of total Japanese production had been shifted abroad by 2000. This rate was

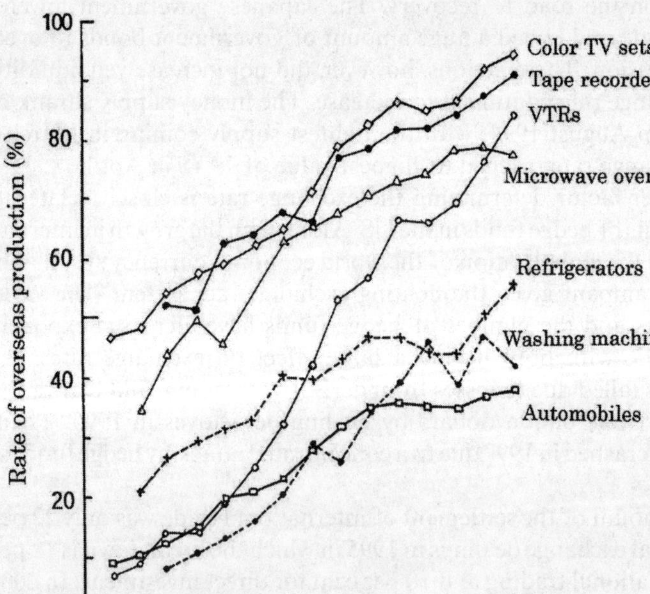

Source: Hirooka (2002b).

Figure 8.10 Rates of overseas production by Japanese companies

significantly higher than the 17–18 per cent shift in the US by the end of the 1960s which caused serious damage to the US. The trend in overseas versus domestic production for color TV sets and video tape recorders by Japanese enterprises is shown in Figures 8.11 and 8.12. While domestic demand remained almost constant, production shifted almost completely overseas.

Purchasing Power Parity and International Competitiveness

The real strength of a national economy can best be evaluated by comparing its purchasing power parity (PPP). The PPP figures published by the OECD are determined on the composition of GDP. Since PPP indicates a parity value (how much must be paid to purchase the same good) it is a proper index for comparing national economies. However, the PPP is not always the same between sectors of the economy. Nevertheless, international comparisons by government are usually carried out using PPP.

The trend of the exchange rate and purchasing power parity between the yen and dollar is shown in Figure 8.13. After the Nixon shock in 1971, the exchange rate under the floating system closely corresponded to the purchasing power parity up until the Plaza Accord in 1985. After the Plaza Accord, however, the exchange rate remained higher than the PPP for almost 20 years, which is unnatural. The higher valued yen produces various anomalies in the economic data. The GDP and trade surplus tend to be overvalued, making the economy look more competitive than it actually is. If the national performance is compared with the purchasing power parity, the actual figures can be estimated. The figures for Japan compared to those of the US, based on the PPP, are shown in Table 8.1. They come from a survey by the Japan Productivity Center and show that Japanese GDP per capita and labor productivity by industry have been consistently lower than those for the US. The lower values in 1975 indicate that Japan suffered more from the oil shock than did the US. The lower labor productivity of the Japanese automobile industry is explained by the fact that total productivity tends to be lower when productivity at subsidiary SMEs is taken into account. Another factor is that a large percentage of the exports have lower profitability. These data indicate that productivity in various Japanese manufacturing industries has never exceeded that in the United States, except in the steel industry.

Another problem for Japan is the development of China. Since its liberalization in 1978, China has developed economically and has made its presence known especially since 1990. This great development of China has shifted the weight of economic development from Southeast Asia to China, and all of the advanced countries have rushed into the Chinese market. The

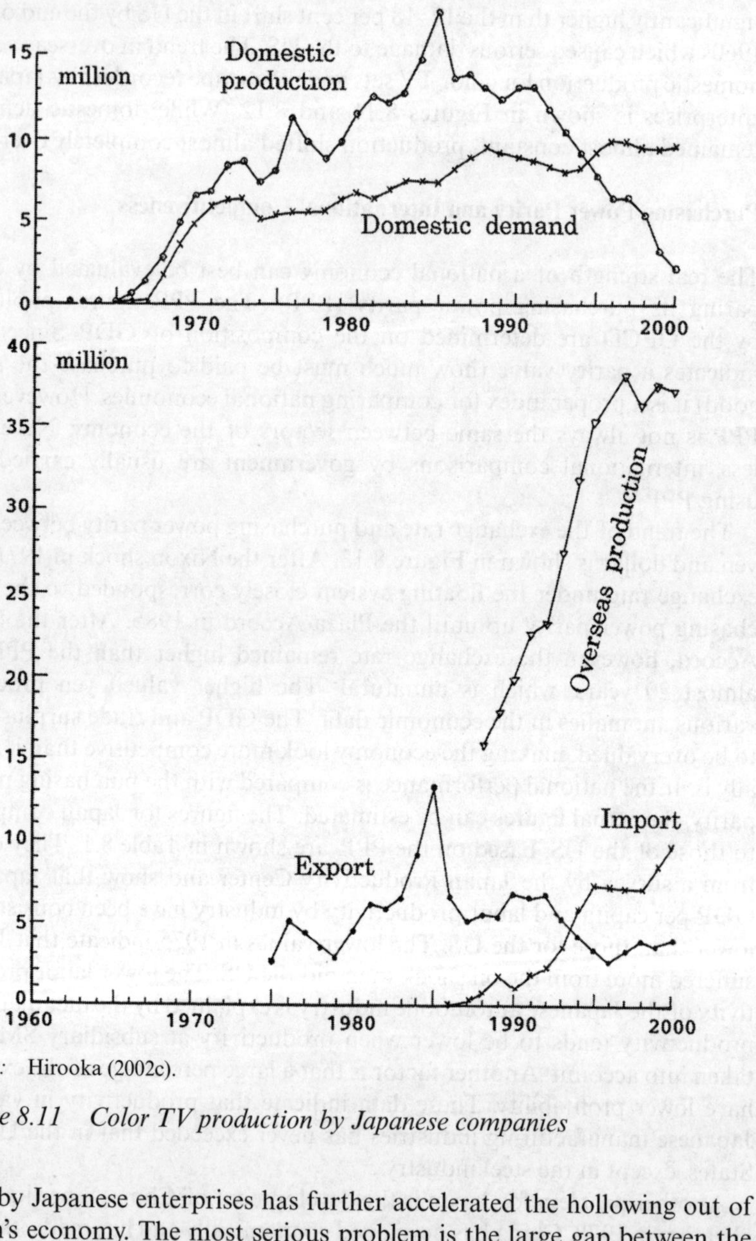

Source: Hirooka (2002c).

Figure 8.11 Color TV production by Japanese companies

rush by Japanese enterprises has further accelerated the hollowing out of Japan's economy. The most serious problem is the large gap between the dollar exchange rates of the two countries. While the Japanese rate remains high at 131 ¥/$ against a PPP of 168 ¥/$, the Chinese yuan has been at an unusually low level of 8.3 yuan/$ against a PPP of 2.1 yuan/$, with a base

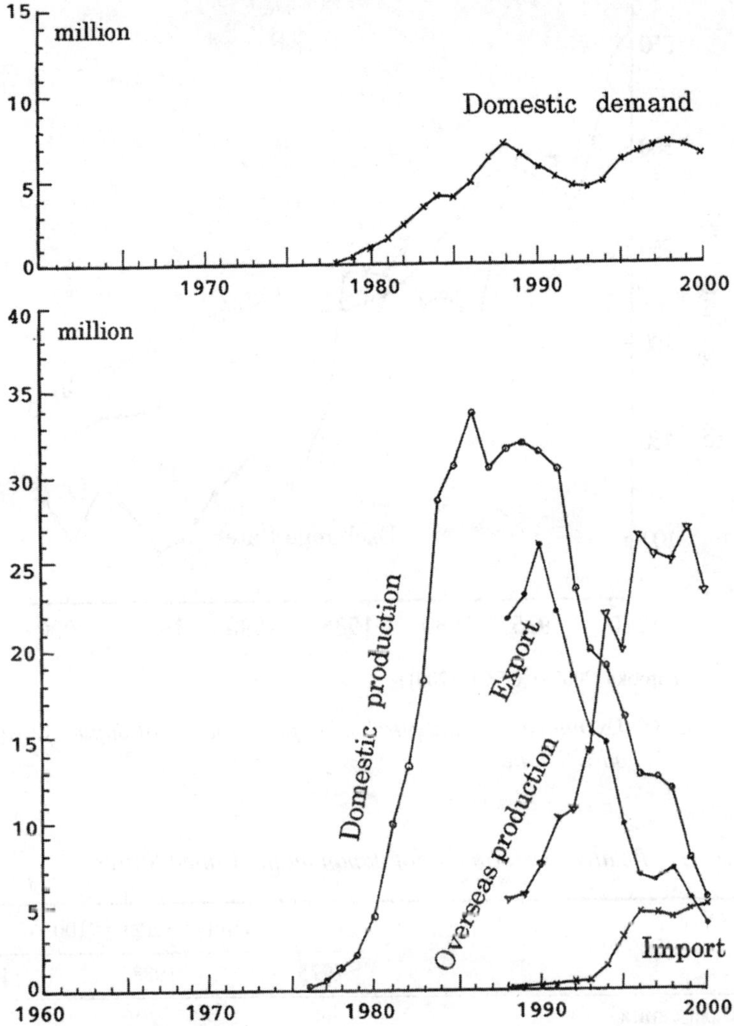

Source: Hirooka (2002c).

Figure 8.12 VTR production by Japanese companies

year of 1998 (World Bank, 2000). In, 2000, the yen was at 108 ¥/$ against a PPP of 156 ¥/$.

According to the analysis by the World Bank, the relationship between income levels and the disparity between a currency's PPP and nominal exchange rates is inversely proportional. While the less developed countries

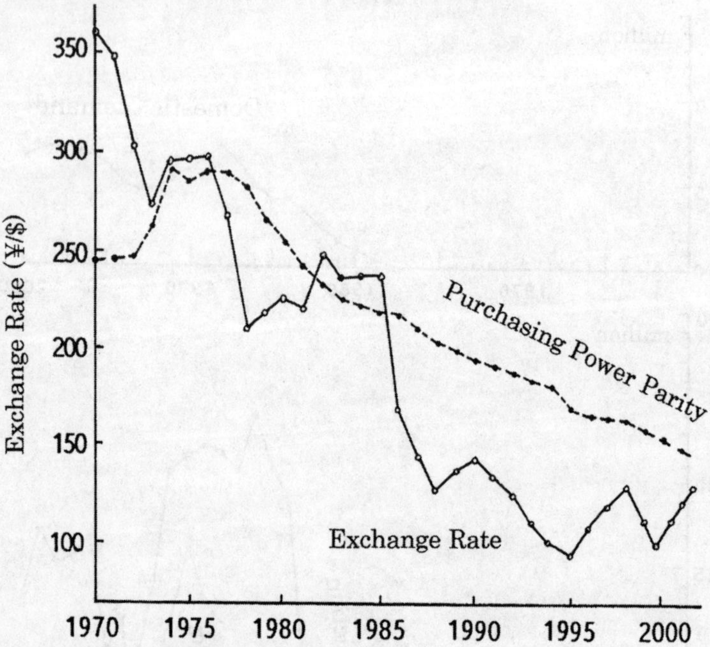

Sources: Hirooka (2002b), OECD (2001).

Figure 8.13 Exchange rate and purchasing power parity of Japanese yen on US dollar

Table 8.1 Relative performance of Japan on the United States

	United States = 100		
	1975	1988	1995
GDP per capita	66	80	80
GDP per worker	56	76	74
Labor productivity by industry			
Total manufacturing	51	61	71
Electric, machinery	23	86	59
Automobiles	42	68	84
Steel	68	113	119

Note: On the basis of purchasing power parity.

Sources: Hirooka (1992c, 2002b); data from Japan Productivity Center (1991, 1998).

are privileged to have lower exchange rates, there is no reason why Japan, an advanced country, should suffer from the highest exchange rate of all advanced countries. This great gap between Japan and China has accelerated the transfer of Japanese operations to China and has worsened the hollowing out of Japan's economy.

The movement of Japanese operations to China has reached a level that cannot be ignored, and the scale of investments to China by Japanese companies leaves no way back. Cheaper goods produced by Japanese factories in China have replaced more expensive ones produced in Japan, resulting in significant deflation in the Japanese economy. Now that Japanese companies have set up shop in China, the disadvantage due to the unusual disparity in currencies is turning into an advantage due to globalization. The only way for the Japanese economy to recover is to develop new businesses. In fact, Japanese electronics companies have been doing this by developing, for example, such electronics appliances as digital cameras, high-density memories and various digital appliances. In these areas, Japanese firms are again world leaders, but this situation is unlikely to last long.

Moreover, while the domestic Japanese automobile industry has suffered from the big gap between the exchange rate and PPP and has been squeezed out abroad, automobile production worldwide has increased from 13.5 million in 1990 to 16.4 in 2000, and 18.9 in 2003. Toyota recently announced that their worldwide production should reach 7 million in 2005, and their overseas production will exceed 50 per cent of their total production.

8.6 SUMMARY

In terms of national systems of innovation, the behaviors of firms and the actions of governments were analyzed using the innovation trajectories of the electronics paradigm, and the international competitiveness of industries was discussed.

I. The behavior of Japanese electronics firms and the fostering actions of the Japanese government, especially by MITI (Ministry of International Trade and Industry) were analyzed.
   1. Japanese electronics firms started operating early during the technology trajectory. Transistor radios and calculators, both typical Japanese export products, came to dominate the world market.
   2. MITI undertook various actions to promote the fledgling electronics industry, starting with the Provisional Act for the

Promotion of Electronics Industries in 1957, which is located early in the first half of the electronics technology trajectory.
3. These early actions taken by firms and the government were the key to Japan's success in achieving international competitiveness.
4. The VLSI collaboration research project was launched in 1976 as a four-year national project and was successful. Its timing correlates with the most effective business opportunities, during the first half of the electronics development trajectory.

II. The national systems of innovation in the United States and Europe were also analyzed.
5. To compete against Japan's success in electronics, the United States launched the VHSIC collaboration project in 1980 and set up the SEMATECH consortium in 1988. The EC governments launched their own collaborative programs: ESPRIT in 1985 and JESSI in 1988. The relative timing of these projects is shown by their positioning on the innovation trajectories.
6. The United States began implementing measures to strengthen its competitiveness, following President Carter's 'presidential message on innovation' in 1979 and the passage of the Stevenson–Wydler Technology Innovation Act and Bayh–Dole Act in 1980. During the 1980s the US organized engineering centers to foster technology transfers for ongoing businesses and science centers to deal with forthcoming industries. These actions are those of a national system of innovation and have worked effectively.

III. The national systems of innovation in several Asian countries were also analyzed.
7 The electronics industry in the Asian region developed along the same trajectories of the electronics paradigm, which means that they shared the same trajectories and the business was simultaneously synchronized in the world.
8. Korea and Taiwan started their electronics industry with strong initiatives from government, and their industries became self-sufficient during the latter half of the development trajectory.
9. Two ASEAN countries, Singapore and Malaysia, invited in foreign electronics companies and developed their electronics industries with the assistance of governmental initiatives.
10. The Indian government implemented a policy for developing its information and electronics industries during the latter half of the development trajectory and was mainly successfully in developing its software industry.

IV. Another factor in international competitiveness and the collapse of the Japanese economy was discussed.

11. The upswing of the Kondratiev cycle is governed by innovation diffusions and international competitiveness is technology oriented. In the downswing, however, different indices determine competitiveness. In the present Kondratiev cycle, the exchange rate has been an important factor determining competitiveness.
12. After the Second World War, the world economy enjoyed an era of prosperity during the upswing of the fourth Kondratiev cycle. At the end of the upswing, however, the competitiveness of the United States had been weakened due to higher valued dollars, which resulted in the hollowing-out of the economy. Japan, on the other hand, was in a strong position for exporting products due to the cheaper yen.
13. This crossover in competitiveness induced trade friction between the US and Japan which began in the late 1960s and has continued for more than 20 years. The history of the frictions corresponds to the rising growth of world trade. This indicates that Japanese industry enjoyed a 'following wind' during the upswing of the fourth Kondratiev cycle and the benefit of a lower yen exchange rate.
14. After the Plaza Accord in 1985, the Japanese yen suffered from a prolonged period with an excessively high-value yen, which led to a hollowing-out of the Japanese economy. Japanese export-oriented industries were forced to advance overseas and internationalize. This movement abroad accelerated the development of the ASEAN countries.
15. Since the crash of Japan's bubble economy in 1990, the yen has remained overvalued. Behind this was Japan's failure to properly control the liquidity of money supply as well as the rise of hedge funds and derivatives. This has caused unusual monetary distortion worldwide.
16. Japan has suffered from a distorted monetary situation and has been forced to move operations overseas, accelerating Japan's globalization. The recent rise of China's economy has further accelerated this trend, and deflation in Japan has been enhanced by the large gap between the cheap Chinese yuan and expensive Japanese yen.

9. Technology fusion and the evolution of innovation paradigms

An innovation develops based on a radical invention or unexpected discovery and eventually reaches maturity. The innovation paradigm is then absorbed into the economy and becomes a part of it. Society consists of accumulated matured innovation paradigms, and daily life is in a kind of equilibrium system, as pointed out by Schumpeter (1939). When an innovation paradigm emerges, a new industry appears, and existing industries absorb the new activities of innovations in an evolutionary way, enhancing the value added to the economy. Kuznets (1959) and Mowery and Rosenberg (1998) pointed out the importance of this adoption of new technologies by mature industries, sparking growth in productivity and even initiating the appearance of new products. Many old industries have experienced a significant boost in productivity as a result of the intersectoral flow of new technologies.

In the previous chapters, the process of innovation was described using three trajectories, and the behaviors of firms and government were discussed. The development trajectory was shown to play important roles, and knowledge transfer among various actors was shown to be crucial for development. This chapter extends the discussion to innovation by existing industries through technology transfer from new industries and to the evolution of the innovation paradigm. Trunk innovations in particular have a pervasive impact on existing industries, reinvigorating them through technology fusion. Furthermore, an innovation paradigm evolves to the next innovation paradigm through a new finding or breakthrough discovery.

9.1 ANALYSIS OF TECHNOLOGY FUSION BY EXISTING INDUSTRIES

Existing industries begin to introduce an innovation if it has versatile and adaptable capacity as a 'trunk innovation' or 'general purpose technology'. Computers and electronics are examples of such versatile innovations. This is the phenomenon of technology fusion, and the locus of technology introduction creates a fusion trajectory.

Origin of Technology Fusion

Kodama (1985, 1986) distinguished two types of innovations, breakthrough and fusion, and showed that 'technology fusion' can be regarded as a special form of cross-sector diffusion of a new innovation paradigm. The expression 'mechatronics' was coined in Japan in 1975 to describe the technology fusion of electronics with machine tools. As described in the previous chapter, the Japanese government (MITI) designed an innovation policy to induce technology fusion and interaction between industries through the Provisional Act for the Promotion of Specified Machinery and Electronics Industries in 1971. As Freeman (1987) pointed out, the Japanese economy has successfully adopted new paradigms appropriate to the needs of various industrial sectors: robotics, NC (computer numerically controlled) machine tools, and FMS (flexible manufacturing systems). Fusion-type innovation is brought about by a concerted effort of the several different industries involved. This is the origin of the international competitiveness of Japanese industries in the 1980s.

As described in the previous chapter, the Japanese government enacted various measures to support the electronics industry and succeeded in making it competitive after the success of the VLSI project in 1976. A more important characteristic of the Japanese electronics industry, however, is its ability to quickly adapt electronics technologies to various consumer products and to industrial processes and products by technology fusion. Gregory (1986) pointed out that the Japanese electronics industry has always been first to industrialize consumer electronics by introducing electronics technology since the transistor radio was introduced in the 1950s. Japan's market shares steadily increased, from 10 per cent for transistor radios in 1955, 50 per cent for color TVs in 1965 and 60 per cent for calculators in 1970 to 90 per cent for video tape recorders in 1980. It is interesting to use the map of the innovation trajectories to study how Japanese industries introduced computers and electronics technologies by technology fusion. This is the main subject of this chapter.

Complex Product Systems and Technology Fusion

One of the interests of technology management is to analyze how to manage complex product systems. These are large sophisticated products and systems composed of various technologies combined in complicated ways, such as flight simulators, aircraft engines, offshore oil drilling platforms and advanced transportation systems, which require more sophisticated management. The Complex Product Systems research project (CoPS) was organized in the Complex Product Systems Innovation Centre, which is the collaboration research group of the Science and Technology Research Unit (SPRU) at the University of Sussex and Brighton University in the UK. This chapter also analyzes complex product systems produced by technology fusion.

There are two types of complex product systems. One is produced by technology fusion which enhances the value-added. For example, the precision and speed of machine tools were remarkably improved by the technology fusion of electronics, and the value-added was greatly enhanced. The other type is a system incorporating additional functions to fulfill some particular demand. While innovations are generally accepted by the market as enhancing the value-added, other demands could be related to safety, amenity or environmental requirements. Fulfilling the demand may increase the cost of the system, but it should be evaluated not by cost but by the functions added due to societal request or environmental pressure. In the first type of complex product systems, the product price falls during the course of innovation diffusion. In the second type, the price increases in response to enhanced requests. For example, while the production cost of general purpose plastics has been decreasing due to economies of scale, the cost of aircraft per unit weight has been exponentially increasing over the last 30 years (Adachi, 1981).

A characteristic feature of every complex product system is that it has a simple purpose and is not so complex. The aircraft engine is a propulsive device used to fly airframes, the oil drilling platform is a structure used to drill oil in the seabed, and the advanced transportation system is used to control traffic. Moreover, the basic operating concept is also simple. The aircraft engine acquires propulsive force from the explosion of kerosene, the oil drilling platform drills using pipes, and the advanced transportation system controls traffic using computers.

Procedure for Analyzing Technology Fusion

A trunk innovation develops in two phases. The first is the formation of the industry itself. The second is the dissemination of the innovation among versatile applications by technology fusion. The latter phase is related to the techno-economic paradigm change and to construction of infrastructures through penetration of the innovation into the economy. The present chapter deals with the technology fusion along trajectories on the innovation paradigm in the cases of computers and electronics.

In this chapter, we study how various industries have introduced computers, that is, mainframe computers. And in the case of electronics, we discuss how existing industries enhanced their value-added by technology fusion with electronics. The timing of the introduction of computers and electronics by various industries is described and we discuss how they were evolved by the introduction, making fusion trajectories.

The process of an existing industry adopting a new innovation can be described as the way in which it innovates its products, creating a fusion trajectory that is comparable with the development and diffusion trajectories of the innovation.

This chapter first describes the introduction of computer technologies by the steel, cement, chemical and railway industries as well as their use in flight simulators. It next describes the introduction of electronics technologies into machine tools, cameras, automobiles and aircraft engines. Then it compares each fusion trajectory with the trajectories of the computer or electronics paradigm. As shown in Figure 9.1, the technology trajectory of the computer paradigm is composed of a series of computer core technologies, and the development trajectory can be described as the development history of mainframe computer products by IBM and of computer operating software. The electronics paradigm has already been described in Figure 7.3.

9.2 TECHNOLOGY FUSION OF COMPUTERS

Existing industries adopted computers to improve their business activities. We studied technology fusion of computers in the steel, cement, chemical and railway industries and in flight simulators. If an industry simply adopts computers as a user, the fusion trajectory must develop along the diffusion trajectory of computers or even later. If the industry adopts computers as a tool for developing their products, the fusion trajectory could appear much earlier than the diffusion trajectory and could be along the development trajectory of computers. It is interesting to identify the timing of the fusion trajectory.

Technology Fusion by Manufacturing Industries

Steel industry

The Japanese steel industry developed during the upswing of the fourth Kondratiev wave and became highly competitive. One of the reasons it obtained such competitiveness is its quick adoption of computer technologies. Figure 9.2 shows the fusion trajectory of the steel industry in Japan. It is plotted using data from Aida (1997) and company histories, especially the chronicles of Nippon Steel (1981). Nippon Steel began using an IBM 7070 for technical calculation in 1961, and planned to use computers to control all production plants in 1966. It started by automating its thick plating, blooming, cold rolling and automatic rolling processes, and by 1970 had

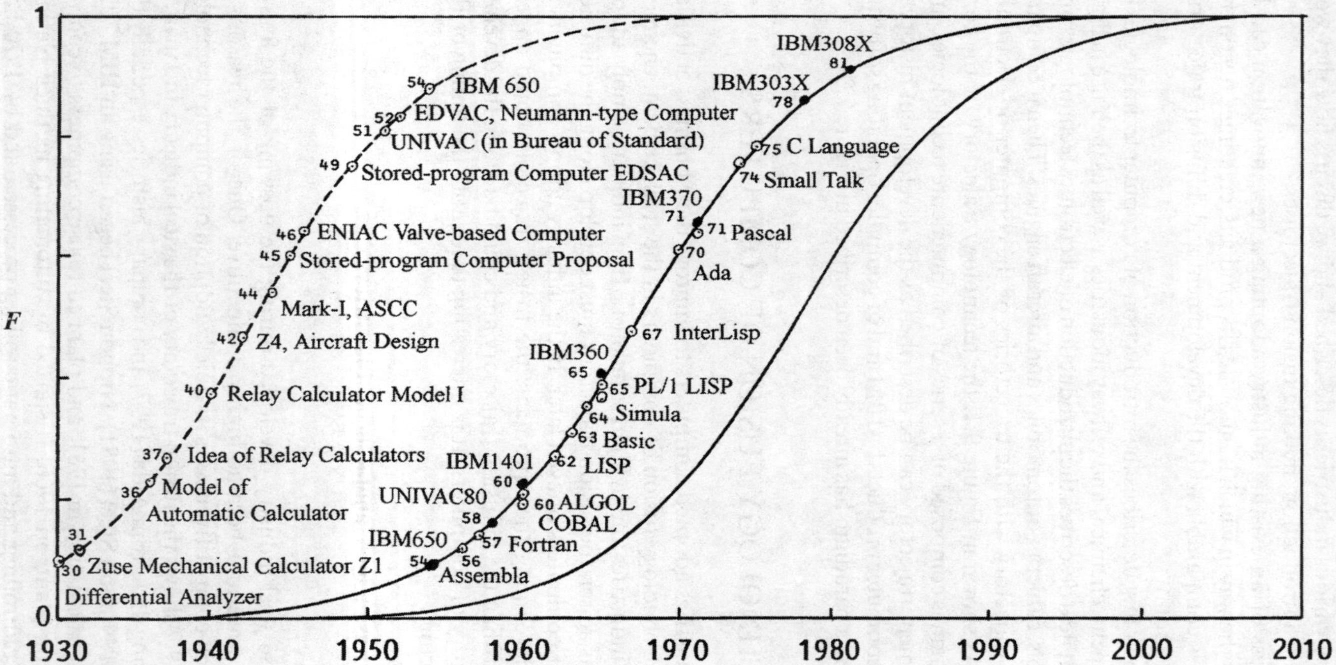

Source: Hirooka (1998c).

Figure 9.1 Innovation paradigm of computers

automated all its processes. It introduced an IBM 360/40 in 1967 and used it to develop a data bank, accounting system and demand model. An automatic design program was started in 1970, and development of a comprehensive information processing system was initiated in 1973. Thus, the fusion trajectory for the steel industry started around seven years later than the development trajectory of computers, and earlier than the diffusion trajectory of computers, as shown in Figure 9.2.

Cement industry

The fusion trajectory of the cement industry in Japan can be traced from company chronicles, especially those for Onoda Cement (1981). The introduction of computers was unbelievably early. Onoda Cement set up an IBM computer in 1957 and used it to process the daily reports of all production factories. It was used to process sales data starting in 1961. The company switched to an IBM 360 in 1967. Starting in 1968, computers were installed for newly constructed kilns to control stability and optimize the operation. In 1970, an XEG system was introduced, and closed-loop perfect control of kilns was fully implemented using an IBM 1800. Online systems for feedstock preparation, analysis and kiln operation, and for production cost accounting, labor management, office management and sales data online control were completed by 1972. These actions formed a fusion trajectory that started only three years later than the development trajectory of computers, as shown in Figure 9.3.

Chemical industry

The adoption of computers by the Japanese chemical industry was comparatively late – almost along the computer diffusion trajectory. The chronicles of five leading chemical companies, Mitsubishi Chemical Corp. (1981), Mitsui Petrochemical Industries Ltd (1988), Sumitomo Chemical Co. (1981), Mitsui Toatsu Chemicals Inc. (1994) and Mitsubishi Petrochemical Co. (1988) were used to gather relevant data. Their adoption of computer technology and the creation of technology fusion is summarized in Figure 9.4 in which firms A to E are those listed above, but in a different order. The introduction of computers started around 1962–66; they were used mainly to rationalize office work. Their use then shifted to technical calculation and factory automation.

Of particular interest is the fact that Sumitomo Chemical succeeded in the development of a new process for Soderberg aluminum smelting using a computer control system in the 1960s and exported the technology to all North American aluminum manufacturers and others in Europe and Japan in the early 1970s. The chemical industry now widely uses computers in offices, factories, logistics and laboratories. Computers are used not only for

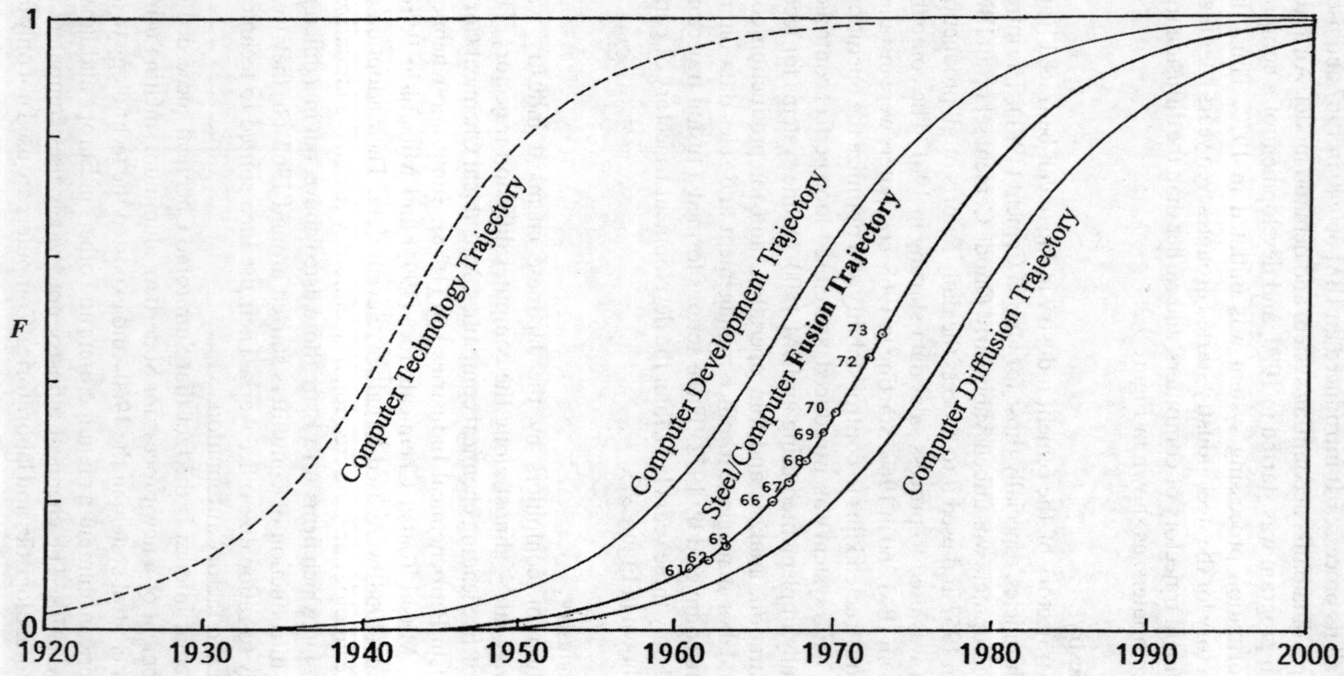

Source: Hirooka(1998c).

Figure 9.2 Technology fusion of computers by steel industry

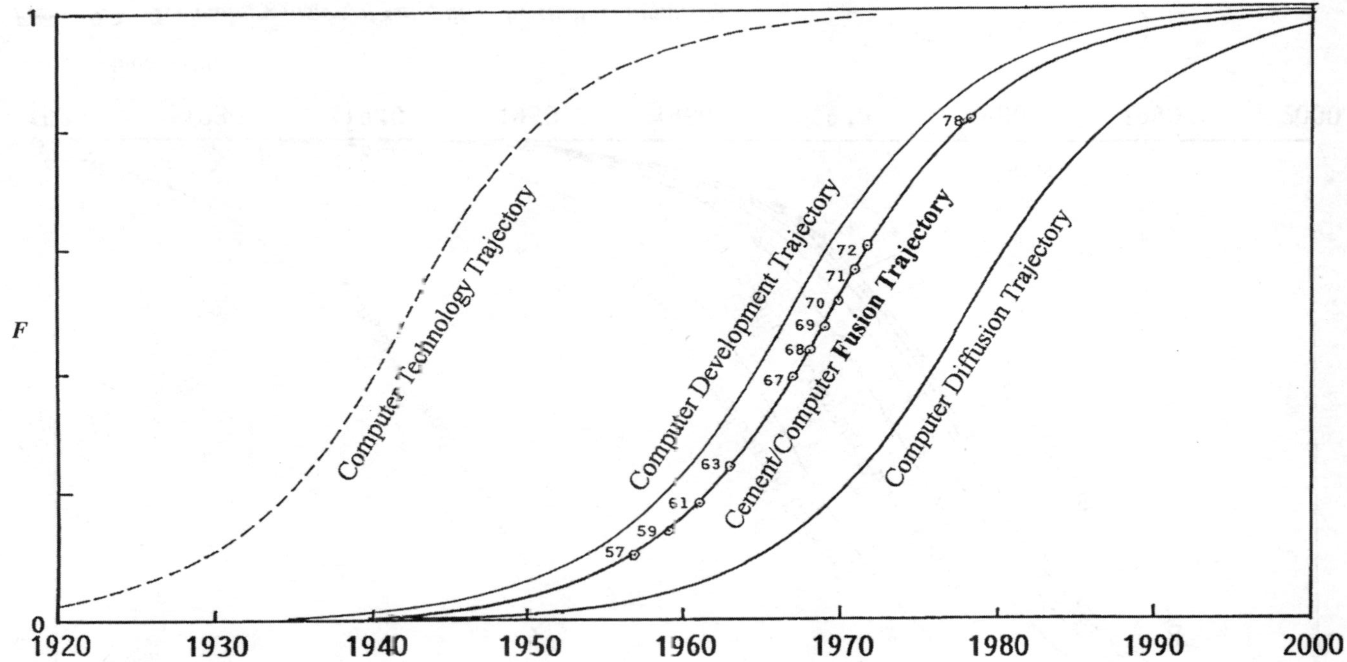

Source: Hirooka (1998c).

Figure 9.3 *Technology fusion of computers by cement industry*

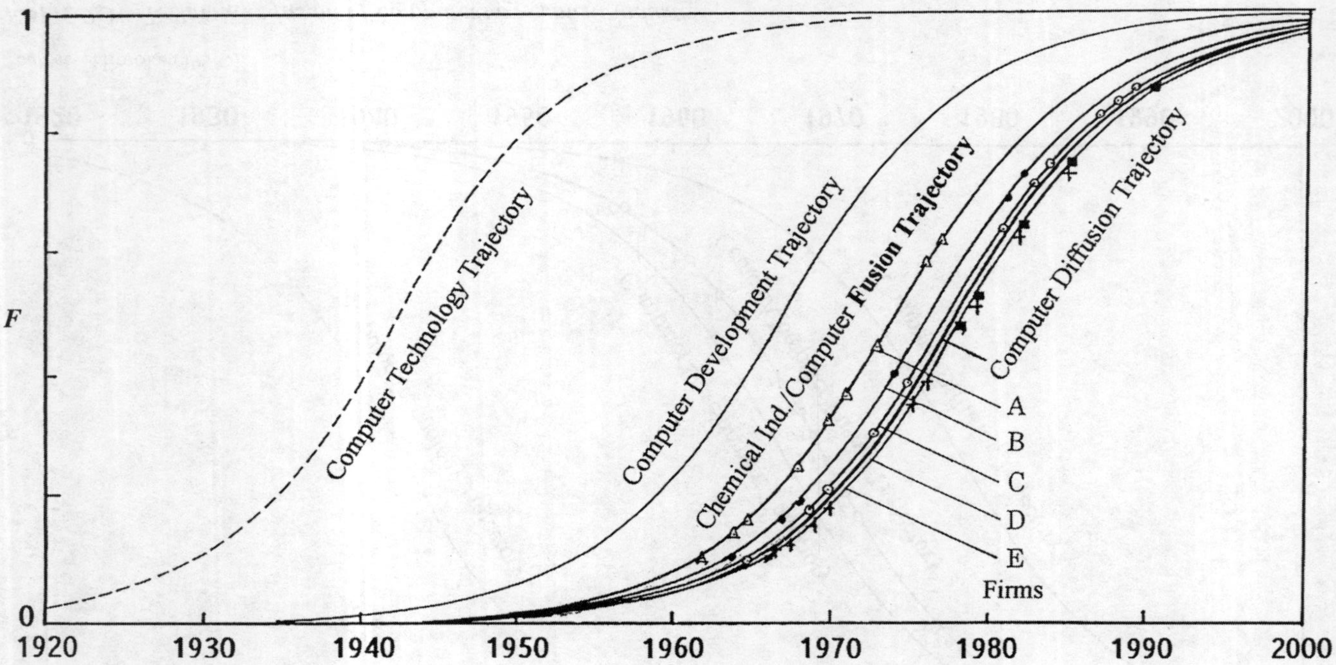

Source: Hirooka (1998c).

Figure 9.4 Technology fusion of computers by chemical industry

factory control and office work, but also for pesticide screening based on computer chemistry, for pharmaceutical development by combinatorial chemistry, for nonlinear analysis of the finite element method in fluidized processing and mold design of synthetic resins, and for automated experiments in the laboratory.

Technology Fusion by Public Sectors

Railways

Japan National Railways (JNR, now several private JR companies) introduced computers and developed various systems. Data collected from the Association of Traffic Signals & Safety, Japan (ATSS, 1987), and Takano (1997), shows that computers were first introduced in the JNR Technology Research Laboratory in 1957, starting with the Bendix G-15 from the US. At that time, fewer than ten computers were in operation in Japan, and this was the only one used for technical calculation. In the 1960s, computers were used for the development of such computer systems as train scheduling systems, train operating systems and a seat reservation system. The seat reservation system, the Magnetic Electronic Automatic Reservation System (MARS), was developed and put into regular use in 1967. The Shinkansen ('bullet train') was put into service in 1964, under the support of two systems: ATC (Automatic Train Control) and CTC (Centralized Traffic Control). JNR developed CTC using relays in 1962, electric CTC in 1967, and electronic CTC in 1984. An ATS (Automatic Train Stop) system was introduced in 1963; S-ATS was introduced in 1966 and H-type ATS in 1987. A COMTRAC (Computer-aided Traffic Control) system was applied to Shinkansen trains in 1972. PRC (Program Route Control) equipment was developed in 1972, PRC using medium-size computers in 1976, and electronic PRC in 1984. An electronic level-crossing controller was developed in 1984, and electronic interlocking equipment was put into use in 1985. Thus, as shown in Figure 9.5, the fusion trajectory is located seven to eight years later than the development trajectory of computers.

Flight simulators

The flight simulator is a typical complex product system as described above. The history of flight simulators was reviewed in detail by Baarspul (1990). Although it is a very sophisticated system, its development began very early in the history of aviation. The Wright brothers made the first flight using motive power in 1903, and a ground-based trainer, the 'Sanders Teacher', was developed by Billing in 1910. Flight simulators were quickly developed after the First World War for training pilots and simulation of bombing. This increased demand accelerated their development, and their systems

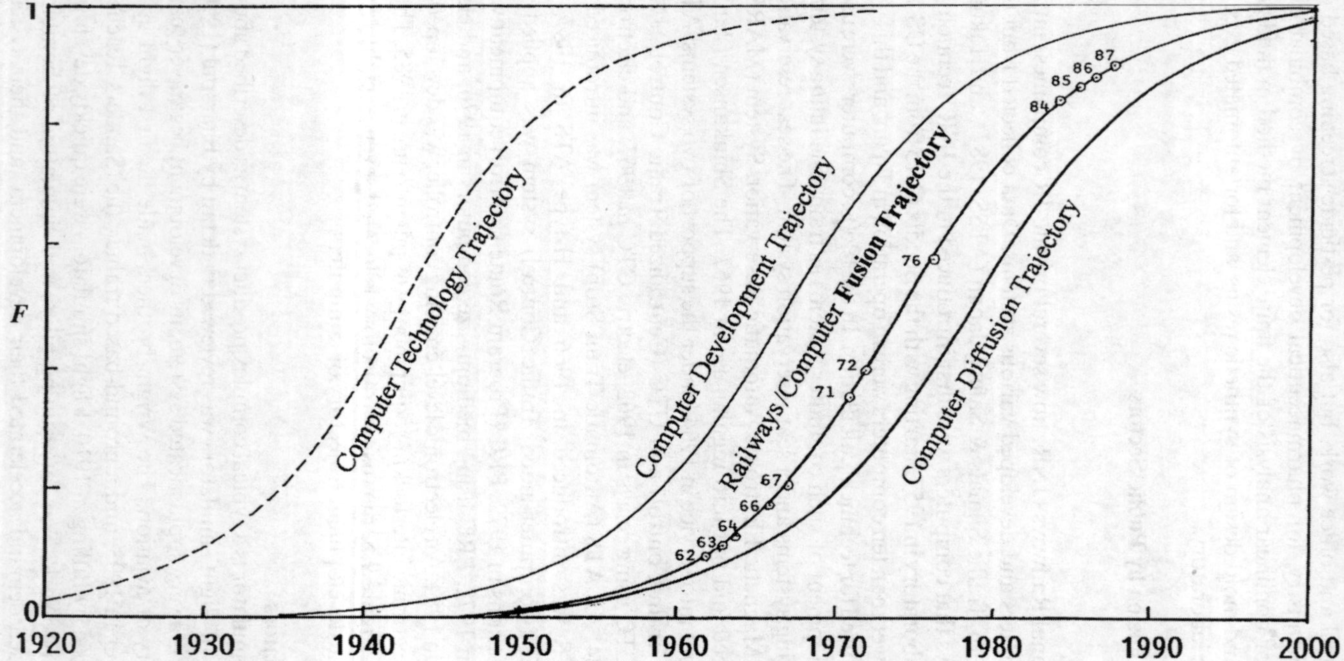

Source: Hirooka (1998c).

Figure 9.5 Technology fusion of computers by railways

were almost completed before the Second World War. The Antoinette trainer was developed in 1919, and Edwin Link developed a high level trainer in 1927, for which a patent was filed in 1930. 'Link Trainers' were soon being fitted with 'flight' instruments as standard equipment. In the early 1930s, blind-flying training was started by the Link Flying School. In 1937, the first Link Trainer was delivered to American Airlines. In 1939, Link was requested by the British to design a trainer that could be used to improve the celestial navigation capabilities of crews and to improve bombing accuracy during night raids over Europe. The first Celestial Navigation Trainer was completed in 1941. In 1945, the Silloth Trainer was put into use as an all-around trainer for basic flying behavior including engines, electrical and hydraulic systems, and sounds. An instructor's panel was also provided.

During the Second World War, a major advance in simulation was the use of a type of analogue computer, the differential analyzer, to solve the equations for aircraft motion. The first known discussion on the application of the computer to flight simulation is that of H.A. Roeder in a German patent in 1929. In 1950, the US Navy initiated a research program on real-time digital simulation at the University of Pennsylvania. The University Operational Flight Trainer (UDOFT) was completed using a Sylvania computer in 1960. In the 1960s, Link developed a special purpose digital computer designed for real-time simulation. In the late 1960s, general purpose computers designed for process control applications were also found to be suitable for real-time simulation. Rediffusion (Redifon) received a contract from BOAC (British Overseas Airways Corporation) in 1958 to develop a pitch motion system for the Comet IV simulator. Later, more complex motion systems, capable of producing motions with two or three degrees of freedom (DOF), were developed. The Delft University of Technology (DUT) started research on flight simulator motion in 1964 and introduced a 3-D motion system (roll, pitch, heave) in 1969. The first commercially available 6-D motion system was developed in 1977.

Visual systems to simulate the out-of-the-window visual scene were proposed. The first visual system achieving widespread use for civil aviation training was based on the CCTV (closed circuit television) method. The first color CCTV visual system was produced by Rediffusion in 1962. The first computer generated image (CGI) systems for simulation were produced by the General Electric Co. for the space program. The first CGI simulation systems economically feasible for airline crew training were of the night-only type using the stroke-writing or calligraphic display method, rather than the raster scan (TV) display method. The first night-only CGI system of this type was the 'Vital II', manufactured by the McDonnell-Douglas Corp. in 1971. The DUT flight simulator, introduced in 1967, provided a so-called 'contact analogue visual display' of the runway outline. It was replaced by a CGI night-only visual system using color CRTs and collimating optics in 1980.

These trends in the development of flight simulators are illustrated in Figure 9.6. The innovation paradigm of flight simulators indicates that the development took off much earlier than that of computers and that the development trajectory of flight simulators coincides with the development trajectory of computers per se. This means that flight simulators were developed simultaneously with the development of computers.

9.3 TECHNOLOGY FUSION OF ELECTRONICS

Several of the industries discussed above incorporated computers into their operation at an early stage. Except in the case of the flight simulator, each industry introduced computers around three to eight years after the development trajectory of computers, but almost before the diffusion trajectory. This is earlier than might be expected to take in computer technologies by technology fusion.

In the case of electronics, microprocessors played an important role and accelerated the fusion behavior because personal computers were easily disseminated in various sectors. Here, the technology fusion of electronics is examined for four cases: machine tools, cameras, automobiles and airplane engines.

Mechatronics of Machine Tools

A typical example of the fusion of electronics is the mechatronics of machine tools. As described in the previous chapter, the concept of mechatronics for machine tools was introduced by Kodama (1986), and the term 'technology fusion' has become widely used in the economics of technological change. Machine tools were first developed during the Industrial Revolution and since then have supported the development of a machine civilization. Machine tools have played an especially important role in the development of continuous production plants with transfer machines. The numerically controlled milling machine invented at the Massachusetts Institute of Technology in 1952 was given instructions using a paper tape with punched holes. It was later replaced by microprocessors. In 1967, the use of IC chips for machine tools was tested, and a direct numerically controlled machine tool using a minicomputer was developed in 1971. Intel's four-bit microprocessor was quickly adopted for use in machine tools, and this became the first step of mechatronics. In 1968, a transfer robot was put into use to automate plants, and a factory automation by FMS (flexible

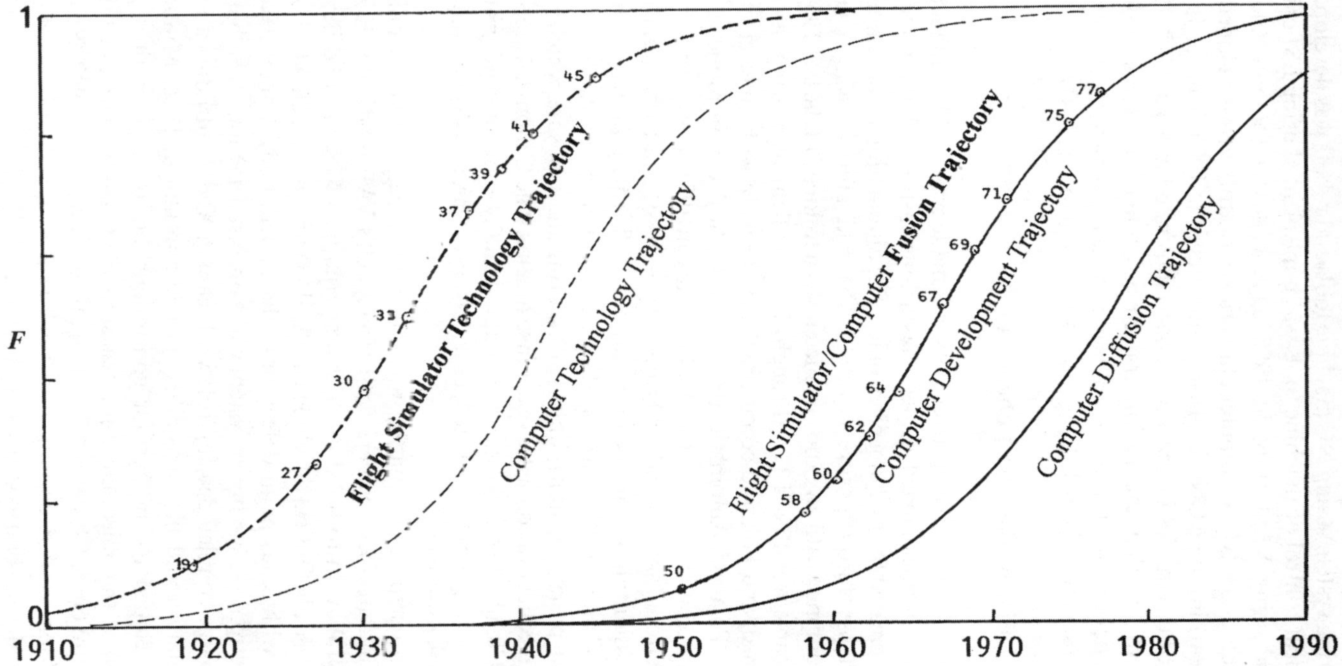

Source: Hirooka (1998c).

Figure 9.6 Technology and fusion trajectories of flight simulators

manufacturing system) using NC machine tools and robots was developed. The numerical control of machine tools was implemented using LSI chips in 1973 and using VLSI chips in 1980. Dialogue-type computer-aided machine tools and CNC (Computer Numerical Control) were developed in 1983, and learning-type CNC tools were introduced in 1985. This technology fusion of electronics by machine tools is illustrated in Figure 9.7. It is interesting to see that the fusion trajectory of machine tools closely matches the development trajectory of electronics. The data for these developments were quoted from Takano (1997) and Fujimoto (1994).

Technology Fusion by the Camera Industry

A photographic system using silver salts was made practical by Daguerre in 1839, and the mechanical camera has a long history. The automatic camera was created by applying electronics. The automation of the camera had two steps: auto-exposure control (AE) and auto-focusing (AF). A camera equipped with an exposure meter was developed in the late 1950s. In 1963, a camera with a TTL (through the lens) light-measuring device was developed, and Polaroid commercialized a camera with an electronic shutter, the 'Polaroid Automatic 100'. Asahi Optical Co. introduced an advanced automated camera, the Asahi Pentax in 1971; it was equipped with an electronic shutter and TTL auto-exposure system using logarithmic compression algorithm memory devices. Full-scale automation was achieved by Canon in 1976 with its AE-1, a sequential control camera with CPU (central processing unit) control of metering and exposure, which was developed in collaboration with Texas Instruments. The Canon A-1, introduced in 1978, had five exposure modes and digitally controlled programming. Sales of this camera rocketed, especially overseas, and exports to the United States rose sharply.

The second step in automation was automatic focusing (AF). Konica developed an auto-focus camera with Honeywell's VAF module in 1977. Canon's Autoboy, introduced in 1979, was equipped with infrared beam AF. In 1982, Asahi Optical developed an Asahi Pentax having an AF zoom mechanism with focus point detection in the camera body. Canon developed an electronic range-finder system with contrast detection and a built-in CCD (charge coupled device) in 1982. Eastman Kodak introduced DX code information and film sensibility into 35-mm film in 1984. Minolta Camera developed technology for controlling the focus point through modification of the object motion forecast in 1988. Canon developed the EOS 5 (Electro Optics System) AF system utilizing eyesight in 1992. Nikon Camera's Zoom 700VR developed in 1994, was equipped with an image stabilization device (blur-compensation device).

The fusion trajectory of electronics by cameras is shown in Figure 9.8. It is clear that the fusion trajectory of cameras is formed in the same locus as the development trajectory of electronics. The description above was quoted by Hatsumei Kyokai, 1997; Ishiyama, 1993; and Nakagiri, 1998.

Technology Fusion by the Automobile Industry: Car Electronics

The performance of automobiles was enhanced to a remarkable extent by the introduction of electronics. Electronic components are now used for engine control, driving control and electrical system control. In engine control, they are used to control the air/fuel ratio, effluent gas backflow, electric spark advance and engine mount. Driving control includes cruise control, traction control, skid control, suspension control, transmission control and steering control. Electrical system control includes electrical and electronic equipment control, peripheral apparatus (doors, windows, mirrors and so on) control, safety function control and self-diagnostic control. Figure 9.9 summarizes these trends in car electronics. The data were adopted from Takano (1997), Eckermann (1981) and JAMA (Japan Automobile Manufacturers Association) (1988).

The automobile industry first considered the use of transistors for controlling car functions, such as ignition, in the 1950s. The actual application of electronics got under way with the development of integrated circuits in the late 1960s to the early 1970s. Specific developments included electronic skid control (ESC) in 1971, electronic fuel injection control (EFI) in 1972, engine automatic start and stop (EASS) in 1973 and electronic transmission control (ETC). In the late 1970s, microcomputers became available, which led to ignition timing control by microcomputers (1976), cars equipped with microcomputers (1977), air/fuel ratio feedback control, engine automatic control system (EACS) (1979), antilock braking systems (ABS), power steering systems, trip computers and keyless entry. In the 1980s, as LSI chips became more advanced, they were adopted for use in various automobile functions: speak-drive in 1980, economy running, electronic suspension control and navigation control in 1981, lean mixing sensors in 1983, digital signal processors in 1986, self-diagnosis connectors in 1987 and active suspension in 1989. Other applications included variable piston control, fuel dilution control, electronic transmission control (ECVT), traction control by wheel sensors (TCS), concerted control of engine and transmission, digital electronic meters and airbags with collision sensors. This period was the most active age for car electronics. In the 1990s, advances in ULSI made the car itself a running vehicle of electronics; the most symbolic innovation was car navigation systems using GPS (global positioning system) in 1991. These developments in car electronics

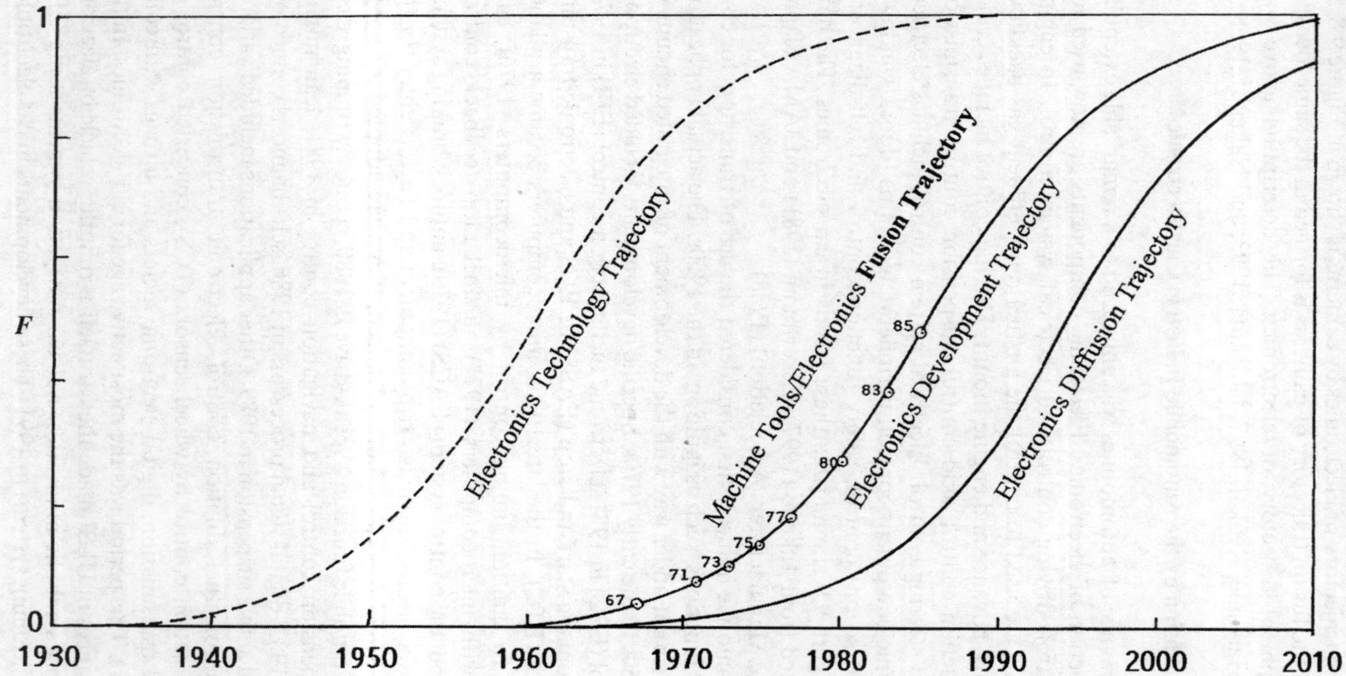

Source: Hirooka (1998c).

Figure 9.7 Technology fusion of electronics by machine tools

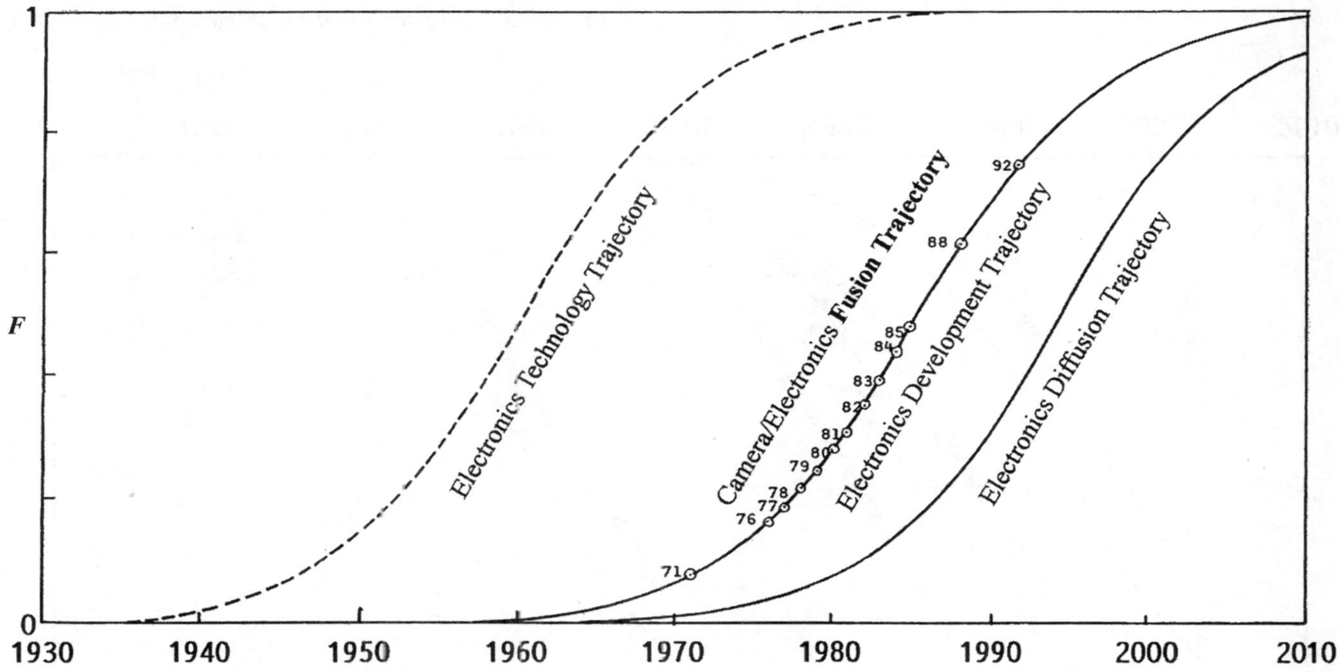

Source: Hirooka (1998c).

Figure 9.8 Technology fusion of electronics by cameras

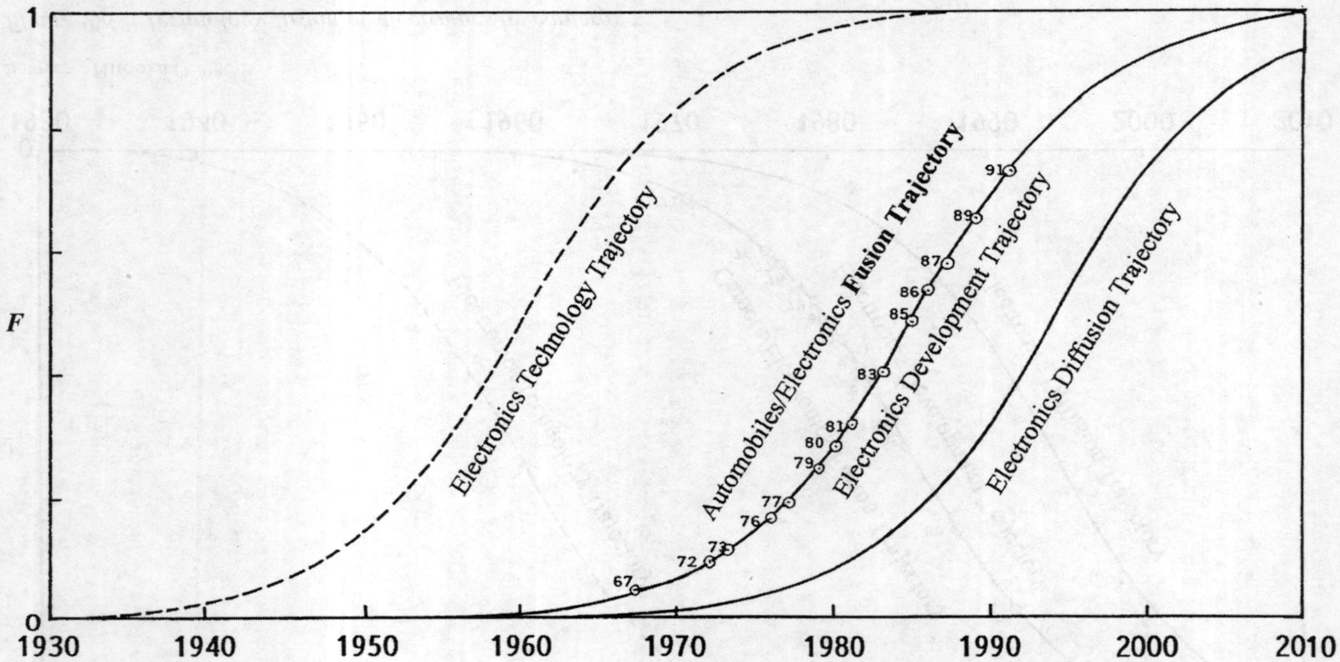

Source: Hirooka (1998c).

Figure 9.9 Technology fusion of electronics by automobiles

are summarized in Figure 9.9. The fusion trajectory of the automobile industry is virtually the same as the development trajectory of electronics.

The demand for clean running, low energy vehicles is now promoting the development of new technologies for energy saving, exhaust gas control engines, and new fuels that do not produce harmful effluents. In other words, R&D in the automotive industry has shifted from the adoption of electronics to improving the environment.

Electronic Control of Aircraft Engines

The jet engines on today's modern aircraft are highly controlled by electronics. Jet engines were developed in the 1930s, and after the Second World War, the technology trajectory of jet engines reached maturity and entered the diffusion period. The history of aircraft engines has been described by McNeil (1990), Yoshinaka (1994), Endo et al. (1986), Kishimoto (1987), Yamazaki and Hiromatsu (1990) and Sugiyama (1996).

The jet engine was independently invented by Whittle in the UK and von Ohain in Germany in the 1930s. Whittle applied for a patent on the jet engine in 1930, prepared a trial model in 1935, and had his engine flight-tested in 1941. In Germany, von Ohain had his engine flight-tested in 1939. In Japan, Nakajima Aircraft Co. developed a jet engine for aircraft in 1945. In 1947, the Bell XS-1 in the US was the first airplane to exceed the speed of sound. In the UK the Vickers was installed with a turboprop engine in 1948. The engine was switched from turbojct to turbofan in 1955.

Jet engines were initially controlled by hydraulic mechanical equipment. As they became more complicated and advanced, the number of output parameters increased. There was thus a growing need for a high degree of propulsion control and for quick acceleration and deceleration control; control devices consequently became more complex. The needs for aircraft jet engine control included engine protection, stability, adjustment of steady performance, response to transitional performance, enhancement of efficiency and actions on starting and stopping. In the 1950s, the most important target was the establishment and enhancement of reliability. Analogue control devices equipped with vacuum tubes were examined, but they did not provide sufficient reliability, so they were deemed impractical. Electronics technologies were just being developed, and the introduction of such devices became an essential factor in the development of modern jet aircraft.

The first digital control system for engines, FADEC (Full Authority Digital Electronics Control), was developed by Hamilton Corp. in the United States in 1965. Independently, Endo et al. of the Aerospace Research Laboratory, Science & Technology Agency, Japan, succeeded in controlling engines using a Toshiba 12-bit microcomputer, which they assembled themselves with control devices, in 1968. The development of FADEC for commercial aircraft jet engines began in 1971. The double FADEC was completed and installed in the engine of the FJR710 in 1980. In the United States, the HDS Co. developed an electric engine control system (EEC) for F-15 fighters in 1974, the JFC 90 supervisory control system in 1979, a digital supervisory controller in 1982, and a double FADEC in 1984. General Electric developed a double FADEC for the Airbus in 1988 and a single FADEC in 1989.

The technology trajectory of aircraft engines and its fusion trajectory with electronics are shown along with the electronics trajectories in Figure 9.10. The technology trajectory of aircraft engines is located some years before the Second World War, and the fusion trajectory with electronics fully meets with the development trajectory of electronics itself, so there is no time lag.

9.4 BEHAVIOR OF FIRMS DURING THE MATURITY PERIOD

The innovation paradigm is composed of three trajectories and the development of the market is described by the diffusion trajectory. During the first half of the diffusion, the market expands and things generally go well. During the latter half, however, the market growth rate falls and various problems occur. As described in Chapter 4, a bubble economy can develop, leading to a sudden drop in the economy, followed by a crash and then a depression. Before reaching a crash, the behavior of firms is changing according to the market maturation. This section deals with the maturation period of the market.

Scenario of Innovation Paradigm

As discussed in Chapter 3, Vernon (1966) explained Leontiev's paradox by using life-cycle theory. During the latter half of the diffusion trajectory, the production technology is exported to the follower countries, and production is started there. Competition switches from a focus on technology to a focus on cost, which means that countries with lower production costs have an advantage. The follower countries then begin to export their products to the original invention country. This is a general trend of the innovation process. The United States in the early twentieth century and Japan during the high growth period after the Second World War had lower production costs and thus had a competitive advantage.

In the last half of the nineteenth century and the early twentieth century, the United States developed its economy by introducing technologies from Europe and then came to dominate the world economy due to its higher

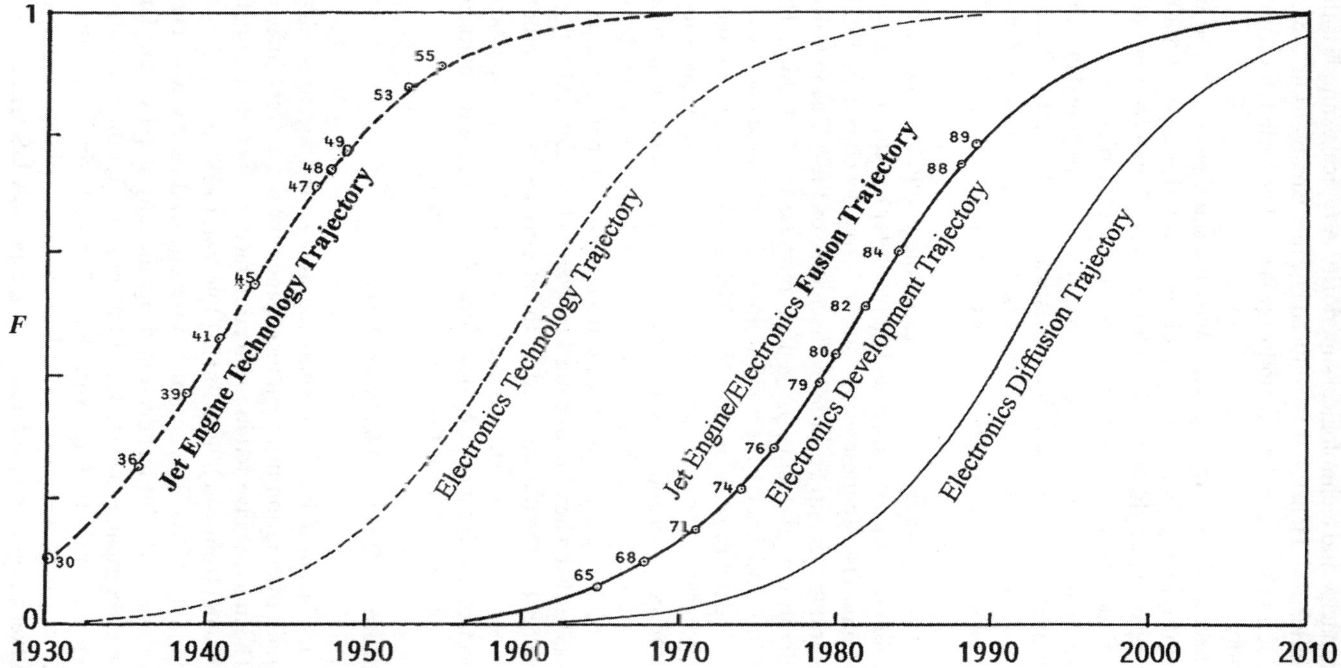

Source: Hirooka (1998c).

Figure 9.10 Technology trajectory of jet engines and fusion trajectory with electronics

production capacity and technological superiority. A typical example is the automobile industry – Henry Ford produced cheaper cars by using a continuous production system. The steel industry is another – the US became the world leader.

After the Second World War, Japan traced the same path followed by the US by introducing Western technologies and developed rapidly. Through the upswing of the fourth Kondratiev cycle, Japan became fully developed by making use of such trunk innovations as textiles, steel, automobiles, petrochemicals, electrical appliances, and then integrated circuits. Although the pursuit by Japanese industries was initially 25 to 40 years behind the Western countries, the time lag was shortened year by year until Japan reached about the same level as the West around 1970, as illustrated in Figure 3.8.

Recent innovation paradigms have been globalized, and the time lag has been almost eliminated, as discussed in Chapter 8 for the case of the electronics paradigm. The Japanese electronics industry developed smoothly and took a world leadership position by the middle of the diffusion trajectory, especially in the production of general purpose DRAM chips. The Japanese situation, however, has since deteriorated, and Korea and Taiwan moved up to leadership roles in the 1990s. Meanwhile, the United States shifted from general purpose products like DRAM to value-added products like microprocessors and software. Intel concentrated on the production of microprocessors, and Microsoft conquered the business world with Windows. The trade friction regarding electronics products between the US and Japan was settled when Japan agreed in 1986 to limit DRAM exports to the US. Since this dispute was settled, the electronics industry in both countries has been changing, which is in the middle of the development trajectory of electronics. What actions have firms taken during the maturity period of the paradigm? This is the subject of this section.

Strategic Alliance of Japanese Electronics Firms

The electronics business has been changing along the trajectories. The degree of circuit integration has been increasing four times every three or four years. The amount invested in production facilities has exponentially increased, growing from ¥40 billion for 1M DRAM, to ¥80 billion for 4M and ¥120 billion for 16M. More seriously, investment efficiency, which was 4 per cent for 64K DRAM, has decreased significantly step by step. This trend has forced companies to accelerate the implementation of countermeasures; one step was to form strategic alliances with foreign firms. International alliances were also formed to avoid complaints about the Japanese market being closed to foreign firms, especially US firms.

A survey of strategic alliances was carried out for the 'Nikkei Telecom' database, which is compiled by the Nihon Keizai Shimbun Inc. (Japan Economic Newspapers). The actual survey data was collected from the Nihon Keizai Sangyo Shinbun Inc. (Japan Economic Industrial Newspapers) from 1980 to 1999 and eight leading electronics companies: Toshiba Corp., Hitachi Ltd., Mitsubishi Electric Corp., Matsushita Electric Industrial Co., Sony Corp., NEC Corp., Fujitsu Ltd., and Oki Electric Industry Co. The survey was carried out for three areas: general purpose DRAMs, custom ICs (specifically, ASIC: application-specific ones, including gate arrays) and microprocessors (MPU). As these areas are related to product development, the strategic alliance events can be plotted on the development trajectory of electronics, as shown in Figures 9.11, 9.12, and 9.13, respectively.

Collaboration on DRAM development

Figure 9.11 shows the trend in strategic alliances formed by Japanese electronics firms with foreign firms, especially US and Asian firms, in the area of general purpose DRAM development.

Hitachi started joint working with Texas Instruments (TI) on the development of 16M DRAM in 1987, just one year after the US/Japan Semiconductor Agreement in 1986. Hitachi licensed SRAM (static random access memory) technology to TI and agreed the mutual OEM (original equipment manufacturing) production in 1989. Hitachi extended the development agreement to 64M DRAM in 1991 and 1G DRAM in 1997. Hitachi and TI jointly established Twinstar Semiconductor in Richardson, Texas to produce 16M and 64M DRAM in 1995.

Toshiba signed a wide-ranging agreement with Motorola in 1986 to jointly develop and produce DRAM; they set up a joint venture, Tohoku Semiconductor, in Japan in 1987. Working with Motorola, Toshiba developed 4M DRAM in 1989 and 16M in 1993 and, working with Motorola, IBM and Siemens, it developed 1G DRAM in 1995. It is said that Toshiba saved ¥20 billion ($200 million) on the cost of developing 256M DRAM in 1992 by collaborating with these companies. Toshiba also licensed DRAM technologies to Siemens in 1986, to Winbond Electronics, Taiwan in 1995 and 1996, and to Samsung Electronics, Korea and Chartered Semiconductor, Singapore in 1997.

NEC Corp. licensed LSI technologies to Standard Microsystems in 1987. NEC tied up with AT&T for the trust production and technology transfer of semiconductors in 1990 and collaborated with them to develop 256M DRAM in 1992. NEC organized a cooperative alliance with Sundisc for producing 256M DRAM. NEC approached Samsung in 1994 on the joint development of 256M DRAM and procurement of 4M DRAM and

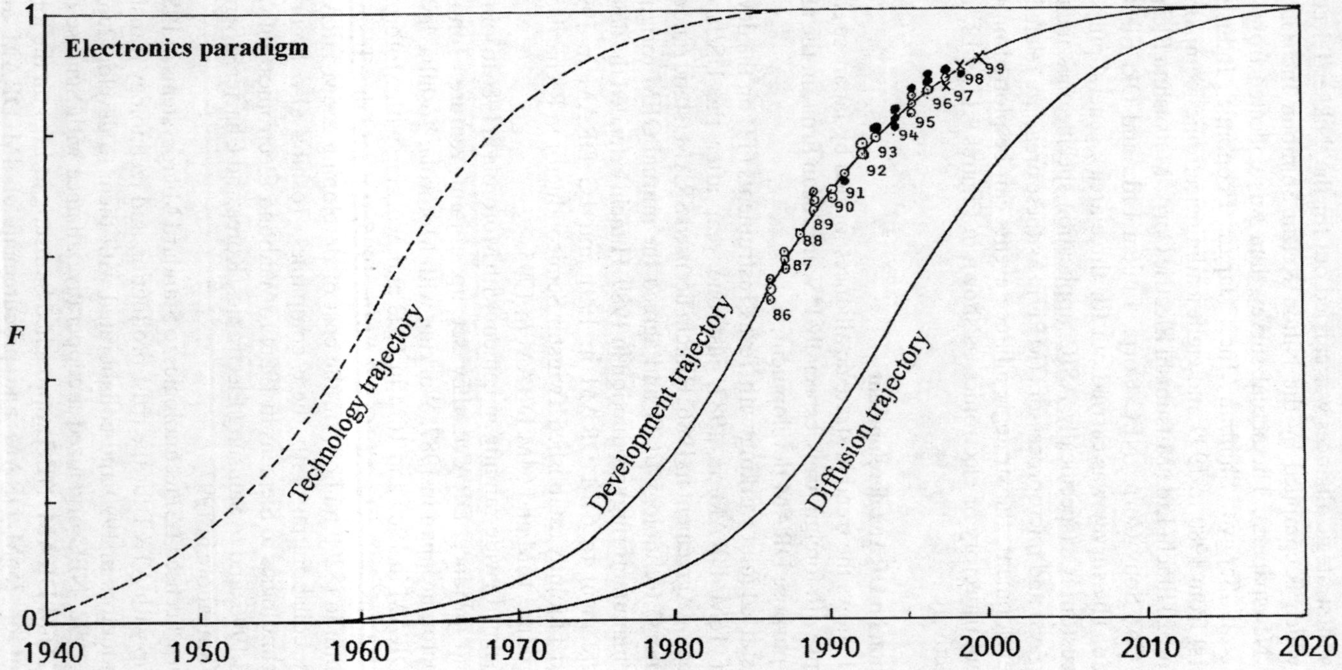

Source: Hirooka and Nitta (2002).

Figure 9.11 Strategic alliances on DRAMs

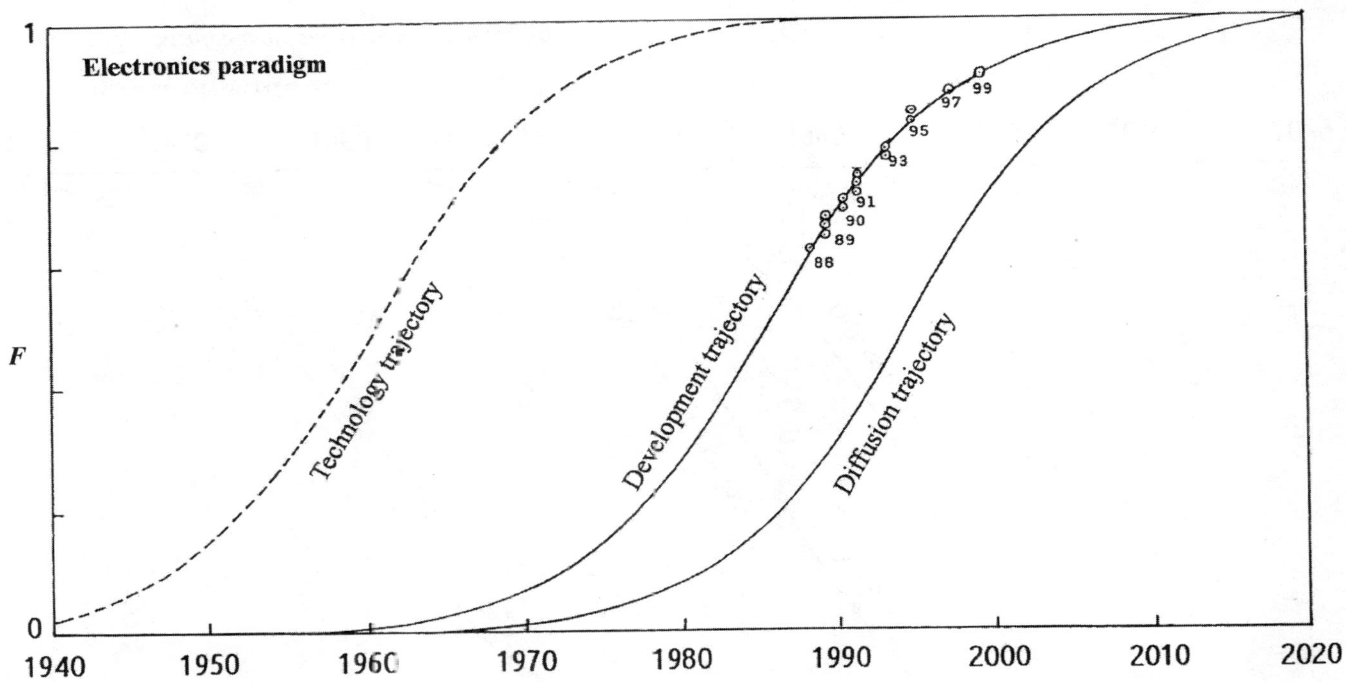

Source: Hirooka and Nitta (2002).

Figure 9.12 Strategic alliances on custom ICs

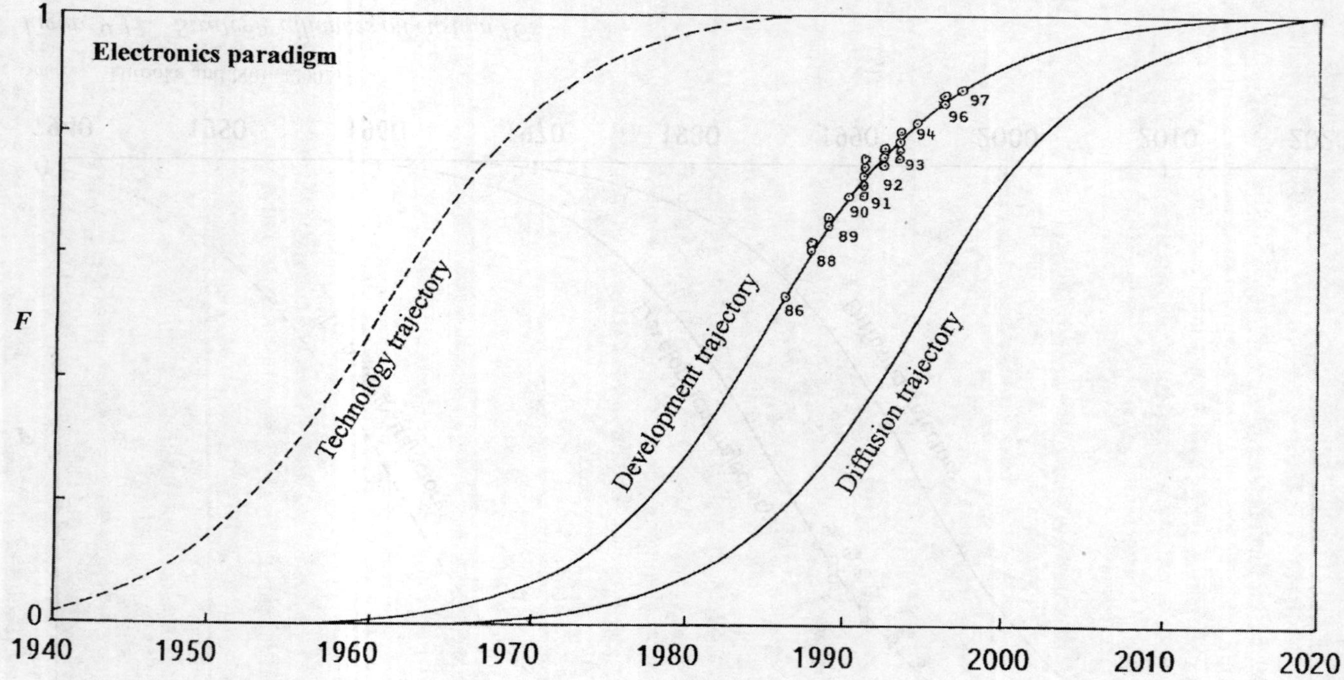

Source: Hirooka and Nitta (2002).

Figure 9.13 *Strategic alliances on microprocessors*

licensed their production technologies to Samsung in 1996. NEC set up a joint venture in Beijing to produce 4M DRAM in 1991 and started production in 1993; full-line production started in 1994. NEC also constructed a joint venture plant in Shanghai in 1974 to produce 4M DRAM and set up an office in Japan in 1998 to promote their China business.

Sony contracted with Advanced Micro Devices to develop SRAM in 1988. Mitsubishi Electric contracted with Texas Instruments to conduct mutual OEM production of DRAM in 1989. Oki Electric Industry formed an alliance with SGS-Thomson in 1990 to jointly produce 4M DRAM. Mitsubishi Electric contracted with Motorola in 1992 for mutual OEM procurement of DRAM and in 1996 for cross-licensing of system LSIs. Mitsubishi Electric also agreed in 1997 to collaborate with TI and Hitachi on the development of 1G DRAM.

In the latter half of the 1990s, the development trajectory of electronics entered into a maturity period, and strategic alliances became obsolete after the development of 1G DRAM. The strategic alliances between the US and Japan thus faced a turning point. Motorola decided to withdraw from the DRAM business in 1997, and the alliance between Toshiba and Motorola was dissolved; Tohoku Semiconductors was transferred to Motorola in 1999. Texas Instruments also gave up the DRAM business and sold the business to Micron Technology in 1998. The tie-up between TI and Hitachi for collaborative development and joint venture production was ended in 1999.

Domestically, however, collaboration between Japanese firms accelerated. Hitachi and NEC agreed to set up a joint venture company in 1998 for designing and developing DRAM, Toshiba worked with Fujitsu to develop memory production technologies and Fujitsu gave up producing DRAM in 1998.

Japanese electronics companies quickly lost their competitiveness because they have no proprietary technologies any more, and Korean and Taiwanese companies became dominant. Another important reason for the weakening of the Japanese electronics industry was the high valuation of the yen following the Plaza Accord in 1985. After the Plaza Accord, Japanese export-oriented industries quickly lost their competitiveness and were forced to move operations abroad by direct investment in foreign production facilities. Japanese electronics companies rushed into Asian countries, especially Thailand, Singapore and Malaysia, as described in Chapter 8. Japanese electronics firms have also become intimate with Korea, Taiwan and China.

The positioning of international strategic alliances can be well expressed on the development trajectory of electronics, as shown in Figure 9.11 in which the white circles indicate alliances of Japanese firms with Western countries, the black circles indicate collaborations with Asian firms, and the crosses show the terminations of alliances.

Strategic alliances for custom ICS

Strategic alliances were formed between US and Japanese companies not only for DRAM but also for much more advanced areas of system LSIs. One was integrated circuits for specific applications and logic ICs. They are represented by specific applications: ASICs (application-specific ICs), semi-specific logic memories, gate arrays, decoder ICs and CMOS ICs. This section focuses on the strategic alliances for ASICs, including gate arrays.

Strategic alliances for custom ICs were first formed in about 1988 and continued to be formed for about ten years. There were many collaborations on ASICs and gate arrays: Hitachi and VLSI Technology in 1988, Toshiba and Integrated CMOS Systems in 1989, Toshiba and Siemens in 1989, NEC and Silicon Computer Systems in 1989, Toshiba and Motorola in 1990, NEC and AT&T in 1990, NEC and Cadence Design System in 1991, Fujitsu and Cadence Design System in 1991, Hitachi and VLSI Technology in 1991, Toshiba and Synopsis in 1993, Matsushita and Actel in 1995, and NEC and AT&T in 1995. Motorola stopped producing DRAM in 1997 and shifted to producing ASICs at Tohoku Semiconductors. NEC began producing ASICs in a joint venture in Beijing in 1999.

Strategic alliances for microprocessors

The epoch-making microprocessor was first developed by Shima, Hoff and Faggin at Intel in 1971. The microprocessor made it possible to make much smaller personal computers, and microprocessors became essential devices in electronic products. Suzuki at NEC soon developed another microprocessor.

Operation software was developed along with the development of computers. The first operating system (OS) for open systems was developed by Thompson at AT&T in 1969 and was dubbed UNIX the next year. A consortium to promote the UNIX OS was set up in 1988, and a corresponding consortium, OSF (the Open Software Foundation) was organized by IBM, Digital Equipment (DEC) and Hewlett Packard, also in 1988. The operating system runs on the central processing unit (CPU) or microprocessor (MPU). Motorola commercialized the CISC (Complex Instruction Set Computer) as an MPU for UNIX in 1979. MIPS Computer Systems also produced an MPU. Patterson at the University of California, Berkley, developed a higher speed MPU, RISC (Reduced Instruction Set Computer) in 1982 and Sun Microsystems commercialized RISC chips, 'SPARC' in 1987. These inventions of software and microprocessors form the development trajectory of electronics, as discussed in Chapter 7.

In the latter half of the development trajectory of electronics, strategic alliances were actively formed for microprocessor units. Toshiba first contracted with Motorola in 1986 and introduced a RISC MPU in 1997. Toshiba introduced RISC technologies from MIPS Computer Systems in 1990 and 1993. Toshiba set up a joint development with Siemens in 1991 to develop RISC-type MPU, and contracted with Integrated Device Technology to jointly develop RISC MPU, and with AT&T in 1992. NEC contracted with MIPS Computer Systems to introduce a RISC MPU in 1988 and introduced RISC technologies in 1991 and 1993. NEC signed a licensing agreement on MPU for PDA with AT&T in 1991. NEC introduced RISC MPU technology from Sun Microsystems in 1996 and collaborated with Philips in 1997. Hitachi worked with Hewlett Packard to develop MPU and workstations in 1989 and RISC chips in 1991. Sony introduced RISC from MIPS Computer Systems for their engineering workstations in 1989 and 1991. Mitsubishi Electric Corp. was licensed to produce RISC MPU from DEC (Digital Equipment Corp.) in 1993 and signed a business collaboration agreement with IBM for RISC MPU in 1996. Fujitsu agreed with Sun Microsystems to be supplied with engineering workstations running UNIX by an OEM in 1988 and extended the agreement for RISC MPU for 32- and 64-bit workstations in 1994. Oki Electric Industry Co. formed a comprehensive tie-up with Hewlett Packard on MPU and computer networks in 1992.

Alliances in Asian Regions

As described above, from the late 1980s, the Japanese electronics industry became less competitive while the Korean and Taiwanese industries made remarkable advances due to their robust competitiveness. More recently, the Chinese electronics industry has also advanced, becoming a large supply base. This section focuses on strategic alliances between Japanese and Asian firms.

The Korean electronics industry

The Korean electronics industry started with the formation of Samsung Electronics Co. in 1969 and grew from there. Samsung on its own established an experimental production line for 1M DRAM in 1985 and became self-reliant. The Korean electronics industry developed 16M DRAM with the same timing as Japan, and Samsung surpassed NEC in 1992 to become the world's leading producer of 16M DRAM; in 1997, Samsung accounted for 19 per cent of the world's production, far exceeding NEC's 12 per cent share. After working closely with NEC in the 1990s, Korean electronics firms then grew on their own, avoiding intimate alliances with foreign firms.

The Taiwanese electronics industry

The Taiwanese electronics industry started with the establishment of the Industrial Technology Research Institute in 1973 and the Electronics Research and Service Organization (ERSO) in 1974. The activities of ERSO played a decisive role in the development of the electronics industry in Taiwan, as described in Chapter 8. Another important event was the establishment of the Hsinchu Science-based Industrial Park in 1980. It became a kind of Silicon Valley, enabling a wide variety of venture firms to function in an active community. The establishment of a foundry plant, the Taiwan Semiconductor Manufacturing Corporation (TSMC), in cooperation with Philips, is particularly noteworthy because it made available a very important and valuable service for Taiwanese electronics firms. Taiwanese electronics firms were too small to construct on their own the huge facilities needed for device production, as the large foreign electronics firms could. The foundry plant made it possible for small firms in Taiwan to produce their devices in small lots. Many venture firms without their own plants have used this common foundry plant. This was a key event in the development of the Taiwanese electronics industry – a large number of firms without their own fabrication facilities were born around the TSMC. This facility is suitable for producing a variety of ASICs in small lots and even for producing general purpose DRAMs as well. Taiwan is now a major producer in the world of electronics products. In 1994, one-third of the personal computers produced worldwide were produced in Taiwan.

During this period, Japan faced a difficult situation due to the highly valued yen, and Japanese electronics firms rushed abroad to build production facilities by making huge investments or forming joint ventures. Toshiba contracted with the Winbond company in Taiwan, provided DRAM and ASIC technologies to them in 1987, and moved on to production on commission at the TSMC plant in 1995. Mitsubishi Electric established a joint venture, Powerchip Semiconductor, in 1994 to manufacture DRAM, joined the foundry business in 1997, and expanded its investment in Powerchip in 1999. Nan Ya Science & Technology signed an agreement with Oki Electric Industry in 1995 for technology collaboration. Oki Electric Industry supplied wafer and DRAM production technologies to Mosel in 1993. Fujitsu contracted with TSMC for foundry services for ASICs and MPUs in 1994. TSMC also provided services to Advanced Micro Devices (MPUs) and NEC (ASICs and gate arrays) in 1994. NEC transferred motherboard technologies to the Elite Group in 1994. Hitachi supplied DVD technology to Acer Inc. in 1998.

US firms also worked closely with Taiwanese firms. TI established a joint venture, Macronix, with Acer in 1990. Macronix introduced RISC technologies from MIPS Computer Systems in 1992. Winbond introduced SRAM technologies from Motorola and RISC technologies from Hewlett Packard in 1996. TI and Acer dissolved their venture agreement in 1999. The collaboration of Taiwanese electronics firms with foreign enterprises is plotted on the development trajectory of electronics in Figure 9.14.

Behavior of Foreign Firms in East Asia and China

After the Plaza Accord in 1985, Japanese export-oriented industries suffered due to the higher value of the yen and turned their business strategy to production abroad through direct investment in various countries. The electronics industry was no exception to this trend, focusing their overseas investment especially on the East Asian countries of Thailand, Singapore and Malaysia. The ASEAN countries and regions had developed quickly, following Korea, Taiwan and Hong Kong. Since countries in this area were ready to develop and eager to accept foreign advanced technologies, Japanese companies enjoyed easy entry. They received favorable tax treatment and found well-organized infrastructures. The Japanese electronics industry thus quickly moved into these countries starting in 1985.

In the previous chapter, the behaviors of firms and governments were discussed, and it was shown that these countries developed through the acceptance of foreign business enterprises and by providing land and labor. The paths they took differed from those taken by Japan, Korea and Taiwan in terms of self-production, as shown in Figures 8.6 and 8.7 for Singapore and Malaysia. During the first half of the development trajectory of electronics, the electronics industries in these latter countries constructed factories to handle the assembly processes during the later stages of production, which require less skill. In Singapore, National Semiconductors, Hewlett Packard, NEC, Matsushita Electric Industry, Intel and Motorola built production facilities from the last half of the 1960s to the last half of the 1970s. In Malaysia, facilities were constructed by National Semiconductors, Hitachi Ltd, NEC, Intel and Motorola.

Starting in 1985, facilities were also built to handle the earlier stage processes, so that there were full production facilities, and foreign enterprises, especially from the US and Japan, increased their operations. Japanese firms especially rushed in during the last half of the 1980s. One reason was that production processes had become highly automated; another was the pressure of the highly valued yen after the Plaza Accord. Japanese firms used funds raised by equity financing and bank borrowing. This was because a bubble economy quickly developed in Japan due to the excess fluidity of

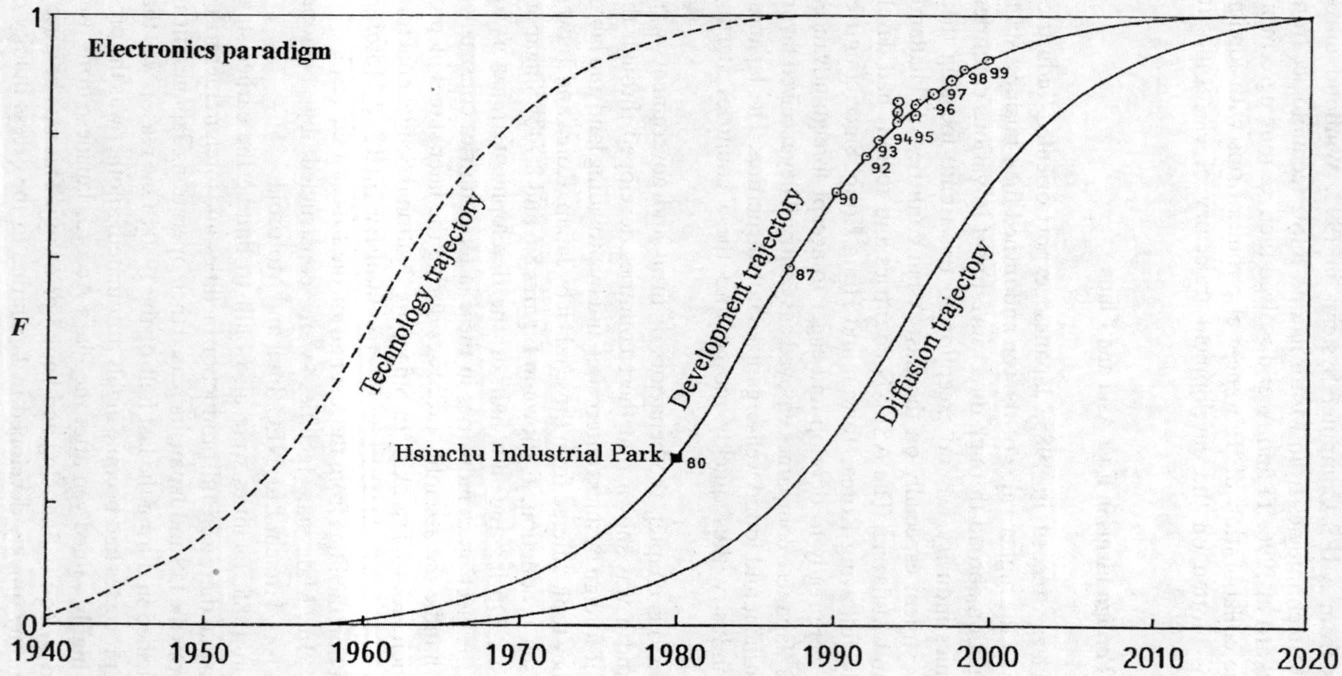

Source: Hirooka and Nitta (2002).

Figure 9.14 International collaboration of electronics firms in Taiwan

money resulting from efforts to stimulate domestic demand by the Japanese Ministry of Finance and the National Bank of Japan. Thanks to these trends in the Japanese economy, the economies of Asian countries and regions enjoyed accelerated development. The number of Japanese firms in Malaysia increased from only 276 in 1985 to 841 by 1995. Most of them came after the Plaza Accord, and most of them were electrical or electronics firms. The situation has again changed, and China's economy has quickly developed.

These trends in the electronics industry indicate that firms behave differently during the latter half of the development trajectory of electronics and that alternation of the business generations has quickly proceeded, as described by Vernon's life-cycle theory (Vernon, 1966). The electronics industry passed through a bubble period (around the year 2000) and is in a turning period according to Perez (2002). We can expect another upswing during the fifth Kondratiev cycle, maybe towards 2040, forming an infra-trajectory of electronics and information technologies.

9.5 EVOLUTION OF INNOVATION PARADIGMS

Innovations in electronics and information technologies include various paradigms, and each paradigm is discontinuously separated. The computer and electronics paradigms are strictly interrelated, and the driving part of the computer is composed of electronics, which evolved from vacuum tubes. The electronics element is composed of semiconductor devices, which were developed to produce microprocessors. Microprocessors made it possible to make computers smaller, that is, to make personal computers. Personal computers thus evolve towards the multimedia paradigm. The innovation paradigms of information technologies nowadays consist of computer, electronics and multimedia paradigms. Here we discuss the evolution of these paradigms and their interrelations.

Formation and Structure of the Multimedia Paradigm

The multimedia paradigm develops accompanying the ISDN (Integrated Services Digital Network) paradigm with internet web structures. Multimedia represents a combination of various media, such as letters, voice and pictures, that are unified by digitization, making it possible to transmit data as a single transmission medium. That is, multimedia information can be sent over an ISDN.

ISDN paradigm

ISDN was defined by the CCITT (Consultative Committee of International Telegraph and Telephone; now the ITU-T) in 1972, as an information system in which digitized information is transmitted over an optical fiber network. The technology trajectory is composed of the idea of data transmission using optical fibers and semiconductor lasers as coherent light sources proposed in the 1960s, and improvements in the transparency of optical fibers by the invention of production processes which achieved fiber transparency of 0.2 dB/Km in 1979; it was completed with the establishment of the basic concept of ISDN in 1980. The development trajectory is a series of information network experiments and the development of transmission processes, a packet communication method using asynchronous transfer mode (ATM), and broadband transmission. The diffusion trajectory started with the construction of optical trunk lines throughout Japan in 1985, continued with the laying of fiber trunk lines across the Pacific Ocean from Japan to the US in 1992, 1996 and 2000, and will finish with the laying of fibers to the home by around 2015. The total paradigm is shown in Figure 9.15.

Multimedia paradigm

The technology trajectory of the multimedia paradigm consists of digitization, protocol development, standardization, ATM transmission process, interface construction and image compression. The trajectory started with the idea for hypertext by Ted Nelson in 1962, continued with the development of the packet communication process by P. Baran in 1964, the development of the Dynabook concept and the NLS interactive computer system in 1968, and the development of internet core technology by ARPANET in 1969, the Ethernet in 1976, the CS Net in 1979, and NSFNET in 1986. The internet protocol developed based on the concepts of TCP (Transmission Control Protocol) in 1973, the TCP/IP (Internet Protocol) protocol principle in 1974, the image compression protocol in 1980, the electronic mailing system in 1982, and the HTML and SGML markup languages in 1986. The development trajectory started with the transfer of NSFNET to Merit Network, Inc. for civilian application in 1987.

During the initial stage of the development trajectory, several important inventions were achieved one after another: www (world wide web) at CERN (the European Laboratory for Particle Physics) in 1990, the Mosaic web browser, MOSAIC in 1993, the JAVA programming language in 1995, and the MPEG-4 standard in 1998. Certain companies and related organizations were deeply involved during the initial stage of the development trajectory: Merit Network, Inc. in 1987, ANS in 1989, GOPHER in 1991, the WWW Consortium in 1994, and Netscape and YAHOO in 1995. Digital TV was a symbol of the digital age and started in the UK and US in 1998. These events compose the development trajectory as the elements. Other groups of elements also compose the development trajectory: the development of electronic money and electronic commerce and the development of mobile telephones and related technologies.

The innovation paradigm of multimedia is as yet unfinished, so its actual make-up is still not clear; however, as the technology trajectory has already finished and its time span was 25 years, the development and diffusion trajectories could each last about 25 years. These analyses strongly suggest that the multimedia paradigm will be finished around 2030. These trends of the paradigm are summarized in Figure 9.16.

Another important element in the multimedia paradigm is the trend of standardization. As standardization is a key to creating communication networks, strategic alliances have been organized from the fairly early stages of the development of multimedia technologies. This differs from the electronics paradigm, which was described above. As multimedia materializes using various communication protocols, standardization of transmission processes is essential. The internet process was established by formulation of TCP/IP during the early stages of multimedia innovation. It became a de facto standard and has been adopted as an international standard. After the basic concept of ISDN was established in 1980, the development of various protocols and standards progressed. The standardization of the image compression process started in 1980; the International Standards Organization (ISO) decided to standardize the image compression protocol, and the MPEG-1 standard was adopted in 1992, MPEG-2 in 1995 and MPEG-4 in 1998.

Computer operating system (OS) software has also been standardized. The first was UNIX, an open system OS released by AT&T in 1969. Because it was open, unlike IBM's closed system, it developed in various ways. AT&T and Sun Microsystems tried to unify the UNIX operating systems by setting up Unix International in 1988 and inviting many firms to participate. A total of 261 companies had joined by 1992. In response, IBM organized another organization, the Open Software Foundation, calling DEC and Hewlett Packard in the same year of 1988, and 335 companies joined by 1992. IBM together with Apple Computer set up a company called 'Kaleida' in 1991 to establish a basis for multimedia systems. Their intent was to develop software related to next-generation multimedia. To counteract their effort, Microsoft organized the Multimedia Personal Computer Marketing Council, MPCMC, along with 11 core firms including AT&T, NEC and Fujitsu. Their aim was to accelerate the dissemination of Windows, Microsoft's OS. Strategic alliances have also been organized to support the development of RISC-type MPU and flash memories.

In 1990, 'General Magic' was set up as a subsidiary of Apple Computer to develop a multimedia apparatus for integrating the functions of letters, voice and images. Six companies – Apple, AT&T, Motorola, Philips, Matsushita and Sony – provided support to this company in 1993, and Mitsubishi Electric, Fujitsu and Oki Electric Industry joined in 1994.

This trend indicates that protocols and standardization play key roles in the multimedia paradigm; work on establishing protocols and implementing standards has progressed, and systems have been constructed from the latter half of the technology trajectory to the first half of the development trajectory, as shown in Figure 9.17.

Evolution of Information Technology Paradigms

The information technologies are covered by the three paradigms of computers, electronics and multimedia. Although they are different paradigms, established within individual disciplines, their close interaction led to the formation of the modern information society. The computer system evolved through the adoption of electronics, and the multimedia system is building on the basis of the computer and electronics technologies. The evolutionary trend of these three paradigms is illustrated in Figure 9.18 using two loci of the development and diffusion trajectories.

The relative positioning of these trajectories indicates that the computer and electronics paradigms are now in a maturity period and the multimedia one is still immature. This gap provides a very important clue to the origin of the IT recession. We experienced a bubble phenomenon in the electronics and information industries around 2000 and have now entered a kind of air pocket, expressed as an IT recession. According to Perez's (2002) explanation, we are now at a turning point, from the 'frenzy' phase to the 'synergy' phase. As the multimedia paradigm is still not far enough along to provide sufficient value-added, the economy cannot yet enjoy the fruits of multimedia. This gap explains the present slackness. Thus, we can expect a synergy phase during the upswing of the fifth Kondratiev cycle towards 2040. The electronics and information infra-trajectory will then most certainly enter the second full swing phase.

9.6 SUMMARY

Existing industries are renovated by adopting new innovation technologies. Technological fusion by existing industries was examined on the basis of computers and electronics as typical trunk innovations. The behaviors of firms during the mature period of innovation were analyzed, and the evolution of innovation paradigms was described.

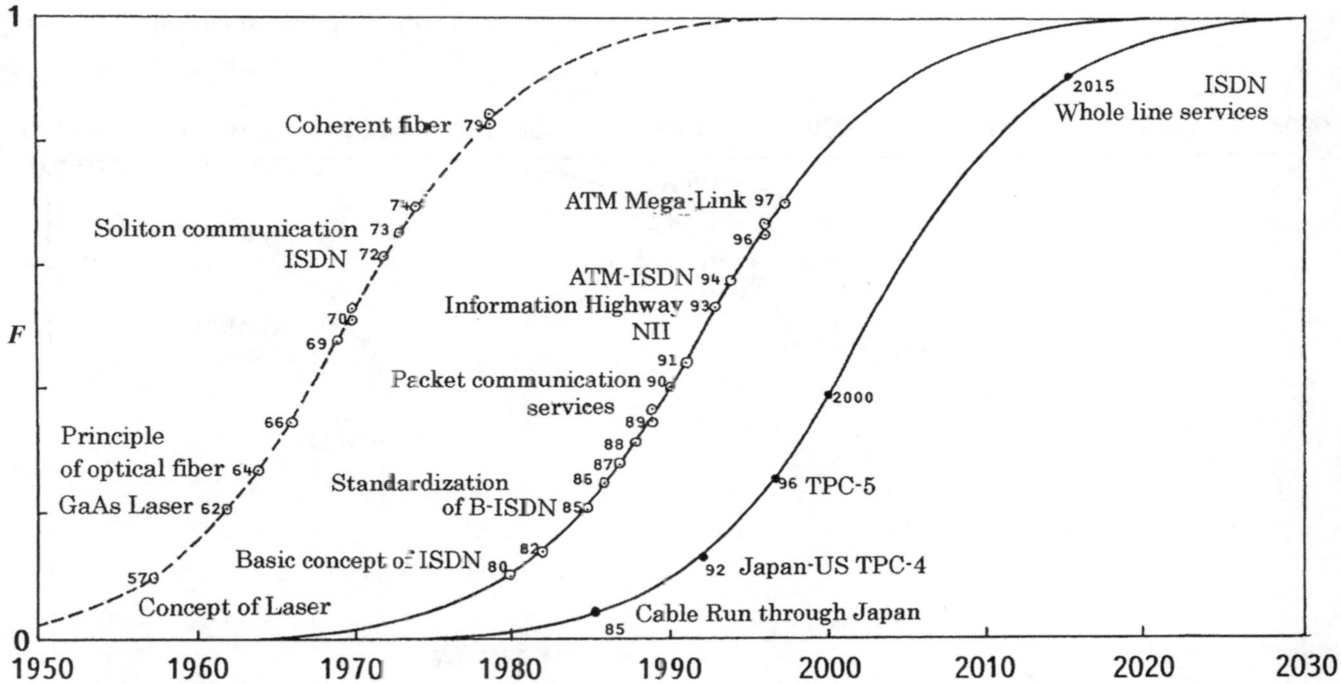

Source: Hirooka and Nitta (2002).

Figure 9.15 Innovation paradigm of ISDN

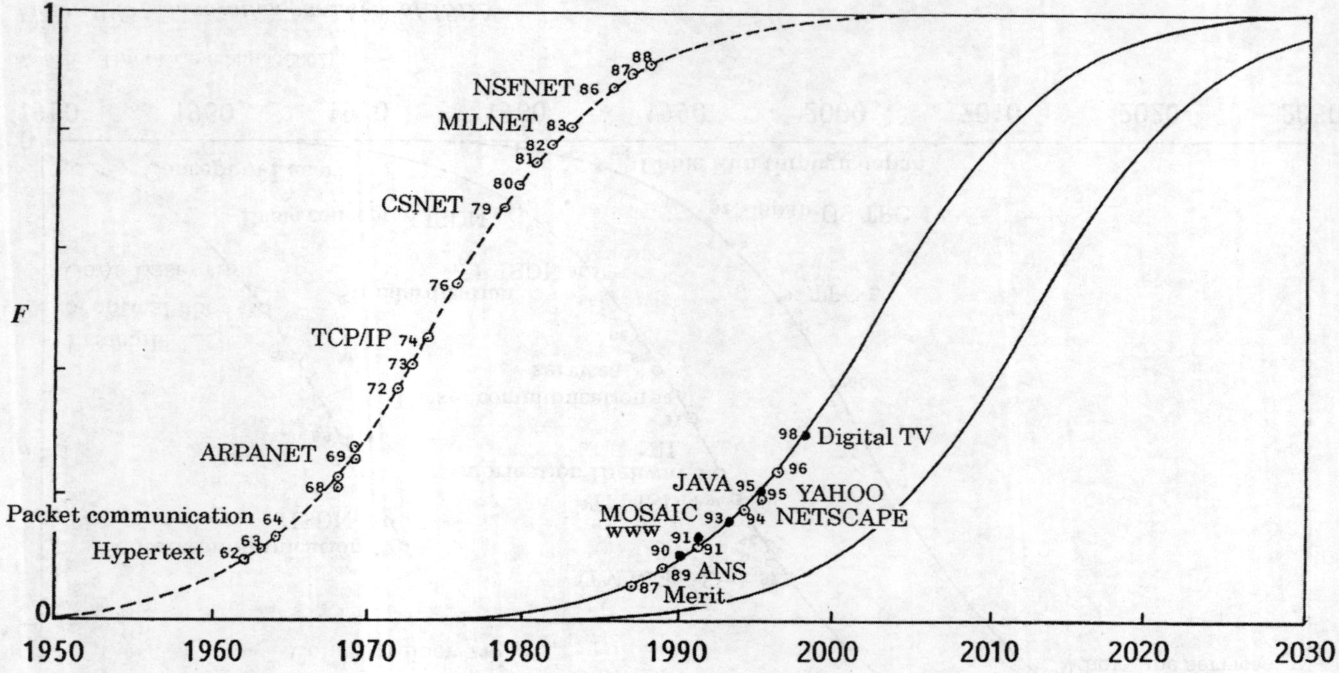

Source: Hirooka and Nitta (2002).

Figure 9.16 Innovation paradigm of multimedia

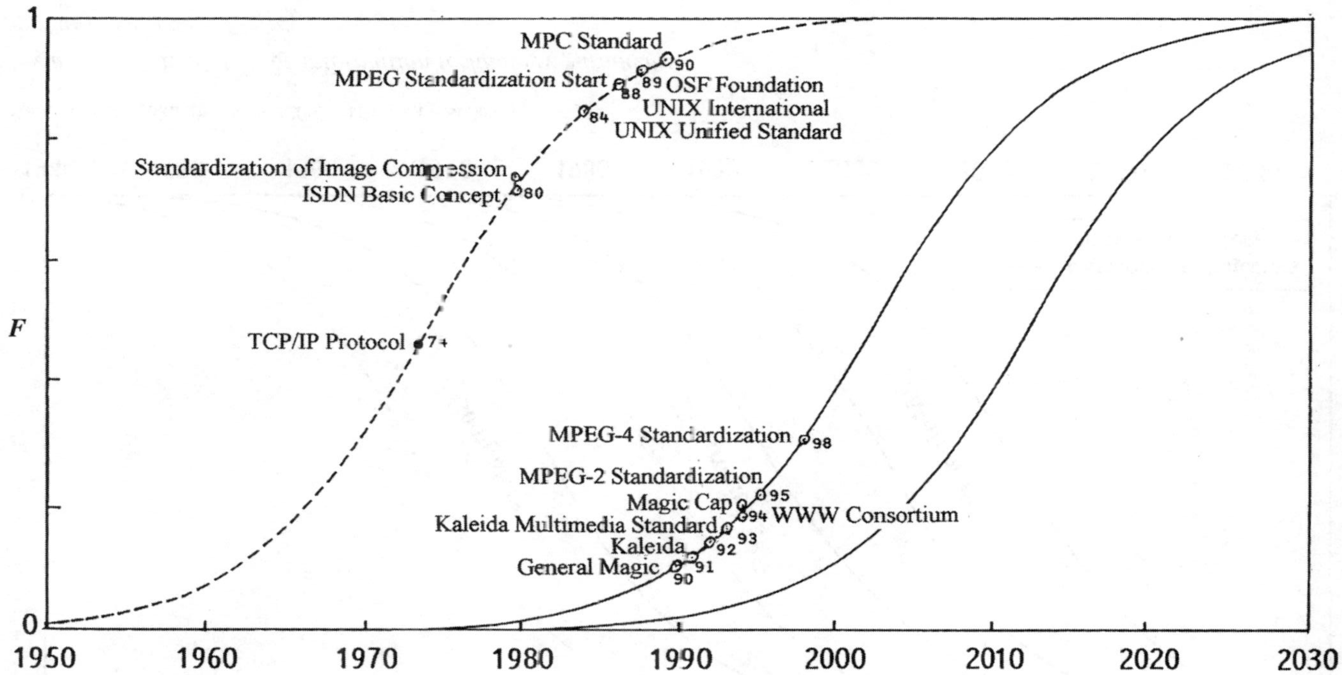

Source: Hirooka and Nitta (2002).

Figure 9.17 Standardization of multimedia and strategic alliances

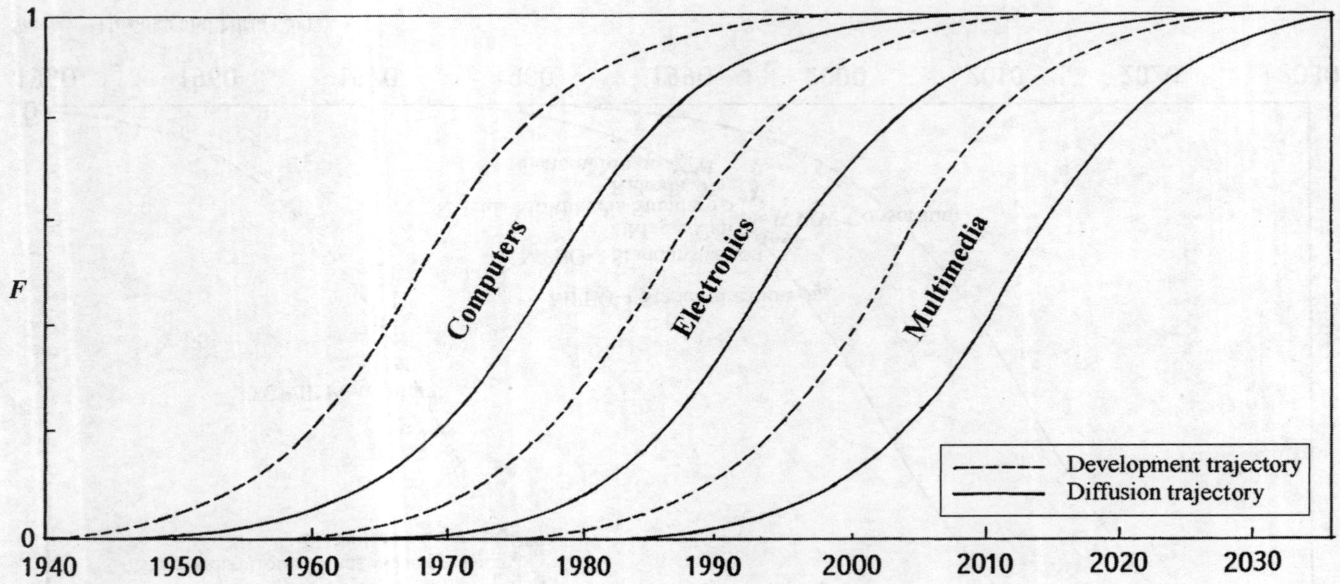

Sources: Hirooka and Nitta (2002), Hirooka (2003a).

Figure 9.18 Evolution of information technology paradigms

I. Technology fusion by existing industries

1. *Technology fusion of mainframe computers* by the steel, cement, chemical, railway and flight simulator industries was analyzed. Flight simulators began to use computers along the development trajectory of the computer paradigm, and the fusion trajectories of the other industries were seven to eight years behind the development trajectory of computers. The chemical industry trailed a bit further behind and, as simple users, did not adopt computers until almost the start of the diffusion trajectory of computers.
2. *Technology fusion of electronics* was analyzed for machine tools, cameras, automobiles and aircraft engines. Electronics technologies were introduced into all of these industries just along the development trajectory of electronics itself, and it was surprising to learn that they introduced electronics technologies at the same time as the development of electronics products per se.
3. *Technology fusion of innovation technologies* by existing industries is very important because it induces institutional change. Technology fusion produces a complex product system. There are two kinds of production systems. One produces cheaper products thanks to the effective utilization of innovations, such as the introduction of computers for higher quality products, labor reductions and higher efficiency for production. The other produces more sophisticated products by adding value: making products safer or more environmentally friendly, adding amenities, and so on; however, the price increases correspondingly. For example, the price of aircraft per weight has been increasing exponentially.

II. Behavior of firms during the maturity period of the innovation paradigm

4. During the latter half of the diffusion trajectory, international strategic alliances are formed to avoid excessive competition due to the reduced growth rate and larger number of competitors.
5. According to Vernon's life-cycle theory, latecomers in developing countries have international competitiveness during the latter half of the development trajectory due to lower costs rather than to technological disparity.

III. Evolution of innovation paradigms

6. Innovation paradigms intermittently evolve to other paradigms. In the information and communications sectors, the computer, electronics and multimedia paradigms are interrelated.
7. The multimedia paradigm has not reached maturity yet. The computer and electronics paradigms are mature, and the recent recession

can be interpreted as the gap between these two paradigms and the multimedia paradigm.

8. The infra-trajectory of information and communication sectors are now at the turning point and could be coming up to the second development period in full play of the three paradigms of computer, electronics and multimedia towards the peak of the fifth Kondratiev cycle in around 2040.

PART III

Nonlinear Dynamics of Innovation and Knowledge Transfer

In the previous chapters we have developed a dynamics of innovation in terms of three interlinked nonlinear trajectories each with a logistic nature. Part III deepens the examination of innovation dynamics to verify its nonlinear nature and expands it to wider knowledge dynamics than innovation per se.

The first chapter in this part explains how the nonlinear nature of innovation can be expressed using various types of sigmoid curves and examines the suitability of one of them – the logistic equation. More precisely, the development of an innovation is a discrete system because knowledge about the innovation is acquired and accumulated through knowledge transfer from person to person. The discrete system of innovation is examined by applying difference equations instead of differential equations. Innovation systems are interpreted in terms of complexity with a dissipative structure. In particular, various types of fractals are examined using empirical data. The phenomenon of self-organization by innovation dynamism is also discussed. Thus, the complexity in innovation systems is empirically clarified. It is noteworthy that a new mathematical derivation of the logistic difference equation has disclosed a unique feature of the logistic phenomenon.

The method of logistic analysis is, then, expanded to examine the nonlinear nature of science per se. The development of modern sciences including economics is outlined and compared with the process of industrial innovation.

Technology foresight is important for estimating the immediate future of industries. Estimating future innovation paradigms by extrapolating nonlinear trajectories becomes feasible through the ongoing development of science and technologies, achieves more accurate forecasts than the Delphi process and indicates more accurately the timing of next-generation industries. The global trends of energy and environmental issues are discussed through the trajectory analysis.

Furthermore, the knowledge transfer analysis beyond innovation is expanded to wider human activities as 'knowledge development dynamics'. Thus, a general analysis of modern social behavior is carried out through knowledge development trajectories and a kind of mega-structure of human activities is considered for a far-reaching time span.

10. Knowledge transfer and complexity in innovation systems

We have previously described the innovation process in terms of three interlinked logistic trajectories and clarified the role of each trajectory. There are, however, many ways to describe a sigmoid model. In this chapter, we compare several mathematical expressions, analyze their suitability for describing innovation dynamism, and determine whether a logistic equation is the most suitable one. Furthermore, we discuss the process of knowledge transfer in the innovation paradigm, which we treat as a discrete system. While there has been much discussion about discrete systems and complexity, no empirical verification has yet to be carried out. This is an experiment to examine the complexity of innovation systems by using empirical data.

10.1 DESCRIPTION OF INNOVATION DYNAMICS

The discussion in this book has used a logistic equation to describe an innovation trajectory. However, many equations other than a logistic one have been presented to describe a sigmoid curve. The present section discusses the appropriateness of a logistic equation compared with other types of equations. The analysis is carried out not only for innovation dynamism but also for biological growth. This is a comprehensive analysis of sigmoid-type dynamic behavior.

Description of Innovation and Growth Functions

Growth in a biological system was first described using a logistic equation by Verhulst (1838). Responding to Malthus' population theory (1798), Verhulst described a decreasing population growth rate by using a logistic equation. Logistic equations have also been used as an 'epidemic model' in the field of pathology to describe the spread of infectious diseases as well as biological propagation. Various equations have been proposed for describing population ecology. Shinozaki (1994) summarized and analyzed various equations that describe biological growth models.

In the field of economics, Schumpeter (1939) recognized the applicability of Verhulst's formula, but did not use it to describe the growth of innovation. He expressed the situation as follows:

> Still more treacherous and pregnant with danger of speculative temerity may be the application of Verhulst's formula, $y = a/(be^{-t} + 1)$, which was intended (1838) to represent certain features of organic or of similar types of growth. Even perfect fit in the least square sense would not prove anything. We are, however, on somewhat safer ground when applying such expressions to the behavior in time of quantities of individual commodities. (pp. 492–4)

Griliches (1957) first showed that the diffusion of innovation products can be expressed using a logistic equation, and many economists confirmed this correlation. Using Griliches's data plus other data, Dixon (1980) demonstrated that the diffusion of hybrid corn is much better expressed by the Gompertz equation than by a logistic equation. Mahajan and Peterson (1985) summarized various equations used as models to express sigmoid growth curves for innovation diffusion.

Metcalfe (1994a) considered that the diffusion of innovation products must have a mechanism similar to that of the growth and propagation of organisms. This is based on the concept proposed by Roughgarden (1979) and Slobodkin (1980). In this concept, the dynamics of organisms is governed by two factors – the breeding rate (r) and carrying capacity (K) in the living circumstance. This is the so-called 'r-K theory'; internal factor r is determined by the species, and external factor K is governed by the circumstances. Metcalfe applied this concept to the diffusion of innovation products into the market, which is governed by two factors – the intrinsic rate of diffusion and market capacity.

Following Metcalfe's argument, the diffusion of innovation products and the growth of an organism can be expressed using a sigmoid (S) curve consisting of two factors: internal factor (g) and external factor (f). The internal factor is related to the intrinsic nature of innovation products, the diffusion coefficients for technological innovations, and the breeding rate coefficient for organisms. The external factor is related to the capacity of the market and to the environmental capacity.

As these factors are interrelated, the diffusion rate of products or the breeding rate of organisms can both be expressed by the product of two factors:

$$dy/dt = g(y)\, f(y)$$

where y is the measure of diffusion at time t.

For both innovation diffusion and organism breeding, the basic prerequisite is a deterministic increasing sigmoid-type function converging towards a specific value of y_0 at the ultimate market or at the ultimate size of the ecological sphere. Thus, both the diffusion rate and the breeding rate can be expressed using a first order differential equation. The general treatment in the case of innovation diffusion is as follows.

For innovation diffusion, if $f(y) = y_0 - y$, the diffusion rate is

$$dy/dt = g\,(y_0 - y) \tag{10.1}$$

where y_0 is the ultimate market determined by the external factor and g is the diffusion coefficient determined by the internal factor.

As Mahajan and Peterson (1985) demonstrated, $g(t)$ is generally expressed as a function of the product y,

$$g(t) = a + by + cy^2 + \ldots \tag{10.2}$$

which is typically expressed as either of three models:

(1) $g(t) = a$,
(2) $g(t) = by$, or
(3) $g(t) = a + by$

(1) External model
In the first model, as the internal factor is a constant, this model depends only on the external factor. The equation can thus be expressed as

$$dy/dt = a\,(y_0 - y) \tag{10.3}$$

where a is the diffusion coefficient (positive real number). The solution is given by

$$y = y_0(1 - e^{-at}) \tag{10.4}$$

or

$$\ln\,[1/(1 - y/y_0)] = at \tag{10.5}$$

This equation was adopted by Coleman et al. (1966) and Hamblin et al. (1973) for describing the diffusion curve. In this case, the diffusion proceeds driven only by external information, without contact with the internal agents. The rate is maximum at the initial stage and then gradually decreases. This model is appropriate when members of the social system are isolated.

(2) Internal model
The model $g(t) = by$ is based on diffusion that develops only through interpersonal contacts. This is the logistic equation as an imitation model.

$$dy/dt = by\,(y_0 - y) \tag{10.6}$$

The solution by integration is

$$y = \frac{y_0}{1 + Ce^{-by_0 t}} \tag{10.7}$$

or

$$\ln\,[y/(y_0 - y)] = C + by_0 t \tag{10.8}$$

If the expression is logarithmically described, the equation is the Gompertz model:

$$dy/dt = by\,(\ln y_0 - \ln y) \tag{10.9}$$

Solving this equation,

$$y = y_0 \exp\,(-Ce^{-bt}) \tag{10.10}$$

or

$$\ln\,(\ln y_0 - \ln y) = C - bt \tag{10.11}$$

(3) Mixed model

Model $g(t) = a + by$ is a mixture of models (1) and (2). As this equation was derived by Bass (1969) based on a hazard rate function, it is called the Bass model.

$$dy/dt = (a + by)\,(y_0 - y) \tag{10.12}$$

Solving this equation,

$$y = \frac{y_0 - aCe^{-(a+by_0)t}}{1 + bCe^{-(a+by_0)t}} \tag{10.13}$$

This model was used by Bass (1969), Webber (1972) and Mahajan and Muller (1979). If this equation is modified by substituting $a + by = Y$ and $a + by_0 = Y_0$, it becomes

$$dY/dt = Y\,(Y_0 - Y)$$

This indicates that the mixed model is a kind of logistic model.

(4) Modified models

Various modified models have been proposed. If a logistic equation is simply expressed as

$$dF/dt = bF\,(1 - F) \tag{10.14}$$

its integration is

$$\ln [F/(1-F)] = bt + c \quad (10.15)$$

This equation is applied to the following modified derivations.
1) von Bertalanffy model
The generalized equation proposed by von Bertalanffy (1957) is

$$dF/dt = [b/(1-\theta)]F^{\theta}[1 - F^{(1-\theta)}] \quad (10.16)$$

This equation is not solvable as is, but if $\theta = 0$, it is an external model, and if $\theta = 2$, it is a logistic equation. When $\theta = 1$, the equation is indefinite but tends to zero, the equation is expressed as $dF/dt = bF \ln (1/F)$, leading to the Gompertz equation.
2) Mitscherlich–von Bertalanffy model
The equation is expressed by

$$dF/dt = 3b\,(F^{2/3} - F) \quad (10.17)$$

Solving it,

$$F = (1 - k\,e^{-bt})^3 \quad (10.18)$$

3) m-order logistic model
A more generalized logistic equation is

$$dF/dt = b\,F\,(1 - F^{\mathrm{m}}) \quad (10.19)$$

Solving it,

$$F^{\mathrm{m}} = 1/(1 + Ce^{-mb}) \quad (10.20)$$

4) Floyd model
Floyd (1968) proposed the relation $dF/dt = bF\,(1-F)^2$. Its solution is close to a Gompertz curve and has an unsymmetrical curve with a flex point at $F = 0.33$.

Development of Logistic Models

The diffusion of innovation products has been discussed a great deal since Griliches (1957) first analyzed the phenomena using a logistic equation. Many economists have confirmed the applicability of a logistic equation and have developed more advanced treatments, especially descriptions reflecting the behaviors of firms from the microeconomics viewpoint. As summarized by Coombs et al. (1987), such treatment of firm behavior was carried out by David (1975), Davies (1979), Metcalfe (1981, 1982), Stoneman (1984) and Cameron and Metcalfe (1987).

David (1975) proposed a profit model in which an innovation is considered as a stimulus for a firm. Each firm has a critical level of stimulus and will adopt an innovation only when the stimulus represented by the innovation itself exceeds the critical level. Davies (1979) considers a model in which inter-firm differences play an important role. Due to uncertainty, firms make decisions in a way that can more appropriately be described as satisfying than as profit maximizing. The most important aspect of inter-firm difference is the maximum acceptable pay-off period, R^*_{it} which represents the equivalent of David's critical level.

Davies (1979) also recognized that there are two types of innovation diffusion. The first type of innovation, called group A, consists of technologically simple and inexpensive processes, which are built off-site. The second type, group B, consists of technologically complex expensive innovations, usually built on a one-off basis and requiring lengthy periods of installation on the adopter's site. Type A innovations might be consumer goods – TV sets, washing machines and so on – and type B innovations are chemicals or steel. According to Davies the different types of innovations lead to different shapes of diffusion curves.

Another important deficiency of the epidemic model was the neglect of the supply side. An innovation can only be adopted if it is produced, and this in turn depends on how profitable it is to manufacture products embodying the innovation. Some trials to involve the contribution of supply side into the diffusion equation of innovation were carried out by Stoneman (1983, 1984) and Metcalfe (1981, 1982, 1994a, 2001, 2003). Cameron and Metcalfe (1987) tried to introduce a mechanism for replacing an old product with a new one by using a logistic equation.

Metcalfe Models: Setting of Demand and Supply Sides

Development of logistic model

Metcalfe (1981, 1982) first examined the issue of the supply side by using a logistic equation to comprehensively describe the market growth rate on the demand and supply sides. The growth rate can be expressed by a logistic equation:

$$dy/dt = b\,y\,(y_0 - y) \qquad (10.21)$$

The appearance of new technology creates an adjustment gap, which is defined by the difference between the equilibrium market demand, $m(p) = y_0$, reaching saturation and actual demand, $y(t)$, at particular instant t during the diffusion process. The growth rate of demand, $g_d(t)$, is proportional to this adjustment gap:

$$g_d(t) = \frac{1}{y} dy/dt = b[m(p) - y] \quad (10.22)$$

where demand at the ultimate market, $m(p)$, is a function of price, p. If the price at time t, $p(t)$, is assumed to be given by the following correlation:

$$m(p) = c - a\,p(t) \quad (10.23)$$

if the price decreases, the demand at the ultimate market increases.

On the other hand, the rate of growth of supply, $g_s(t)$, is assumed to be proportional to the profitability of the innovation for suppliers. The profitability increases when the price of the innovation product increases but is negatively affected by increasing costs of all the inputs used in the production of the innovation (labor, materials, machinery and so on).

$$g_s(t) = \frac{p(t) - h_0 - h_1 x(t)}{k} \quad (10.24)$$

where $x(t)$ is the scale of production, h_0, h_1, and k are constants, and k reflects the supply of capital funds and investment requirements. The profitability of suppliers increases with the price, and a reverse relation to (10.23) holds. If we hold constant the fundamental parameters of the diffusion process relating the supply of finance, the innovation technology, the equilibrium demand curve, and the supply conditions of the composite non-capital input, we can trace out the balanced diffusion paths for output, prices, costs and profits as follows. Along the balanced path, we have $g_d(t) = g_s(t) = g(t)$, and $y = x(t)$.

Substituting equation (10.24) into (10.23), and the outcome into (10.22) we have

$$g(t) = B\,[C - x(t)] \quad (10.25)$$

where

$$B = \frac{b(1 + ah_1)}{1 + abk} \text{ and } C = \frac{c - ah_0}{1 + ah_1}$$

B corresponds to diffusion coefficient *b*, and *C* corresponds to saturated market y_0.

These correlations of demand and supply indicate that in the course of innovation the price decreases and profitability decreases gradually. The change in demand is expressed by an S-curve, and the size of the saturated market is estimated to increase due to the improvements brought by innovation products.

In this treatment, however, the expression of equation (10.23) is as a first order approximation, linear, so alternate treatments are likely to provide further insights into the process.

Development of the Gompertz model

The Gompertz model was developed by Gompertz in 1825 to describe the trend in the death rate by age:

$$dy/dt = by\,(\ln y_0 - \ln y) \tag{10.26}$$

Solving this,

$$y = y_0 \exp\,(-k\mathrm{e}^{-bt}) \tag{10.27}$$

This equation, which also describes the death rate of animals, indicates that the death rate increases exponentially with age. It is thus a model of aging and indicates that damage to genes progresses along with aging. This can be interpreted in terms of deterioration of homeostasis (the function that maintains the body's internal conditions against changes in the external environment). The rate of reduction in life expectancy due to radiation and the change in the rate of machinery failure due to wear can also be expressed using the Gompertz equation.

Using the *r-K* theory of Roughgarden and Slobodkin, Metcalfe (1994a) asserted that economic development must be governed by a mechanism similar to that of growth and retardation. Various economists have discussed the retardation of industry. Using the principal retardation theories of A.F. Burns and S. Kuznets, Abramovitz (1989) identified structural change and a tendency towards retardation in the process of economic growth. Three major and distinct mechanisms are identified to describe the development of an economy. The first is the dynamics of demand for a new product, or, more precisely, the dynamics of preference formation for a new commodity. The second is the growth in capacity to supply a new commodity, in which profitability plays a central role. The third involves the development of the technology due to product improvements. These three mechanisms provide the basis for the ecological dynamics of industry growth driven by technical progress.

Metcalfe asserted that there is room for selection between candidate equations for explaining innovation diffusion. He selected the Gompertz equation because of its convenient properties in explicating the role of key economic elasticities and incorporating time dependent changes in the environment of the industry. The Gompertz process is an asymmetric sigmoid curve with an inflection point at approximately 37 per cent of the carrying capacity or long-run value of the dependent variable.

The Gompertz process is expressed by

$$\frac{d}{dt}\log y_d = \beta(\log y_0 - \log y_d) \tag{10.28}$$

where y_0 is the ultimate level of market demand at which all learning of preferences has been completed, y_d is the demand at time t, and β is a constant.

In this innovation process, price $p(t)$ of the innovation product at time t and the long-run level of demand, y_0, are given by

$$y_0 = m(p) = \eta_0[p(t)]^{-\alpha} \tag{10.29}$$

where $m(p)$ is demand in the ultimate market as a function of price p, α is the elasticity of long-run demand, and η_0 is a constant.

Firms set prices relative to unit costs in order to finance growth at a particular rate, so that capacity growth is proportional to current profitability. The innovation analogue to the demand side learning process is

$$\log p(t) = \kappa \frac{d}{dt}\log y_s + \log h(t) \tag{10.30}$$

where $d\log y/dt$ is the growth rate of capacity, coefficient κ depends on rules about the funding of investment, on the capital requirements per unit of output, and on the conventions determining the rate of depreciation. The industry has a long-term supply curve given by

$$h(t) = h_0 y_s^{\varepsilon} \tag{10.31}$$

where $h(t)$ is the long-run unit cost level, and ε is the (inverse) elasticity of long-run supply and positive so that, as the industry expands, unit costs increase and profits are squeezed.

It follows that for any arbitrary $p(t)$ we have two Gompertz curves giving different paths for demand and capacity. If considered a balanced path by setting $y_d = y_s = y$ in equations (10.28) to (10.30), this balanced process satisfies the equation:

$$\frac{d}{dt}\log y = B(\log C - \log y) \tag{10.32}$$

where coefficients B and C are given by

$$B = \frac{\beta(1+\alpha\varepsilon)}{1+\alpha\beta\kappa},\ \log C = \frac{1}{1+\alpha\varepsilon}(\log \eta_0 - \alpha \log h_0)$$

Solving for the path of prices, as a function of current output,

$$\log p(t) = (\kappa B \log C + \log h_0) + (\varepsilon - \kappa B)\log y \tag{10.33}$$

Provided the elasticity of supply is not too low, relative to $\kappa\beta$, $p(t)$ declines over time and always falls relative to the unit cost level, $h(t)$. Since profits and growth are closely linked, it follows automatically that retardation and a squeeze on profit margins are two sides of a coin.

Balanced expansion is a useful ordering device but it is clearly a strong requirement; for entrepreneurs to keep capacity growth in exact step with demand growth would require a considerable degree of foresight and a tranquil environment. Now while the rules adopted in relation to pricing and investment behavior can well tolerate small departures or self-correcting cyclical departures from balanced expansion, they are not likely to cope with major shocks which are bound to lead to a fractured process of secular change. Figure 10.1 illustrates what might occur in the face of an unforeseen reduction in demand or an unforeseen increase in costs, the immediate consequence of which is to create excess capacity in the industry and to reduce the long-run niche to C'. Up to time t_1, expansion is balanced but the shock drives output, log y_2, below capacity, log y_1, and destroys the basis for investment. Investment and capacity expansion drop to zero, or close to it, depending upon precisely what one assumes about the effect of excess capacity on investment behavior, until demand has grown again to fully absorb capacity. From then on balanced expansion can resume but is directed towards a new niche that reflects the changed long-run fundamentals. Thus, the path a–a is fractured and ultimately resumes as path b–b. Similarly, an unforeseen increase in demand will generate a capacity shortage and a price above its balanced value. Growth is again fractured until the capital stock has caught up with market demand.

In all such cases, the growth curve is fractured and not well summarized by a single Gompertz or any other curve. Indeed, Hirooka and Hagiwara (1992) have shown such shocks in an oil crisis.

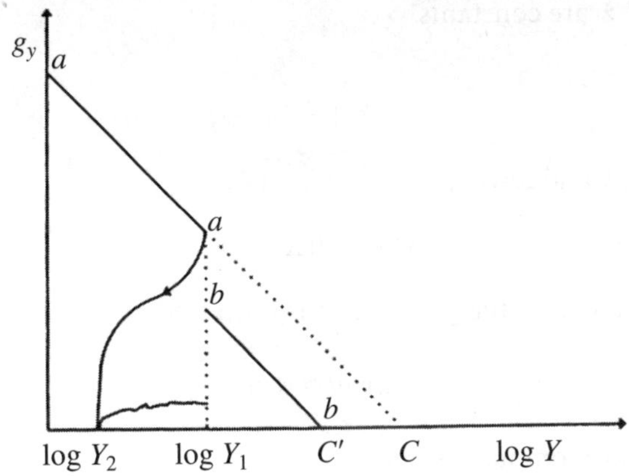

Source: Metcalfe (1994a).

Figure 10.1 Fracture due to economic turbulence and resumption

Growth Models in Biological Systems

The above treatment makes it possible to discuss the interrelation between demand and supply by introducing an economic elasticity. Economic behavior seems similar to the ecological behavior of organisms, as pointed out by Metcalfe. Now, let us examine the growth theory of organisms compared with economic development.

Shinozaki (1994) set forth six preconditions for the growth of organisms.

1) There is a homogeneous closed system without interaction with the outside.
2) The amount of growth, y, is a positive real number and changes continuously within a deterministic region.
3) The growth is described by a first order differential equation on time, t, the same as for a chemical reaction.
4) Ideal growth is expressed using an exponential function, as shown by population breeding.
5) The growth rate, $s = 1/y \cdot dy/dt$, never exceeds a positive finite value. This is reasonable as the capacity of organisms.
6) There is an upper limit on y so that it never reaches infinity.

Thus, we can consider a first order linear differential equation for the rate of growth on time t from 3) and 6):

$$dx/dt = a_0(X - x) \tag{10.34}$$

where a_0 and X are constants.
Solving this,

$$x = X(1 - ke^{-a_0 t})$$

Setting $k' = kX$ and rewrite,

$$k' = e^{a_0 t}(X - x) \tag{10.35}$$

From requirement 3), the growth rate is expressed as

$$dy/dt = v(y) \tag{10.36}$$

which is an autonomous system.
Now, we consider an appropriate function of y:

$$x = G(y) \tag{10.37}$$

We can determine $G(y)$ as the linear differential equation of (10.34) is satisfied, that is,

$$dx/dt = a_0[X - G(y)] = dG(y)/dy \cdot dy/dt = v(y)dG(y)/dy$$

Thus,

$$dG(y)/dy + [a_0/v(y)]G(y) = a_0 X/v(y) \tag{10.38}$$

From this linear differential equation, $G(y)$ is obtained. This $G(y)$ is a linearized function of growth function (10.36). The integral of (10.38) is naturally equivalent to the integral of (10.36). When $G(y)$ is obtained, the solution of (10.36) can be obtained from (10.35):

$$k' = e^{a_0 t}[G(y_0) - G(y)] \tag{10.39}$$

where $y_0 = y|x \to X$, and $v(y_0)$ is expressed by G:

$$v(y) = \frac{dx}{dt} / \frac{dG}{dy} = a_0[G(y_0) - G(y)] / \frac{dG}{dy} \tag{10.40}$$

Add condition 4) and tend to $t \to -\infty$.
As $e^{a_0 t} << 1$, and $G(y_0) = \text{const.}$,

$$k' \cong -e^{a_0 t}G(y)$$

The initial growth should be expressed by an exponential function: $y \propto e^{at}$.

As the constant part should not be counted from equation (10.39),

$$G(y) \fallingdotseq 1/y^m, \quad m = a_0/a$$

That is, in order to ensure the initial growth is exponential, at least during a smaller region of y,

$$G(y) \fallingdotseq 1/y^m$$

The growth curve corresponding to the linearized function is

$$1/y \cdot dy/dt = a\,[1 - (y/y_0)^m]$$

that is,

$$y^m = y_0^m/(1 + Ce^{-mat})$$

$$G(y) \fallingdotseq 1/y^m$$

That is, this is an mth degree logistic equation, and when $m = 1$, it is the simplest logistic equation having reversal linearity.

For the Gompertz and Mitscherlich–von Bertalanffy equations, $G(y)$ is $\ln y$, and $y^{1/3}$, respectively. These correlations indicate that the logistic equation is the most suitable one among growth functions to satisfy the requirements of 4) and 5).

Descriptive Equations for Innovation Diffusion

The above examination indicates that versatile models expressing a sigmoid curve can be classified into three categories, as shown by Mahajan and Peterson (1985). The external models can be excluded as being insufficient because there are no interactions between persons concerned. Although the mixed models show a variety of modifications, there is no apparent special reason for their adoption. Thus, the internal models should be the most appropriate; they include logistic equations in a broad sense and the Gompertz equation. A common feature of these models is that they start at a slow rate, accelerate gradually, and, at a flex point, begin to decelerate gradually to a reduced rate until they reach a saturated level; that is, they form an S-shaped curve. As the actual diffusion is affected by various types of economic turbulence, it is quite difficult to determine which equation should be adopted.

Most importantly, however, is that while the actual diffusion of innovation products often suffers from various types of economic turbulence, causing the logistic curve to become deformed, the diffusion resumes at the original rate after sound economic conditions are restored. This means that the diffusion of innovation has an intrinsic diffusion coefficient; that is, the diffusion phenomenon has an intrinsic logistic nature. The logistic equations fulfill the six prerequisites defined by Shinozaki (1994). Although any logistic *m*th order equation fulfills the six prerequisites, a first order logistic equation ($m = 1$) having reciprocal linearity is considered to be the most suitable and fundamental equation. In short, among the various types of equations showing a sigmoid curve, a first order logistic equation is the most appropriate one, not only for ecological models but also for innovation diffusion systems.

10.2 EMERGENCE OF INNOVATIONS BY KNOWLEDGE TRANSFER

The development of innovations has been discussed using a logistic equation. Not only innovation diffusion but also the technological development period can be described using a logistic equation. Because the essential nature of innovation lies in knowledge transfer from person to person, it follows that this process is closely related to the emergence and progress of innovations.

Logistic Paradigms of Innovation

While the logistic nature of innovations has been more systematically clarified by the above examinations, it should be more clearly demonstrated that innovation is a phenomenon of knowledge transfer in the human community.

Mahajan and Peterson (1985) discussed the cause of innovation having a sigmoid curve and summarized the following six items:

1. A function of the extent of economic advantage of the innovation: the amount of investment required to adopt the innovation and the degree of uncertainty associated with the innovation (Mansfield, 1961).
2. Supply and demand rationale as a diffusion explanation (Griliches, 1957; Robinson and Lakhani, 1975; Brown, 1981).
3. Learning perspective when explaining diffusion patterns (Casetti and Semple, 1969; Sahal, 1981).
4. Information transfer explanation (Hagerstrand, 1967; Bernhardt and MacKenzie, 1972).
5. Technological substitution frame of reference when interpreting diffusion processes: innovation typically replaces an existing product, service or technology (for example, the substitution of synthetic fibers

for natural fibers); the dynamics of this replacement process account for both the rate of diffusion and the shape of the diffusion curve (Blackman, 1974; Sharif and Kabir, 1976).

6. Communication-based theory for interpreting the diffusion of diverse phenomena (Rogers, 1983).

Among these interpretations, the learning, information transfer and communication theories indicate that information transfer is essential to innovation. In this context, the epidemic model for the spread of infectious disease explicitly describes a typical transfer phenomenon; if the disease is replaced by knowledge, this model becomes an explicit knowledge transfer equation for innovation. Marchetti (1985, 2002) and Modis (1992) observed that the logistic equation can be used to describe the life history of knowledge creation by both artists and scientists. This means that human intellectual activities have a logistic nature.

Innovation Creation and Tacit Knowing

The creation and progress of innovation can be described by a logistic equation for the technology and development trajectories:

$$dy/dt = by\,(y_0 - y) \tag{10.41}$$

where y is the population that has acquired the knowledge and $(y_0 - y)$ is the potential population to have the knowledge at time t. The product of the two indicates the existence of a kind of feedback. This is the simplest equation for showing the interactions of knowledge transfer.

Polanyi (1966) postulated the existence of 'tacit knowing', which is latent accumulated knowledge and plays a crucial role in the emergence of new science and technology. The creation of new knowledge emerges through knowledge transfer from person to person in a field of human community as if it were in an electromagnetic field. This is the hierarchical heuristic field.

Thomas Kuhn (1962) proposed the concept of 'paradigm', which was introduced into the economics of technological change by Dosi (1982) and developed into the concept of the techno-economic paradigm by Perez (1983). Such concepts led to the field of knowledge transfer in which new knowledge is created to give an innovation. Polanyi (1966) explicitly recognized the concept of paradigm and described how the paradigm shift after Copernicus lasted 144 years, until the *Principia* by Newton.

In the field of complex systems, an important concept is 'emergence', which relates to the phenomenon of self-organization. Self-organization was recognized as a concept of dissipative structure by Prigogine (1976). In a non-equilibrium system, a dissipative structure emerges accompanying the production of entropy, and this is a nonlinear thermodynamic system. Living organisms are typical dissipative structures, and termination of the dynamic system means death. Development of the economy and creation of innovations are nonlinear dynamic systems having dissipative structures. The logistic dynamism of innovation exhibits a typical dissipative structure with knowledge transfer in a heuristic field.

The Science of Complex Systems

The science of complex system has been receiving much attention. Figure 10.2 illustrates the paradigms related to complex systems. The fact that such deterministic systems as that found by Lorenz in 1963 and May in 1973 exhibit chaos phenomena was unexpected. Their findings have led to studies on the deterministic chaos leading to the development of complex systems. An important finding was the Li–Yorke theorem; Li and Yorke coined the term 'deterministic chaos' in 1975 (Li and Yorke, 1975). Another approach, by von Neumann, started from cellular automata in 1957. This concept was developed into the life game by Conway, Holland and Langton. It was made into an artificial life system by Langton's introduction of parameter λ in 1983 and by Kaufman's 'edge of chaos' in 1985. The Santa Fe Institute pioneered the concept of complex systems in 1986. These studies on chaos and complex systems form a trajectory for the idea of a complex system; it has a time span of 25 years, as shown in Figure 10.2.

Before then, unsolved problems were found in studies of Newtonian dynamics. Weierstrass tried multi-body analysis for the movement of planets and identified non-solvable cases in 1875. Poincaré proposed and established a concept of topology in 1895. The series of these studies form the topology trajectory lasting until the KAM (Kolmogorov, Arnold and Moser) theory in 1954; its time span was about 90 years. One more important trajectory is the fractal paradigm starting from the snow pattern by Koch in 1904, through Sierpinski's gasket in 1915, and lasting until Mandelbrot established the concept of 'fractal' in 1980. The time span of the fractal trajectory is estimated to be 80 years. These three trajectories are related to complex systems and are illustrated in Figure 10.2.

10.3 COMPLEX SYSTEMS IN INNOVATIONS

An innovation system is a discrete and dissipative system; it is created by knowledge transfer from person to person as a steadily changing nonlinear stream. Since the diffusion of an innovation was found to have a logistic

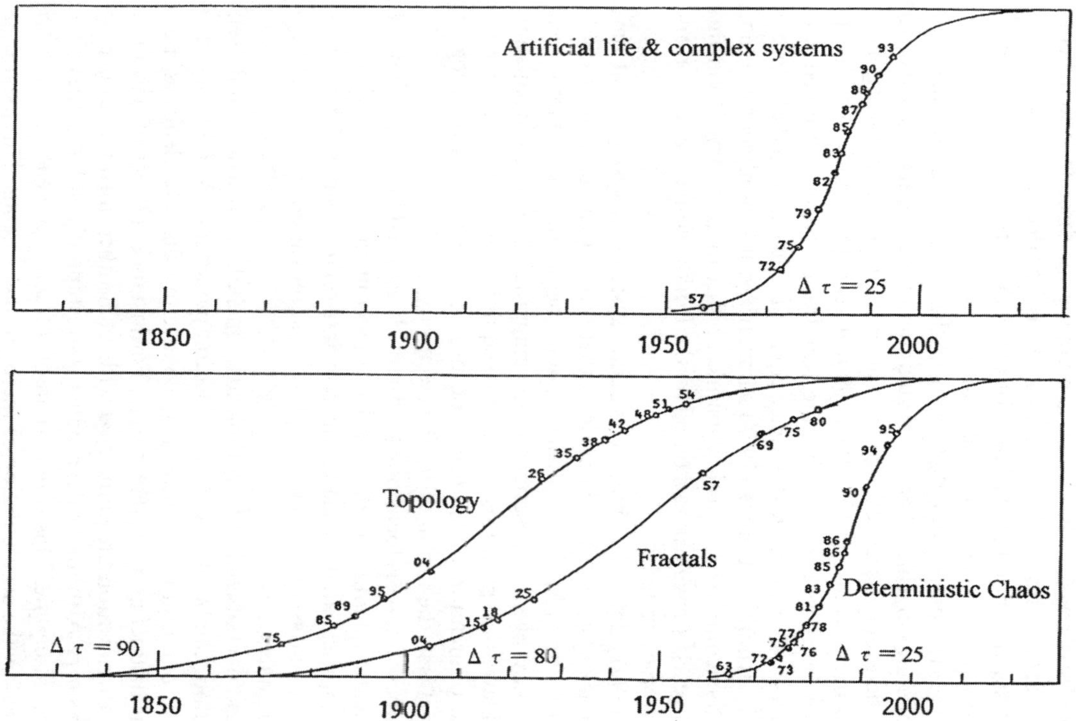

Source: Hirooka (2001b).

Figure 10.2 Development of complex systems and related sciences

nature by Griliches (1957), many economists have confirmed this phenomenon. This means that the system is nonlinear. Analysis of this nonlinear dynamism has been ongoing. One of the most important findings was the discovery of deterministic chaos in the logistic equation by May (1974, 1976). This was revealed in a logistic mapping for discrete systems connected to the naming of chaos by Li and Yorke (1975), and became an important starting point in the study of complex systems. This concept of complex systems has been introduced into economics.

Peters (1991) pointed out that, even in the chaotic region, there is a window of chaos in which there is a regular iterative structure and existence of a specific order. Based on this, he argued that logistic mapping has a fractal nature. Goodwin (1950, 1982, 1990) analyzed Schumpeter's concept of complex business cycles, which are Kondratiev cycles combined with Juglar cycles, using a logistic difference equation and clarified the nature of the chaotic dynamics. Lorenz (1993) summarized his 'nonlinear economics', which encompasses chaotic dynamics, bifurcation theory, Goodwin's model, the class struggle model based on the Lotka–Volterra equation and logistic complex systems.

Peters (1991) defined nonlinear dynamic systems as systems that have (i) a feedback system, (ii) a critical level that produces at least two equilibria, and (iii) fractals. The feedback system means that a previous event affects the next event as described in the mapping of the logistic equation. A critical level that produces at least two equilibria indicates the region in which the bifurcation phenomenon occurs: in the case of May's logistic mapping, it is in the region of $a>2.5$ (seems to be >3).

All of these analyses focus on the discrete logistic equation and indicate that innovation should be discussed in this context. However, there has been no empirical analysis using concrete data on innovation. This section is an experiment to investigate complexity in innovation systems based on empirical analysis.

Various phenomena of complexity, including fractals and self-organization, can be observed in innovation paradigms. A trajectory in an innovation paradigm is identified by connecting discrete elements. Each element is distinct as a discrete system. For instance, a series of new products in an innovation are connected with each other, forming a development trajectory. Although each product is substantially isolated, they share a common discipline. This is the origin of a fractal, and each trajectory is constructed by the accumulation of these product fractals. This means that a trajectory is formed by a series of fractals. Two kinds of fractals have been identified: chain fractals and bundle fractals. The former are connected to each other, as in a chain, and the latter are fractals making a bundle, like a bouquet of flowers.

Chain Fractals of IC Devices

The development trajectory of innovation technologies consists of a series of developments of innovation products. Each innovation product is distinct and has its own development history with a shorter development trajectory. This is the origin of fractals. For instance, devices in the electronics paradigm are developed step by step, following Moore's law. That is, the degree of device integration quadruples every three to four years from one level to the next. The development is not continuous but is a stepwise process. Each process has its own trajectory, which is an element of the fractal, and the accumulated fractals compose the parent development trajectory. The resulting products diffuse, forming a diffusion trajectory with a chain fractal structure, as shown in Figure 10.3. The development trend of integrated circuits is located on the development trajectory; a few years later, the developed product is sold along the diffusion trajectory. The sale of an IC chip continues for several years until it is replaced by a more advanced one. This process is described by a triangle on the diffusion trajectory, as shown in Figure 10.3. Thus, the fractals form a chain structure.

Bundle Fractals in Application Fields

Another fractal structure can be seen in the application fields. The development trajectory comprises various developed products and also various applications. The development histories of innovation products compose the development trajectory, and the products diffuse to be used in various applications, forming a 'fusion trajectory', as described in Chapter 9 (Hirooka 1998c, 1999c, 2003a).

The adoption of innovations by existing industries was termed 'technology fusion' by Kodama (1985) and Freeman (1987) Thus, a trajectory based on technology fusion is called a 'fusion trajectory'. This means that an innovation eventually expands beyond its own products towards wider application through the techno-economic paradigm change. For example, electronic devices not only form an electronics development trajectory, they were also adopted by various industries for use in a wide variety of products, including aircraft engines, machine tools, automobiles and cameras. Each industry creates its own fusion trajectory, as discussed in Chapter 9. These fusion trajectories occur along the electronics development trajectory, meaning that the adoption of electronics innovations by other industries takes place at the same time as the original technological development to form bundle fractals.

The time span of these fusion trajectories is almost the same as that of the original development trajectory because the development of technology

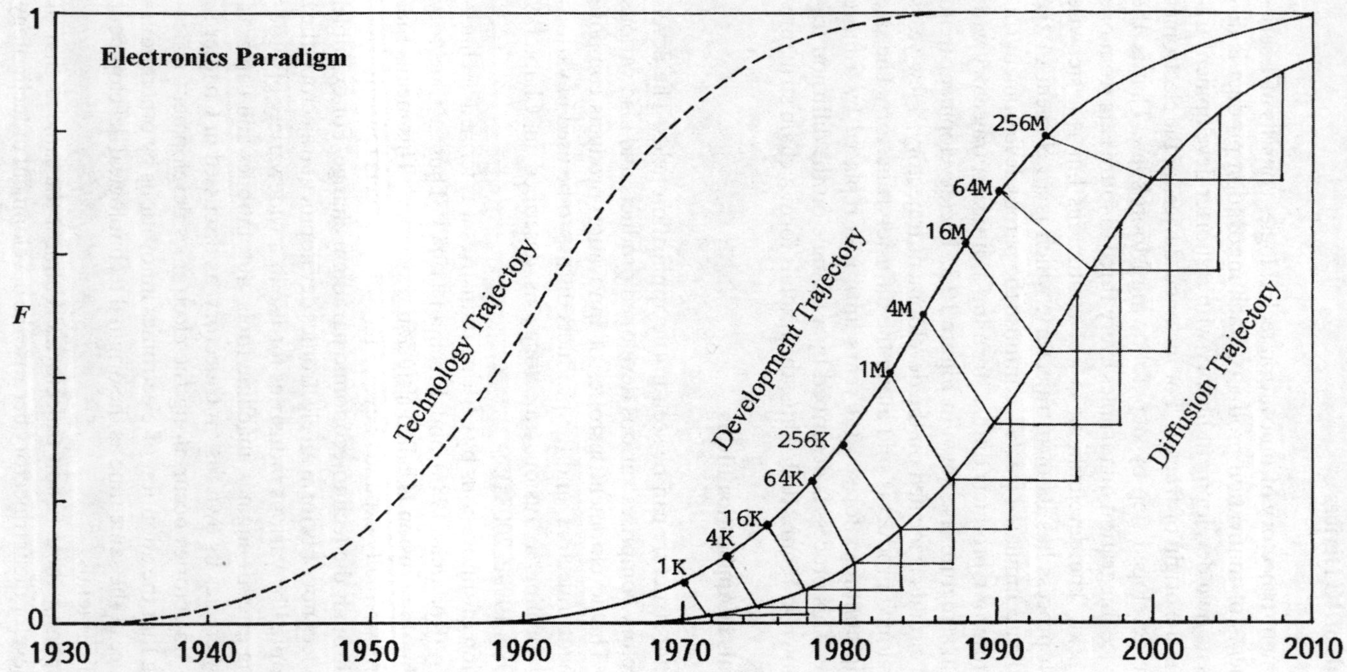

Note: Development and production of IC chips in Japan.

Source: Hirooka (2004b).

Figure 10.3 *Chain fractals of IC chips in electronics paradigm*

fusion takes place along the development trajectory. These fractals form a bundle of fusion trajectories along the electronics development trajectory, as shown in Figure 10.4. It is amazing to see how quickly other industries adopt innovation technologies. This expanded use of innovation technologies enriches the development trajectory and expands the market, contributing to institutional change in the economy.

System Fractals in Innovation Paradigms

The information technologies can be grouped into three paradigms: computers, electronics and multimedia. While these are different paradigms established within individual disciplines, they closely interact and together form the modern information society. The computer system has evolved by adopting electronics, and the multimedia system has been constructed on the basis of computer and electronics technologies. The evolutionary trend of these three paradigms is illustrated in Figure 10.5 using two loci: the development and diffusion trajectories.

After the completion of the core technologies composing the technology trajectory, a series of new products are developed composing the development trajectory. Furthermore, various elementary technological systems are developed along the development trajectory combining the core technologies with some software. These intermittently invented systems support the development of new products or the application of these products. These are more sophisticated and complex systems than simple products directly prepared using core technologies. They sometimes comprise combinations of various devices and software. These are intangible fractals which can be called 'system fractals'.

The development trend of various elementary technological systems is illustrated in Figure 10.5 by bars on the development trajectory. The starting point of each bar indicates the timing of the invention of the technological system, and the end of the bar corresponds to the practical use of the system along the stream of the diffusion trajectory; however, this point does not always correspond to the timing of commercialization. Each bar can be substantially considered to correspond to a small fractal S-curve because each technological system development can be expressed using a small logistic curve. That is, the development trajectory is considered to be an accumulated cluster of relevant elementary technological systems.

In the computer paradigm, computers were first used for scientific and military purposes. They were later (around 1960) put to use for business purposes (accounting and so on). The package home delivery system in Japan began using computers in 1968. Information management and decision support systems were developed around 1970. Office automation systems

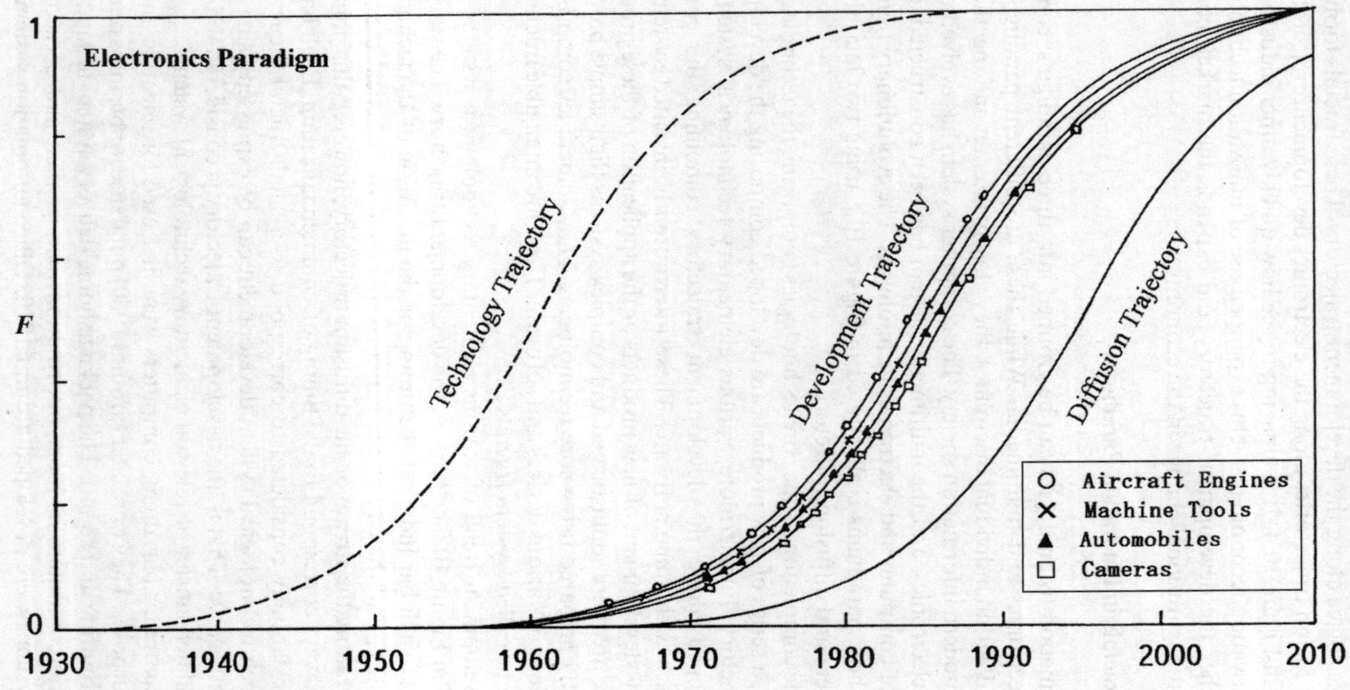

Source: Hirooka (1998c, 2004b).

Figure 10.4 Bundle fractals of electronics fusion trajectories by various industries

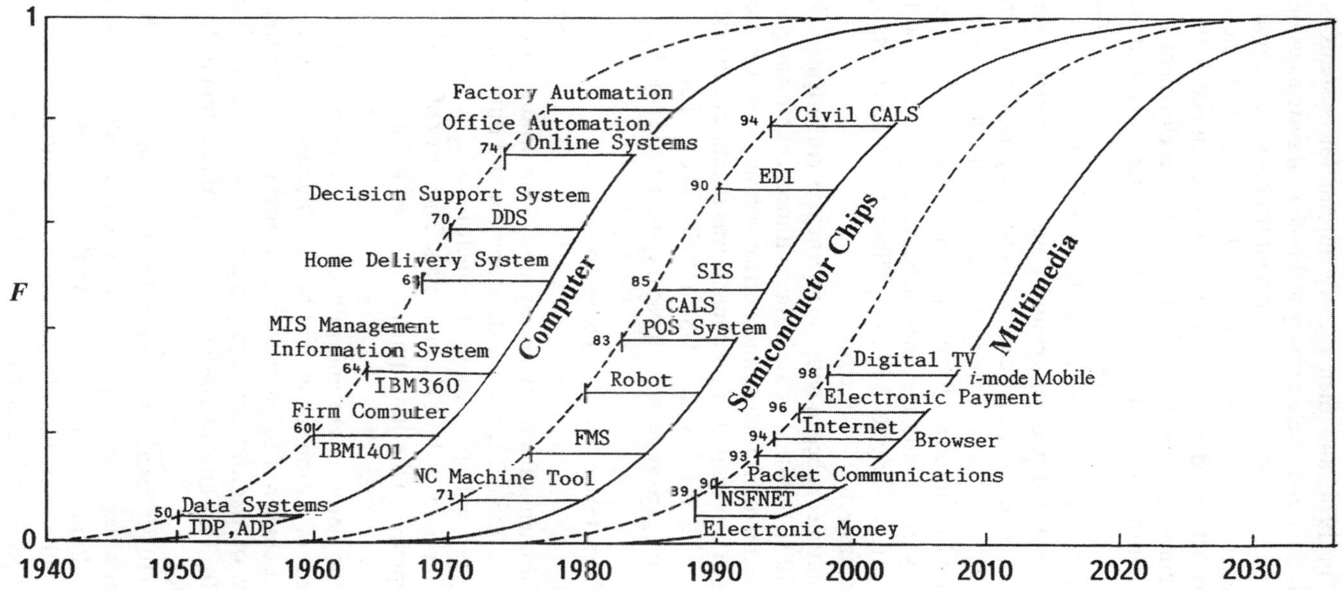

Source: Hirooka and Nitta (2002), Hirooka (2004b).

Figure 10.5 System fractals in information technology paradigms

came into use in 1974, and then factory automation systems became operable. This is the trend of elementary technological systems using computers, forming fractals along the development trajectory of computers.

The development of microprocessors in 1971 initiated the electronics development trajectory. Machine tools were first equipped with microprocessors for numerical control. This led to the development of FMS (flexible manufacturing systems) comprising computer controlled machine tools, robots and unmanned carriers. POS (point of sales) systems were developed in 1983 and soon became widely used in retail businesses. The CALS (Computer-Aided Acquisition and Logistics Support) system was developed as a computer-aided procurement system by the US Department of Defense in 1985. At the same time, strategic information systems (SIS) were adopted by leading firms for comprehensive management of their entire business. Various industries began developing civilian-use CALS (commerce at light speed) systems in 1994. These form chain fractals along the development trajectory of electronics.

The age of multimedia began with the internet system, originally known as NSFNET, in 1990. Now that we are in the multimedia paradigm, electronic commerce has begun to take shape, and electronic money has been commercialized. An important symbol of multimedia was the start of digital television broadcasting in the UK and US in 1998. These chain fractals of elementary technological systems compose the multimedia paradigm.

Composition of Fractals in Innovation Systems

In the case of the chemical industry, many innovation products were developed, as shown in Table 2.1, and the development and diffusion trajectories of the polymer industry paradigm, which both have a time span of 35 years, are composed of these chain fractals. The series of developments of pesticides depicted on the development trajectory in Figure 6.2 also form chain fractals.

In general, the development and diffusion trajectories are composed of an assortment of various chain and bundle fractals. Furthermore, we have identified a cluster of infra-trajectories across two Kondratiev cycles as an extension of the innovation paradigm, as described in Chapter 5 (see Figure 5.10). This kind of cluster also seems to be a bundle fractal of infra-trajectories.

Schumpeter (1939) described the complex dynamism of three business cycles: the Kondratiev, Juglar and Kitchin cycles. This description can be interpreted as a kind of fractal theory. As shown by the above analysis, there is no doubt that fractals compose the innovation paradigm. Interestingly, Table 2.1 indicates that the time spans of innovation products of chemicals, crude steel and automobiles are gathered in a time period of around 10 years, ranging from nine to 19 years. This is substantially the same as the time span of a Juglar cycle. Both semiconductor integrated circuits and electric appliances have a time span of around four years (ten at the most), which corresponds to that of a Kitchin cycle. It can thus be said that Schumpeter implicitly recognized the existence of fractals.

Self-organization of Innovations

The coincidence of innovation diffusion and the upswing of Kondratiev business cycles described in chapters 3 and 5 indicate that Kondratiev business cycles are induced by the clustering of innovations. A direct reason for the clustering of innovations is rather simple – new products resulting from innovations are difficult to introduce in a depressed market during a downswing; the innovations stagnate and then simultaneously take off as the economy recovers, forming the upswing as a cluster. This is a simple self-organization phenomenon. It is evidenced by the delay in full-scale diffusion of railways in the UK until the upswing of the third Kondratiev cycle, starting around 1850, even though railways were initially introduced in 1825. Further evidence is found in the retarded diffusion of automobiles in the UK due to the Great Depression and the resumed diffusion during the upswing of the fourth Kondratiev cycle, as shown in Figure 3.2. Similar retardation behavior was found in 17 examples, which are described in Chapter 2.

Goodwin (1990) ascribed the cause of Kondratiev cycles to a swarm of technological innovations. He explained that a disparate group of technical innovations becomes ordered into a common upsurge by self-organization.

Three trajectories of the innovation paradigm have been well identified. Formation of these trajectories should be a kind of self-organized action because they are sharp and well-defined. The fractal formation of fusion trajectories along the development trajectory, illustrated in Figure 10.4, can also be interpreted as a kind of self-organization phenomenon.

Complexity in Innovation Systems

In short, innovation systems are composed of various fractals and formed by a self-organization mechanism. These characteristics indicate that innovations are a kind of complex system. The self-organization mechanism has been observed on various occasions, and its attractor should be the value added by the innovation. When an innovation is pervasive and widely disseminated in the economy, various fractals are induced by the value added by the innovation and attracted to the trajectories. This is the driving force creating trajectories and expanding the techno-economic paradigm.

The primary products of innovation form the fractals of the development trajectory, for example, electronic devices construct the electronics development trajectory as chain fractals. The devices are introduced by existing industries forming bundle fractals of fusion trajectories. These are the secondary fractals induced by the value added by electronics devices spreading to outer industries. System fractals are formed by the application of electronic devices to create various elementary technological systems, such as FMS, POS and CALS. These systems are introduced by various industries, resulting in the creation of the third fractals. These pervasive effects drive institutional change in the economy. Since the value added by electronics is pervasive, versatile fractals are induced by various industries. The value added by electronic devices works as a self-organizing activity, attracting these fractals towards the electronics trajectories at the same time.

Discrete Knowledge Development Dynamics

Given the above consideration, we can term the behavior of discrete and nonlinear innovation systems 'knowledge development dynamics'. That is, innovation systems are composed of four elements and three trajectories. Innovation develops through relevant actors creating innovation products. Innovation develops through knowledge transfer from one person to another, and its products bring about value-added to the economy. Therefore, four elements – actors, products, knowledge, and value-added – are involved. As the innovation systems are discrete, they are composed of discrete elements. Product elements are actually inventions to produce patents and products. Patents are intangible, but product elements also include intangible products such as software and systems. Innovations develop by various actors along three trajectories. 'Actor' is another discrete element, and each trajectory has its own actors. Knowledge is an essential component of innovation and is transferred among actors and trajectories. An innovation product brings about value-added to the economy, and its pervasiveness and quality is important in determining its impact.

Knowledge development dynamics can be expanded to much wider disciplines beyond innovation dynamics. It covers all human knowledge activities such as knowledge development dynamics of science, knowledge development dynamics of economics, knowledge development dynamics of ideologies, and knowledge development dynamics of arts. These wider activities are discussed in Chapter 11.

10.4 LOGISTIC EQUATION IN DISCRETE SYSTEMS

The logistic equation has a long history, and versatile applications have been developed in the fields of biology, ecology and medicine. After discovery of the deterministic chaos phenomenon by May (1974, 1976) and Li and Yorke (1975), much attention has been paid to logistic mapping, and a new field of complex systems has been developed. This is a phenomenon in discrete systems. This section discusses the relationship between continuous systems expressed by a differential equation and discrete systems expressed by a logistic mapping.

Historical Background of Logistic Equations

This review of the historical background of the logistic equation comes mostly from Hutchinson (1978) and Yamaguchi (1997).

As pointed out above, the logistic equation was first introduced by Verhulst in 1838 to describe population dynamics. A debate had been started by Malthus in 1798 when he warned in his essay, 'The Principle of Population', that a catastrophe was inevitable because while the population tends to increase geometrically, its means of subsistence increases only arithmetically. Sadler recognized in 1830 that the population growth rate decreases as the population density increases. Quetelet and Fourier argued that the decreasing rate of growth is proportional to the square of the growth rate. Verhulst derived the logistic equation as described below for the population dynamics based on the analysis of the growth of organisms in the few years following 1838.

$$dN/dt = rN\,(K - N)/K \tag{10.42}$$

where N is the number of individuals, K is the upper asymptote, and r is a constant.

Solving this for N,

$$N(t) = \frac{CKe^{rt}}{Ce^{rt} + 1} \tag{10.43}$$

Initial value N_0 at $t = 0$ is

$$N_0 = CK/(C + 1) \tag{10.44}$$

The logistic equation has since been used for studies on biological proliferation. Robertson used a logistic equation in 1909 to explain the self-reproduction of vital organs. In 1911, Mckendrik and Pai described the proliferation of *colon bacillus*, and in 1913 Carlson used a logistic equation to analyze the reproduction of yeast. The logistic equation has been used in the field of pathology as an 'epidemic model' to describe the spread of infectious diseases.

Biological proliferation is a discontinuous subject in a discrete system, and there is no basis on which a logistic relation holds. For example, insect reproduction cannot be treated as a continuous function because insects have a breeding season. Yamaguchi (1997) introduced such examples. Professor Uchida of Kyoto University studied the breeding of an insect, *mamezomushi*. This insect dies about 20 days after laying eggs. The eggs hatch and become adult for about 25 days. This means that each generation is isolated from the previous one – they never overlap and form groups having members in different phases of the life cycle. A mapping of logistic equation (10.42) is given by

$$N_{n+1} = \left(\frac{1}{b + cN_n}\right) N_n \tag{10.45}$$

This is a monotone increasing function and cannot explain the above correlation.

An alternative difference equation was thus introduced,

$$N_{n+1} = \left(\frac{1}{b_0 + cN_n} - \sigma\right) N_n \tag{10.46}$$

that well expresses the actual behavior of the *mamezomushi*, in which the breeding has a periodic oscillation of a swarm converging to a specific number of population.

For continuous breeding of insects, however, a logistic equation can be applied even for a discrete system under a condition in which a mathematical convention of continuity holds. Lotka (1924) discussed such mathematical treatment and published *Elements of Mathematical Biology*. During the same period, Volterra (1926, 1927) studied the struggle for existence and introduced a mathematical expression for it. This is the background of the Lotka–Volterra equation.

Logistic Mapping and Relation to the Actual Diffusion

The mapping of the logistic equation has been discussed in various ways. May's mapping in particular has received much attention because he was the first to identify deterministic chaos. May (1974, 1976) precisely analyzed logistic mapping:

$$N_{n+1} = \beta N_n (1 - N_n) \tag{10.47}$$

In this equation, β can take any number between 0 and 4. If it is less than 1, the value tends to 0 by increasing n, and when β is between 1 and 2, N tends to $1 - 1/\beta$. When $2 < \beta < 3$, it tends to $1 - 1/\beta$ with repeated up and down oscillation. When β exceeds 3, it has multiple solutions with a periodicity of 2^n. When β is larger than 3 and less than $1 + \sqrt{6}$, it has oscillation with a periodicity of 2. Periodicity increases further together with the value of β. It is noteworthy that when β exceeds 3.57, the system enters a chaos condition. This is an extraordinary phenomenon that has never been seen before in a deterministic system.

The mapping (10.47) by May can be derived from the logistic equation

$$dF/dt = a\,F(1 - F) \tag{10.48}$$

When the left side of the equation is replaced by the forward difference and the parameter β,

$$\beta = 1 + \delta\,a \tag{10.49}$$

May's difference equation is obtained by mapping. When F tends to 1, this difference equation reaches $1 - 1/\beta$.

Comparison of (10.47) and (10.48) shows that the two equations are interrelated by (10.49). While the actual diffusion is carried out in a discrete system of knowledge transfer from person to person, the system has actually been analyzed using a logistic differential equation. This is based on the assumption that the mathematical convention of continuity assuming a quasi-continuous system holds. The diffusion coefficients derived from the slope of the logistic curve using a Fisher–Pry plot are shown in Table 2.1 of Chapter 2. The smaller the δ, the closer the logistic mapping comes to the original differential equation. If δ is small enough, $1 > \beta > 2$ (because δa is small enough and less than 1), and in this region, N_n tends to $1 - 1/\beta$. This is a regular region in May's mapping before the ergodic and chaotic regions.

An important issue is the emergence of fractals in the course of an innovation paradigm. The logistic mapping has ergodic and chaotic phases, and bifurcation takes place for multiphase defined by the Feigenbaum number (Feigenbaum, 1978). This indicates a kind of fractal, as shown by Lauwerier (1987), Lorenz (1989) and Peters (1991). They have, however, not identified actual fractals and no empirical evidence has been given to prove the existence of fractals as a practical phenomenon. Furthermore, we would like to point out that May's mapping is not suitable to describe the actual innovation process because the mapping does not hold monotone increasing property but is randomly changeable upon the value of a.

Acceptable Logistic Mapping and its Unique Feature

A series of difference equations holding a monotone increasing property can be derived from the logistic differential equation. This is achieved by deriving a bilinear equation, which has been developed in the theory of soliton, and by holding a gauge invariance. The gauge invariance can endow a strict solution fitting the solution of the original differential equation by its perturbation. Actual derivation was demonstrated by Hirota (2000).

In order to make it bilinear, the dependent variable N in a logistic equation (10.50) is replaced by new variables f and g as $N = g/f$, and thus a bilinear equation (10.51) is formulated as follows:

$$\frac{dN}{dt} = aN\left(1 - \frac{N}{K}\right) \tag{10.50}$$

$$\frac{dg}{dt}f - g\frac{df}{dt} = ag\left(f - \frac{g}{K}\right) \tag{10.51}$$

Then, the left side is replaced by forward difference and three kinds of bilinear difference equations holding a gauge invariance are obtained:

$$\text{I. } g(t+\delta)f(t) - g(t)f(t+\delta) = \delta a g(t+\delta)\left[f(t) - \frac{g(t)}{K}\right] \tag{10.52}$$

$$\text{II. } g(t+\delta)f(t) - g(t)f(t+\delta) = \delta a g(t)\left[f(t+\delta) - \frac{g(t+\delta)}{K}\right] \tag{10.53}$$

$$\text{III. } g(t+\delta)f(t) - g(t)f(t+\delta) =$$

$$\tfrac{1}{2}\delta a[g(t+\delta)f(t) + g(t)f(t+\delta)] - \frac{\delta a}{K}g(t)g(t+\delta) \tag{10.54}$$

The left side of each equation can be separated into two linear difference equations which are easily solved.

Thus, bilinear functions I, II, and III are transferred, by the dependent variables $N(t) = g(t)/f(t)$, into the following nonlinear difference equations

$$\text{I. } N(t+\delta) - N(t) = \delta a N(t+\delta)\left[1 - \frac{N(t)}{K}\right] \tag{10.55}$$

$$\text{II. } N(t+\delta) - N(t) = \delta a N(t)\left[1 - \frac{N(t+\delta)}{K}\right] \tag{10.56}$$

$$\text{III. } N(t+\delta) - N(t) = \delta a\left\{\tfrac{1}{2}[N(t) + N(t+\delta)] - \frac{N(t)N(t+\delta)}{K}\right\} \tag{10.57}$$

These difference equations are rewritten into recurrence formulae by putting $t = n\delta$,

$$\text{I. } N_{n+1} = \frac{N_n}{1 - \delta a\left(1 - \dfrac{N_n}{K}\right)} \tag{10.58}$$

$$\text{II. } N_{n+1} = \frac{(1+\delta a)N_n}{1 + \delta a\dfrac{N_n}{K}} \tag{10.59}$$

$$\text{III. } N_{n+1} = \frac{(1+\frac{1}{2}\delta a)N_n}{1 - \frac{1}{2}\delta a + \delta a\dfrac{N_n}{K}} \tag{10.60}$$

These difference equations derived from the logistic equation hold the monotone increasing property.

These solutions are given by

$$\text{I. } N(t) = \frac{(1-\delta a)^{-(t-t_0)/\delta}}{c_0 + \dfrac{(1-\delta a)^{-(t-t_0)/\delta}}{K}} \tag{10.61}$$

$$\text{II. } N(t) = \frac{(1+\delta a)^{(t-t_0)/\delta}}{c_0 + \dfrac{(1+\delta a)^{(t-t_0)/\delta}}{K}} \tag{10.62}$$

$$\text{III. } N(t) = \frac{\left[\dfrac{1+\frac{1}{2}\delta a}{1-\frac{1}{2}\delta a}\right]^{(t-t_0)/\delta}}{c_0 + \dfrac{1}{K}\left[\dfrac{1+\frac{1}{2}\delta a}{1-\frac{1}{2}\delta a}\right]^{(t-t_0)/\delta}} \tag{10.63}$$

When the solutions are rewritten by

$$\text{I. } \frac{1}{1-\delta a} = \exp(\delta\alpha) \tag{10.64}$$

$$\text{II. } 1+\delta a = \exp(\delta\alpha) \tag{10.65}$$

$$\text{III. } \frac{1+\frac{1}{2}\delta a}{1-\frac{1}{2}\delta a} = \exp(\delta\alpha), \tag{10.66}$$

all of the solutions are interestingly expressed by a single equation:

$$N = \frac{g}{f} = \frac{\exp\alpha(t-t_0)}{c_0 + \frac{\exp\alpha(t-t_0)}{K}} \tag{10.67-1}$$

or

$$N = \frac{g}{f} = \frac{K}{1+c_0K\exp(-\alpha t)} \tag{10.67-2}$$

which is the same formula as the solution of the original differential equation (10.50). That is, all solutions of these difference equations are equivalent to the solution (10.68) of the original differential equation and hold monotone increasing property.

$$N(t) = \frac{K}{1+c_0K\exp(-at)} \tag{10.68}$$

It is noteworthy that the solution of difference equations always satisfies the solution of the original differential equation independent of the time difference of δ, in spite of α still being a function of δ, that is, in the case of I (10.64), $\alpha = (1/\delta)\ln[1/(1-\delta a)]$. When δ is getting smaller, α tends to a. Furthermore, even if δ is large enough, the solution (10.67) always satisfies the solution of the original differential equation (10.68).

These results strongly suggest that every fractal, either chain and bundle, as the elements of the logistic S-curve, mathematically fulfills the solution of the original differential equation describing the locus of trajectory. This is noteworthy because the difference equation derived from a differential equation is not always the same as the original – there is some gap. That is, this unique feature of the logistic equation suggests that a fractal of any length can be mathematically included as an element of the parent logistic curve.

Yamaguchi (1997) also introduced a similar interesting phenomenon of a logistic equation. That is, a difference equation derived by mapping the solution of a logistic equation gives exactly the same value as that of the original differential equation irrespective of the time span of the difference.

The logistic equation and its solution are expressed by (10.69) and (10.70), respectively.

$$\frac{dN}{dt} = r\frac{(K-N)}{K}N \tag{10.69}$$

$$N(t) = \frac{KN_0e^{rt}}{N_0e^{rt} + K - N_0} \tag{10.70}$$

When N at time $n\delta$: $N(n\delta)$ is expressed by N_n, the difference expression of (10.70) can be described by

$$N_{n+1} = \frac{N_n}{e^{-r\delta} + \frac{1}{K}(1 - e^{-r\delta})N_n}. \tag{10.71}$$

That is,

$$N_{n+1} = \frac{1}{b + cN_n}N_n \tag{10.72}$$

where $b = e^{-r\delta}$, $c = \frac{1}{K}(1 - e^{-r\delta})$

It is again interesting to see that the N_{n+1} value obtained in (10.72) always satisfies (10.70) irrespective of the time span, δ, as shown in Figure 10.6.

We have first disclosed fractals and self-organization phenomenon in the practical innovation systems. The above mathematical treatment gives an

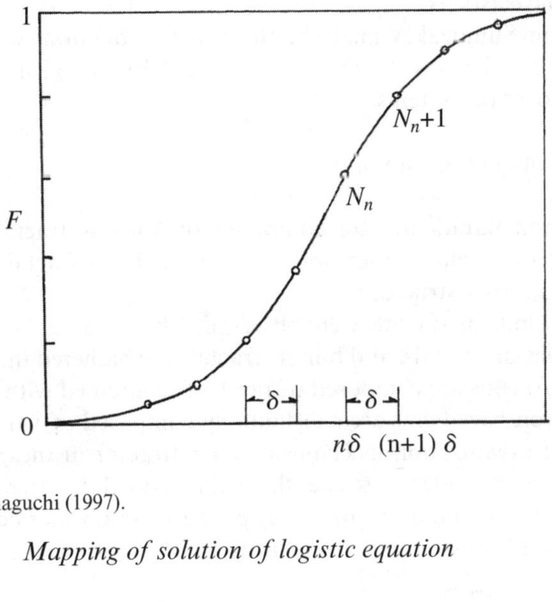

Source: Yamaguchi (1997).

Figure 10.6 Mapping of solution of logistic equation

important background for the emergence of such fractals which can be mathematically involved into the trajectory. That is, the fact that the data of the logistic mapping with monotone increasing property strictly coincides with the solution of the original differential equation, as described above, indicates that any fractal unit within the parent trajectory can be properly included as an element of the trajectory.

10.5 SUMMARY

This chapter dealt with fundamental aspects of innovation. Various sigmoid curves were analyzed, and their suitability for expressing the trend in innovations was examined. As the innovation process is a discrete system formed by knowledge transfer from person to person, the logistic mapping was discussed and the complexity in innovation systems was considered.

I. Mathematical treatments of sigmoid curves and logistic equation

1. Various sigmoid curves for describing innovations were analyzed, and the contact-type diffusion models represented by the logistic and Gompertz models were found to be the most suitable.
2. The Gompertz model by Metcalfe was found to be suitable for describing the supply side as well as the demand side with flexible explicating economic elasticity.
3. Based on Shinozaki's analysis, the logistic equation was ultimately identified as the most suitable growth model for biological proliferation and innovation systems.

II. Complexity in innovation systems

4. Innovation paradigms are composed of various fractals and constructed by a self-organization mechanism. Each fractal has its own small trajectory structure.
5. Several kinds of fractals were observed. Chain fractals are composed of a series of fractals, and bundle fractals are gathered in a trajectory. System fractals are composed of products combined with software.
6. Innovation paradigms are substantially composed of various fractals by a self-organization mechanism. The attractor in these innovation systems is considered to be the value added by the innovation. Pervasive trunk innovations deeply penetrate into wider fields forming fractals and contribute to the institutional change of society.

7. Innovation is a discrete system in which knowledge is transferred from person to person to produce discrete products. In that sense, it is termed 'knowledge development dynamics'.
8. Discrete innovation should thus be described using a difference equation instead of a differential equation. May's logistic mapping was indicated to be a complex system having ergodic and chaotic regions, suggesting the existence of fractals. May's mapping, however, is not the actual case of our innovation systems because it has no monotone increasing property.
9. Instead of May's mapping, a logistic difference equation having monotone increasing property has been derived by bilinear approach in the condition of gauge invariance and found a unique property that any solution of the difference equation fulfills the solution of the original differential equation irrespective of the width of the difference. This unique property seems to be the mathematical origin of the inclusion of any fractal within the trajectory.

11. Technology foresight and long-term perspectives

Part II investigated the innovation paradigm with a nonlinear framework, and discussed the characteristics of the three trajectories forming the paradigm. The relationships between the trajectories can even be used to estimate the future trend of forthcoming industries by extrapolation, once a science or technology trajectory can be identified. It is especially important to estimate when new high-tech industries will emerge. This chapter explores the estimation of next-generation industries on the basis of science and technology in progress.

Although global issues such as energy and environmental problems are crucial, sufficient methods have not been established for estimating future trends in these areas. This chapter concentrates on means for forecasting such issues. Every five years since 1972, the Institute of Science and Technology Policy of the Science and Technology Agency (now the Ministry of Education and Science) in Japan has carried out technology foresight by using the Delphi process. Forecasting technology using logistic analysis is examined, and the results are compared with those of the Delphi process.

Finally, innovation analysis with logistic dynamics is expanded to a general analysis of discrete knowledge transfer phenomenon in human society and to a clarification of the cycles of human activities. The existence of a mega-structure of society is discussed.

11.1 IDENTIFICATION OF SCIENCE PARADIGMS

As discussed in previous chapters, the technology trajectory is often a science trajectory in academia, and a series of scientific developments can be described by a logistic trajectory. Thus, the development of scientific disciplines is examined based on the process of logistic dynamics. It is especially interesting to see how scientific disciplines evolve from one to another.

Establishment of Modern Sciences

A series of core technologies developed in an innovation paradigm form a technology trajectory, and the spread and timing of this trajectory can be determined based on its nonlinear nature. In many innovation paradigms, the technology trajectory is composed of a series of scientific discoveries as shown in those for synthetic dyestuffs and biotechnology (Chapter 7). Furthermore, we have observed that every science discipline is composed of a series of core scientific discoveries, forming a well-defined cluster within a definite time span, the same as for technological innovations. This means that science paradigms are also nonlinear and that the procedure for determining a technology trajectory can be applied to the science disciplines to determine a science trajectory.

Thus, the development of modern sciences is examined by identifying science trajectories. The dawn of the modern scientific approach came during the Renaissance in the fifteenth century. It arose from Copernicus' work on his heliocentric system. The publication of his theory in 1543 led to a series of scientific observations and discoveries regarding celestial movement. In the 1570s Tycho Brahe collected extensive data on celestial movement. His accumulation of knowledge was used by Kepler in his finding that planetary orbits are elliptic and to formulate his three laws. Galileo Galilei made the first astronomical observations using a telescope in the early seventeenth century, and his findings supported Copernicus' heliocentric theory. Newton summarized the laws of universal gravitation in his *Principia Mathematica* in 1687. These events form a trajectory of Newtonian dynamics with a time span of around 150 years.

When the eighteenth century dawned, the main scientific disciplines had been established in the fields of chemistry, biology and physics. The categorization of substances based on the 'the theory of four elements', which originated in Greek philosophy, was put to an end by Lavoisier's interpretation of the phenomenon of burning as a chemical reaction in 1777. Lavoisier established the modern science of chemistry in his *Theory of Elements and Chemistry* published in 1789. The atomic theory of chemistry by Dalton, the discovery of the Avogadro number, and the protocol of the congress at Kahlsruhe finally established the modern discipline of chemistry. The periodic table of elements, developed by Mendeleev in 1869, was a comprehensive summary of the chemistry trajectory. Thus, the trajectory for fundamental chemistry has a time span of 90 years.

Electromagnetics originated with Volta's invention of the electric pile in 1800, by which electricity first became available, and built upon a series of discoveries by Oersted, Faraday, Ohm, Ampere, Kirchhoff and Maxwell. Maxwell established the electromagnetic discipline with his four fundamental equations in 1864, and Hertz verified the existence of electromagnetic waves in 1888. Thus, the trajectory for electromagnetic physics has a time span of 90 years.

Embryology in biology emerged with the postulation of the theory of organisms in 1759, and findings in the areas of insemination, cell differentiation and embryonic breeding led to the establishment of the discipline of embryology in 1861. Thus, the trajectory for embryology has a time span of 100 years. Bacteriology developed in the latter half of the nineteenth century and its trajectory has a time span of 70 years. These trajectories are illustrated in Figure 11.1. It is interesting to see that these three main disciplines of natural science (physics, chemistry and biology) were established at the same time as the Industrial Revolution.

Evolution of Science Paradigms

It is interesting to note that science and technology paradigms develop intermittently. This evolutionary behavior is similar to that in biology. Various innovations have been observed to evolve intermittently, as described for information technologies in Chapter 9 (Figure 9.18) and the chemical industry in chapters 6 and 3 (Table 3.1). Science paradigms have also been observed to evolve in the same way.

Evolution of theoretical physics

Although Newtonian dynamics was established in the seventeenth century, something was observed in the late nineteenth century for which it could not account. In 1881, the speed of light was observed always to be constant, irrespective of the relative speed of the observer. Einstein explained this with his theory of relativity in 1905 and 1915. He postulated that gravitational force is explained by a time–space geometry. A unified theory of electromagnetic force and gravity was established in 1929. Thus, the time span of relativity theory trajectory is estimated to be 50 years, as shown in Figure 11.2a.

While Newtonian dynamics belongs to Euclidian physics, relativity theory is non-Euclidian. Since Bohr established the concept of atoms composing electrons and nuclei in 1913, the ultimate structures of elementary particles have been continuously elucidated, starting in the 1930s from Yukawa's meson theory and Tomonaga's renormalization theory through quark theory in 1964 and the establishment of elementary particle theory. The time span of elementary particle theory is estimated to be 50 years. After the postulation of QCD (quantum color dynamics) and the unification of the interplay between the weak force and electromagnetic force in 1967, theoretical physics entered into a new paradigm, one step closer to an ultimate theory of the universe.

To explain the scattering amplitude of hadrons, Nambu (1970) and Goto (1971) proposed string theory, and Ramon and Schwarz et al. postulated a super-string theory inclusive of Fermi and Bose particles in

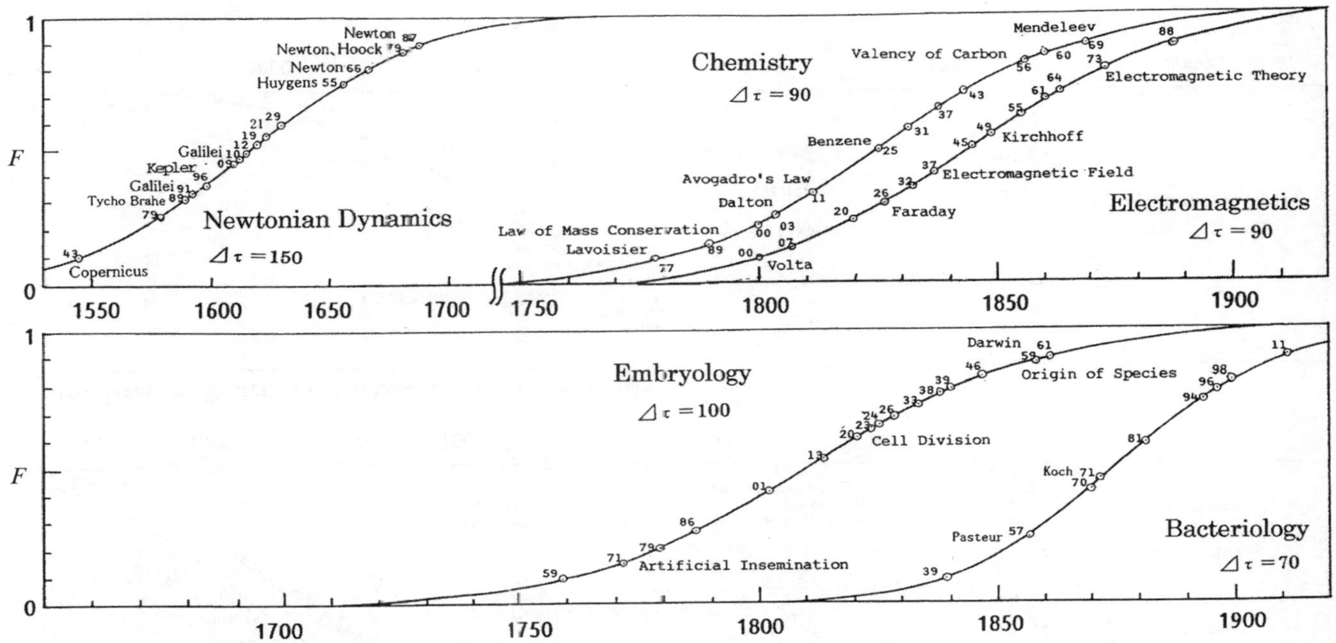

Source: Hirooka (2001b).

Figure 11.1 *Development of modern sciences*

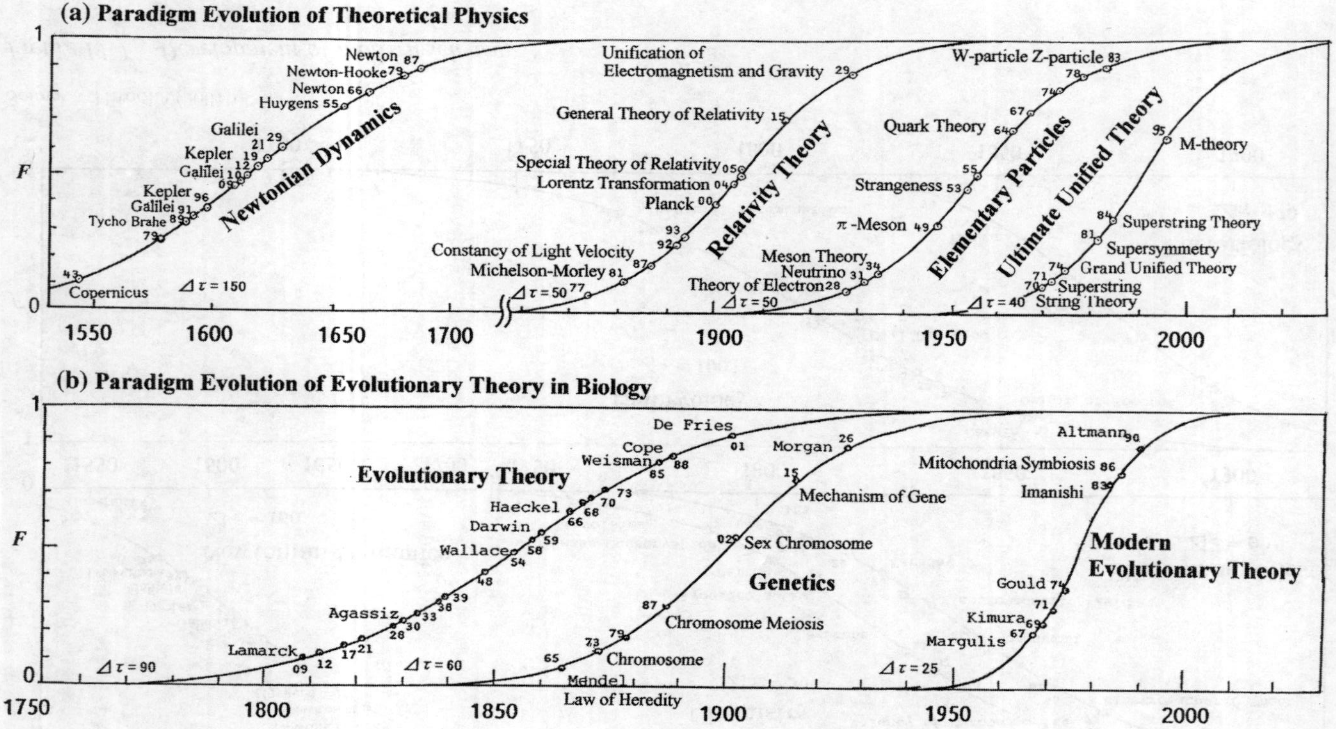

Source: Hirooka (2001b).

Figure 11.2 *Evolution of science paradigms*

1971. The grand unified theory (GUT) was completed in 1974, giving a comprehensive figure across QCD and the unification of weak force with electromagnetism. Introduction of the super symmetry concept in 1984 enabled the use of five super-string theories having ten dimensions to solve the gravity anomaly by incorporating quantum mechanics. A theory with 11 dimensions, the 'M theory', was estimated in 1995 to be the ultimate theory for describing all aspects of the universe. While the M theory systematically has to fulfill the requirements of the five super-string theories, it has not actually been identified yet. It may take a decade or more to get the final equation of the M model, and the trajectory could have a time span of at least 40 years, which means by around 2010. These evolutionary transitions of theoretical physics are illustrated in Figure 11.2a.

Evolution of biological evolutionary theories

The theory of evolution started with the use and disuse theory of Lamarck in 1809 and continued with the field of paleobiology by Agassiz in 1830, the concept of natural selection by Wallace in 1854, the origin of species theory by Darwin in 1859 and the theory of mutation by de Fries in 1901, a time span of about 90 years. This is the story of so-called Darwinism and neo-Darwinism by inclusion of the concept of mutation. After that, genetics developed to form another trajectory based on the law of heredity by Mendel in 1865. The trajectory was followed by the discoveries of chromosomes, the reduction division of chromosomes, sex chromosomes and gene theory by Morgan in 1926. These discoveries form the genetics trajectory, which has a time span of 60 years. On the basis of genetics and molecular biology, modern evolutionary theories have developed, starting with continuous symbiotic theory by Margulis in 1967 and followed by the development of neutral evolutionary theory by Kimura, virus evolutionary theory, punctuated equilibria theory by Gould and Eldredge, Imanishi's evolutionary theory, the mitochondria symbiosis concept in 1986, and symbiosis theory by Altman in 1990, a time span of 25 years. These three trajectories of evolutionary theories compose an evolutionary chain (Figure 11.2b).

Evolution of Economic Disciplines

The paradigms of natural science are easily identified, and their time spans have shrunk from about 150 years during the seventeenth century to about 25–30 years nowadays. Furthermore, they intermittently evolve towards other more advanced paradigms. In this section, the trend in economic disciplines is examined as one of the social sciences. The economic disciplines can be clearly identified, as shown in Figure 11.3. While neoclassical economics and Keynesian economics considered the system to be closed and equilibrium to be important, more recent theories of economics deal with the behavior of the economy as dynamic, as in the institutional economics initiated by Veblen, Schumpeterian economics and evolutionary economics. Here, typical economic theories since classical economics are examined to identify the paradigms. (On the classification of economics see Gill, 1967; and Kobayashi and Sugihara, 1989.)

The development of economics as an academic discipline began at the same time as that of the natural sciences. During the Renaissance (in the fifteenth to seventeenth centuries), merchants gained power and mercantilism emerged, although it did not become a systematic discipline. This timing corresponds to the age of Newtonian dynamics. Another economic precursor was physiocracy in the eighteenth century. The 'tableau économique', as summarized by Quesney in 1758, led to the concept of 'laissez-faire' by Adam Smith. Classical economics originated with Adam Smith's *The Wealth of Nations* in 1776. This timing corresponds to the Industrial Revolution, during which the spinning jenny was invented by Hargreaves in 1765 and the steam engine was patented by Watt in 1769. It also closely corresponds to the emergence of natural sciences. Lavoisier first clarified the actual nature of chemical reactions by an oxidation reaction in 1777, and Ingenhousz elucidated the photosynthesis mechanism by carbonic acid assimilation in 1779. The electric pile was invented by Volta in 1800. This simultaneity does not seem to be solely coincidental. This suggests that a civilization revolution occurred, as well as an Industrial Revolution.

Adam Smith's 'invisible hand' came from Quesney's 'tableau économique', and labor-value theory came from James Steuart. Thus, the trajectory of classical economics runs from the ideas of Quesney in 1756 to those of J.S. Mill in 1848 and includes those of Smith, Malthus, Say and Ricardo, with a time span of 90 years. Along this stream, Darwin's theory of evolution was triggered by Malthus's *Essay on the Principle of Population*. Mill used the logic of natural science to examine economics with a historical methodology in his *A System of Logic: Ratiocinative and Inductive*. Ricardo discussed the interrelations among landowners, capitalists and laborers, and Mill considered the basic issues in the areas of labor, distribution theory and socialism. Owen built a 'new harmony village' in 1826 based on his idea of utopian socialism. Classical economics was based on the labor-value theory. Marx studied classical economics intently, especially the contributions of Ricardo, and developed his own theory of labor value, Marxian value theory. Marxian economics was completed with the theory of surplus value in 1910, so that the time span of Marxian economics is estimated to be around 90 years.

Gossen postulated a new theory on value in 1854, which is the origin of neoclassical economics, but it was not acknowledged. 'Marginal utility theory' was simultaneously proposed by Menger, Walras and Jevons in the 1870s and then by Marshall in 1890. The general equilibrium theory was summarized by Walras. There were three main schools of neoclassical economics: the Cambridge school, the Lausanne school, and the Austrian school. Hicks, Samuelson and Solow further contributed to building neoclassical economics into the 1950s. Thus, the trajectory of neoclassical economics had a time span of about 90 years.

Keynes learned economics from Marshall, and in his *The General Theory of Employment, Interest and Money*, published in 1936, he departed from the previous theories of economics since mercantilism and repudiated Say's law. Well before Keynes's general theory, however, Wicksell contributed *Interest and Prices* in 1898, and it affected Keynes's theory. Keynesian economics developed from short-term macroeconomics to long-term theories of economic development by Harrod and Domar, and to business cycle theory by Kalecki and Kaldor. Thus, the trajectory of Keynesian economics also had a time span of about 90 years.

The neoclassical approach using a general equilibrium theory has been criticized as unrealistic because the actual economy must be a dynamic system. Recent theories of economics have taken this view, and three schools have been classified by Gill (1967): institutional economics, Schumpeterian economics and imperfect and monopolistic competition theory.

Veblen criticized existing economics and from his first publication in 1899 developed an institutional economics with a kind of evolutionary theory, introducing the concept of Darwinian thought. Commons, Mitchell and Myrdal inherited his ideas and expanded them into the sociological region. Polanyi introduced an analytic process for a realistic social system on the basis of institutional analysis in 1944. After World War II, Myrdal, Galbraith and Buchanan developed a new theory of institutional economics, focusing on the collaboration between government and the main economic groups. These institutional movements form a trajectory having a time span of about 90 years.

Schumpeter was a pupil of Böhm-Bawerk and his work developed out of a deep understanding of existing economics. He proposed a *Theory of Economic Development* in 1912 and then developed the economics of technological change accompanying the creative destruction of the economy. His ideas were inherited by neo-Schumpeterians, and various empirical findings accumulated that were used to build various disciplines that formed a trajectory with a time span of about 90 years up until now. Starting with the ideas of Alchian, Nelson and Winter an evolutionary theory of economics developed based on a background of Schumpeterian

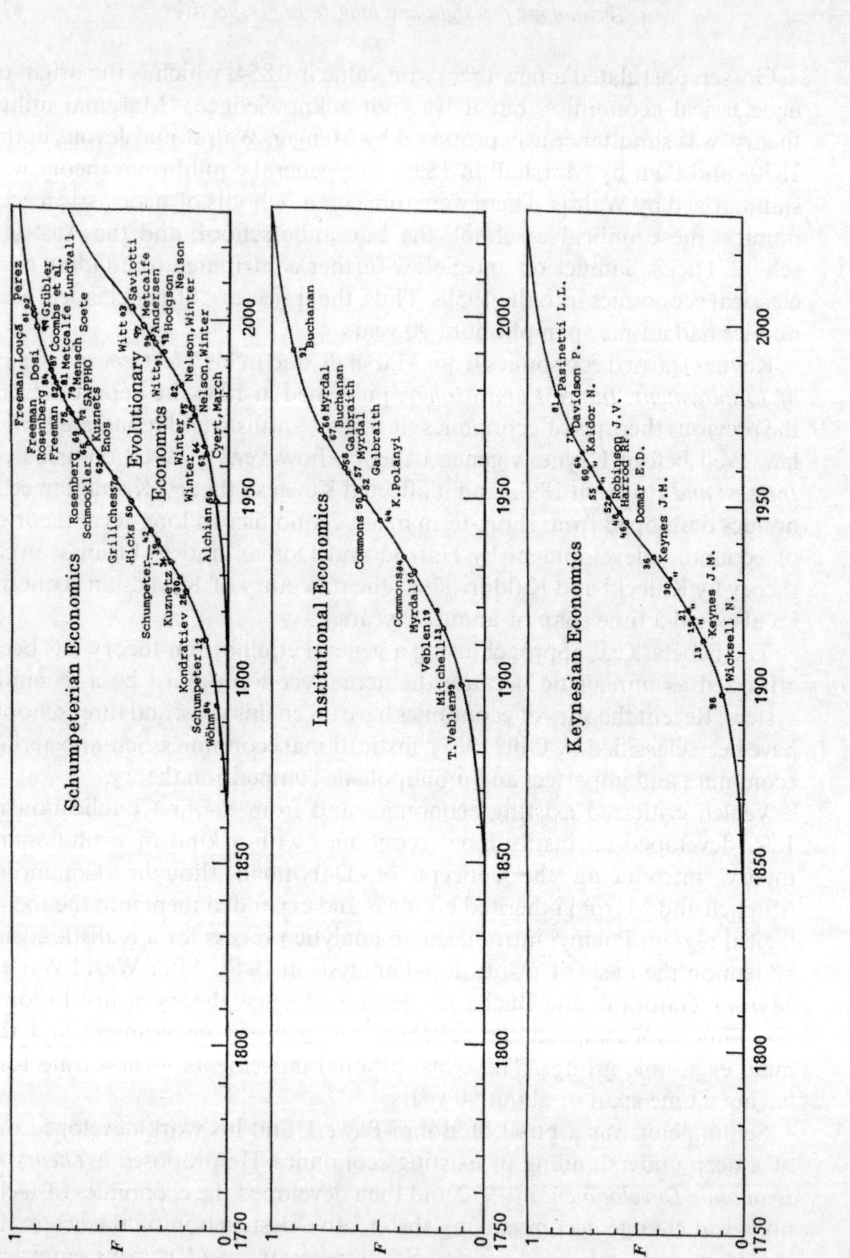

Schumpeterian Economics
Evolutionary Economics
Institutional Economics
Keynesian Economics
F
1
0
1750
1800
1850
1900
1950
2000
Wöhm
Schumpeter
Kondratiev
Kuznetz
Schumpeter
Hicks
Griliches
Schmookler
Rosenberg
Enos
Kuznets
SAPPHO
Freeman
Rosenberg
Freeman
Mensch
Metcalfe
Coombs et al.
Dosi
Freeman, Louçã
Grübler
Perez
Soete
Lundvall
Alchian
Cyert, March
Winter
Nelson, Winter
Witt
Andersen
Hodgson
Metcalfe
Saviotti
Nelson
T. Veblen
Mitchell
Veblen
Myrdal
Commons
K. Polanyi
Galbraith
Myrdal
Buchanan
Wicksell N.
Keynes J.M.
Domar E.D.
Harrod RF.
Robinson J.V.
Kaldor N.
Davidson P.
Pasinetti L.L.

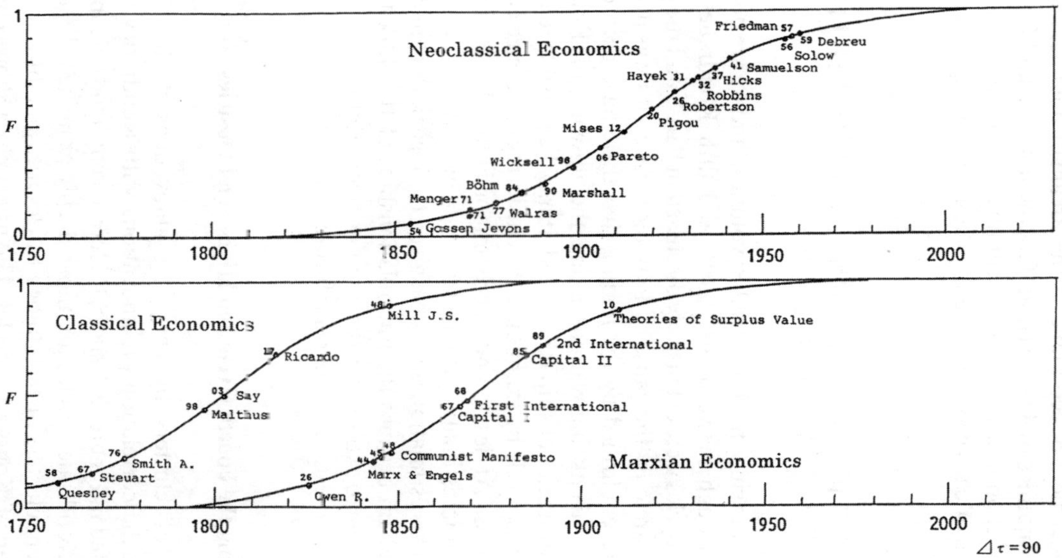

Source: Hirooka (2001b).

Figure 11.3 *Evolution of economics paradigms*

thought. These studies could form a trajectory independent of Schumpeterian economics.

A novel branch of economics was developed by Sraffa, Chamberlin, Robinson and von Neumann. A trajectory is not readily apparent in this case. Another movement of economics is the stream of nonlinear economics and chaos economics by Day, Arthur, Goodwin, Peters and Lorenz.

This outline of economic paradigms is summarized in Figure 11.3. Note that the time span of each economic school was about 90 years, even for the most recent ones.

11.2 INNOVATION CLUSTERS AND THE FIFTH KONDRATIEV CYCLE

It has been 230 years since the Industrial Revolution, and we are now at the bottom of the trough between the fourth and fifth Kondratiev cycles. Kondratiev long cycles were extensively discussed in Part I, and the premise that innovations are only the cause of long business cycles was discussed. This premise was confirmed by the finding that there have been concentrated clusters of trunk innovations along the upswing of Kondratiev cycles. If this causal relationship is accepted, it may be possible to estimate the coming business cycle by examining the clustering of science and technology in progress. At least, and more importantly, this examination should be helpful in estimating the trends in the near future of emerging new industries. This estimation is feasible given the logistic dynamics of the innovation paradigm because the interrelationships of the three trajectories make it possible to extrapolate the ongoing science and technology trajectories in order to estimate the near future development and diffusion trajectories.

Innovations during the Fourth Cycle and High-tech Industries

The formation of clusters by emerging innovations is illustrated in Figure 6.4, and the correlation of innovations with Kondratiev business cycles is illustrated in Figures 3.1 and 5.1. Here, recent trends in innovation clustering and Kondratiev cycles are more deeply examined. Figure 11.4 summarizes recent clusters of technology trajectories and diffusion trajectories on the map of Kondratiev cycles. The upper chart shows clusters of technology trajectories, and the lower chart shows clusters of diffusion trajectories. The upswing of the fourth cycle was driven by the clustering of leading innovations, including electrical appliances, petrochemicals, aircraft and computers. Japan, a latecomer to modern society, introduced

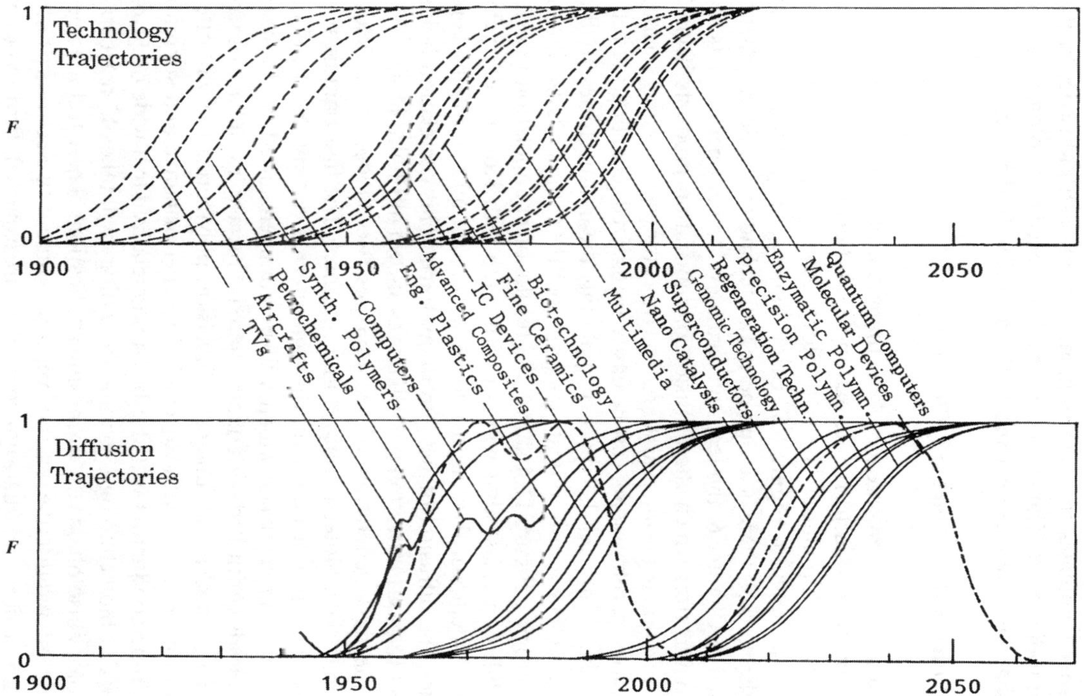

Sources: Hirooka and Nitta (2002), Hirooka (2002a, 2003c).

Figure 11.4 Innovation clusters and the fifth Kondratiev cycle

Western technologies and its economy grew during this upswing, so the production of steel and automobiles is also included in this upswing. The second cluster of innovations includes integrated circuits, engineering plastics, advanced composites, fine ceramics, and biotechnology. This second cluster stimulates the development of high-tech industries in the fourth Kondratiev cycle. The third cluster of advanced technologies can be observed that will drive the upswing of the fifth Kondratiev cycle.

Forthcoming Innovations and the Fifth Cycle

The estimation of the emergence of forthcoming industries depends on how the innovation trajectories are extrapolated. Fortunately, we have an empirical relationship among the technology, development and diffusion trajectories, as shown in Figure 6.7. Before innovation products start to diffuse in the market, there is a long period of technological development. As discussed in Chapter 6, the technological development period lasts for about 30 years, after which the diffusion of innovation products begins, reaching saturation in another 30 years. The trajectories of ongoing science and technology developments can be easily recognized, enabling the estimation of the future market by extrapolating these trajectories to the diffusion phase. The results indicate the timing of the commercialization of these technologies. Figure 11.4 shows the results of extrapolation to estimate the diffusion phase that will constitute the fifth Kondratiev cycle, using historical empirical relationships. After the completion of the core technologies, the diffusion of new products starts. This correlation holds with respect to past innovation paradigms, as shown in Figure 11.4.

As for the next generation of technologies, we choose three fields: advanced materials, biotechnology and electronics. The first paradigm is multimedia which is now taking off through internet systems. Next will be nano-materials and catalysts, which will be commercialized in the near future. Materials controlled with nanometer-size precision will have superior performance and novel functions. Precision polymerization will offer next-generation products with extremely high performance and superior functions. Superconductors will also be commercialized in a wide range of applications. In biotechnology, genome technology will take off; regeneration technology also offers very promising applications. These technologies have already entered into the realm of venture business. In electronics technology, nano-devices and quantum computers are expected. Prototypes of single-electron devices have already been made using carbon nanotubes. Quantum computing has already been demonstrated.

As a whole, these future technologies form a cluster, and this trend indicates that the upswing of the fifth cycle will take place between 2000 and

2040. Hopefully this upswing will be large enough to sustain a prosperous economy because serious problems will be encountered in the not so distant future, such as damage to the environment and shortages of energy resources. These problems are discussed below.

Advanced Technologies in the Twenty-first Century

The economy in the twentieth century developed through a series of trunk innovations, and the frequency of important innovations has been increasing. We can expect even more, and more frequent, innovations in the twenty-first century. The important thing is how to manage these innovations because these determine the international competitiveness of the country.

The United States launched Clinton's 'National Nanotechnology Initiative: Leading to the Next Industrial Revolution' in 2000. Its objective was to strengthen the international competitiveness of the United States and an annual sum of $500 million was committed to priority projects in a wide range of interdisciplinary fields, from nano-materials to biotechnologies, including genomic technology and regeneration technology, and to electronics and information technologies, such as nano-devices. The impact on the economy and society of innovation in these fields should be comparable to that of information technology and of cellular, genetic, and molecular biologies in the twentieth century.

In Japan, the importance of nano-technology was recognized early. In 1979, Professor Namba of Osaka University launched a nano-electronics project supported by a national fund for nanometer technologies, which had the aim of promoting the development of nanometer electronic devices. Following the development of the tunneling microscope in 1983, the Yoshida nano-mechanism five-year project was launched by the Science and Technology Agency in 1985 to promote the development of molecular manipulation for molecular-level electronic devices. The fourth World Congress of the CHEMRAWN (Chemical Research Applied to World Needs) by the policy proposal committee of the International Union of Pure and Applied Chemistry (IUPAC) was held in Tokyo in 1987. The subject of the conference was 'Advanced Materials for Innovations – Energy, Transportation and Communications', and the future action committee prepared a protocol of the conference entitled 'Perspectives and Recommendations' in which they stressed the importance of structural control of materials at the atomic and molecular level in coming innovations. This recommendation was farseeing and their perspective has been revalued by the recent trend of nanotechnologies.

Forecasting Next-generation Technologies

A brief summary of some typical next-generation technologies is presented in this section, and the timing of innovations is examined. Forecasting paradigms can be formulated as follows: once the technology trajectory is almost completed, the time span can be estimated. This is usually about 25 years, based on those for electronics and biotechnology. When the technology trajectory reaches the 90 per cent level, $F = 0.9$, the development trajectory begins, $F = 0.1$. The diffusion trajectory begins just after the technology trajectory is almost finished which is about ten years later than the development trajectory. This is a standard scheme for the forecasting paradigm. If everything goes smoothly, the timing of the trajectories can be predicted accurately. However, commercialization sometimes begins later than expected, depending on comparisons with competing technologies, especially in terms of cost/performance.

Polymer science towards precision polymerization

The polymer industry is now mature, and a variety of polymers have been commercialized. The structure of polymer molecules, however, is not perfectly ordered, and they have many dislocations. If we could completely regulate the polymer's structure, its performance would be greatly improved, and new functions could be discovered. Using a gel spinning process, Smith and Lemstra succeeded in 1980 in arranging a high-density polyethylene with a linearly stretched structure that was well crystallized. This extended-chain-structure polyethylene fiber without defects performed remarkably well – the specific strength was 35 g/denier, which is 200 times that of conventional polyethylene and higher than the 4 g/denier of steel piano wire. While plastics are usually insulators, an electro-conductive polyacetylene developed by Naarmann in 1987 has the same level of electro-conductivity as copper, 10^5 S/cm, which was achieved by regulating the conjugation of double bond in the polymer. This is an example of revealing a new function by regulating the structure of polymers. These examples indicate that the perfect regulation of polymer structures is a very promising subject for future innovations.

Nano-catalysts

One of the promising nano-materials is a high-performance catalyst with its structure controlled at the nanometer level. For example, a zeolite-type mesoporous catalyst having a regular pore size of 2–50 nm was recently developed. 'Meso', as defined by IUPAC, means a pore size of 2–50 nm; 2–10 nm is considered nano-size. Zeolite catalysts were first developed in the 1950s, and the pore size has been artificially controllable since the

ZSM-5 catalyst, which is formed by intercalation with surface active agents like quaternary ammonium salts, was developed by the Mobil Oil Company in 1971. This process was developed to achieve a regular pore size of 2–50 nm, especially within nano-size. Pores with a regular size can be used as meso-size reactors with specific catalytic activities. These pores exhibit specific quantum size effects, and various physical properties are different from those in the normal condition. This results in enhanced catalytic activity and stereospecificity. Aida et al. synthesized extended chain structure polyethylene fibers in mesoporous silica including titanocene polymerization catalyst. This was the first direct production of extended chain structure polymer which was the result of a special performance of the mesoporous structure. The paradigm for nanoporous catalysts is illustrated in Figure 11.5. The paradigm is now in the key timing of the development trajectory, and various new processes using them are expected.

Nano-electronics

Nano-electronics has been studied for more than 20 years, and various trials have been carried out on post-silicon devices.

Molecular devices have been under development since the tunneling microscope was invented in 1982, making molecular manipulation feasible. IBM succeeded in writing atomic letters on a solid surface in 1990, and the technology trajectory is now almost in the final stage. When will the development trajectory become visible? It could be around 2010. Single-electron devices have developed since Likharev first proposed them in 1986. The success of the Coulomb blockade in 1988 made it possible to encapsulate an electron. A single-electron device operable at room temperature was developed in 1993, and single-electron logic gates and single-electron flash memories were developed in the 1990s. Room-temperature single-electron transistors were realized within individual metallic single-wall carbon nanotube molecules in 2001. These inventions form the technology trajectory, which has a time span of 25 years. The development trajectory should start around 2010, and commercialization could begin around 2020, if no fatal problems appear.

The development of quantum computers, on the other hand, has proceeded following a completely different algorithm. The basic concept of the quantum Turing machine was proposed by Deutsch of the University of Oxford in 1985. This is the starting point of the technology trajectory. Shor at AT&T demonstrated that high-speed factorization can be easily performed by quantum computers. This came as a big shock to people in charge of conventional computers because it meant that coded messages could be easily broken. Shor's discovery stimulated the study of quantum computing. Chuang at IBM demonstrated basic calculation using quantum computers in 1997, and a complex mathematical problem was solved using quantum computers at IBM's Almaden Research Center in 2000. In 1999, Takeuchi of Mitsubishi Electric Co. performed basic calculations using photons, and Fukushima of the Osaka Industrial Research Institute did the same using atomic nuclear spins. Nakamura of NEC developed solid-state devices in 1999 and collaborated with the Riken Institute in Japan on the development of basic and logic circuits using superconductors in 2003. If the paradigm proceeds without any trouble and follows a trend similar to that of recent paradigms, the technology trajectory will become mature and the development trajectory will start around 2010, and products will be commercialized around 2020, which means a timing of $F=0.1$, as shown in Figure 11.6.

Development of genome technologies

The biotechnology paradigm, which started with Avery's discovery that DNA is the origin of genes, has proceeded and is now on the diffusion trajectory. This paradigm has evolved step by step, and now we have two important evolved paradigms: post-genome and regeneration technology.

The full decoding of the human genome was completed in 2001, and now the genome paradigm is on the development trajectory. The technology trajectory of genomics started in 1975 with Sanger's sequencing method, which became the standard decoding process. Wada of the Riken Institute in Japan proposed a DNA sequencing project in 1981 and contributed a paper to *Nature* on a high-throughput automatic sequencing method in 1987. Vender of Celera Genomics Corp. developed a 'shotgun sequencing method' in 1995. The international project to sequence the human genome, which started in the United States in 1990, was accelerated by this development. In 2000, Vender announced that three billion genomes had been analyzed, and in 2001, the international human genome project ended with the completion of the analysis. These steps in genome sequencing analysis form the technology trajectory. We are now in the development trajectory of the genome technology paradigm. This means that the venture opportunities for post-genome businesses will be concentrated in the years 2000–2015.

Regeneration technology paradigm

Embryonic stem cells, ES cells, were discovered by Evans and Kaufmann at Cambridge University in the UK in 1981. Their finding that a specific cell can produce any type of human cell was shocking. The discovery of the human embryonic stem cell in 1998 opens up the possibility of producing or regenerating any type of human cell and thus any type of internal organ. Although regenerating a human heart or cerebral nerves has previously been considered impossible, stem cells may change this. Recently, human stem cells, but not embryonic ones, were discovered in the marrow of bone, and these pluripotent progenitor cells can be used for human regenerative therapy. Various internal organs, including blood vessels, skin, eyeballs, kidneys, livers and nerve cells, have been experimentally regenerated so far. The technology trajectory is almost completed, and the development trajectory has just started. This means that we are now in the most promising period for technological opportunities for various venture businesses. The diffusion trajectory will likely begin around 2015, as shown in Figure 11.7.

11.3 DEVELOPMENT OF ENERGY-RELATED TECHNOLOGIES

The above are some observations of forthcoming typical innovations in the fields of materials, electronics and biotechnology. On the other hand, global problems will become serious in the fields of resources, energy, population, food supply and environmental issues. While advanced technological innovations will be needed to solve these problems these are not yet evident. The energy supply situation in particular is not clear. In 1972, the Club of Rome warned of a global crisis due to the explosion of the population, the shortage of energy resources, the gap in the food supply and the increase in severity of environmental problems (Meadows et al., 1972). After the oil crises in 1973 and 1979, much work went into developing renewable energy resources, but no substantial solution has been obtained so far. In addition to its warning in 1972, Meadows et al. (1992) issued a report entitled *Beyond the Limits* describing how the situation had worsened. More recently, various problems caused by increasing populations have become evident, and countermeasures against the build-up of carbon dioxide have been actively considered. Solar energy, fuel cells and wind power generation have been positively promoted for clean energy generation. Although nuclear energy is a candidate for supplemental resources until the ultimate source for clean energy is developed, it has various problems, such as safety and waste disposal. There is no decisive solution to the future energy problem at the moment. As we have learned in this study, development of a technology after a key discovery or basic invention takes a long time to be commercialized. We next survey energy-related technologies and consider the problems to be solved.

Solar Cells

A solar cell was first proposed, as a single crystalline cell, in 1954, and the technology trajectory ended around 1985, a time span of 30 years. The actual transition in conversion efficiency from light to electricity is shown in

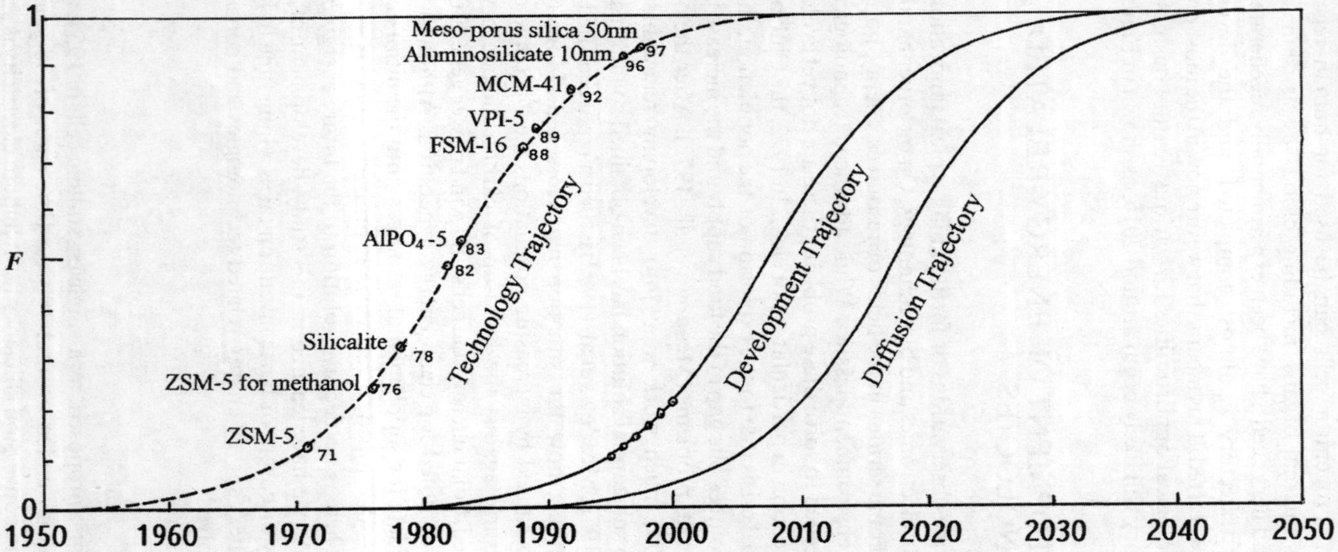

Source: Data from Iwamoto (2000).

Figure 11.5 Innovation paradigm of nano-porous catalysts

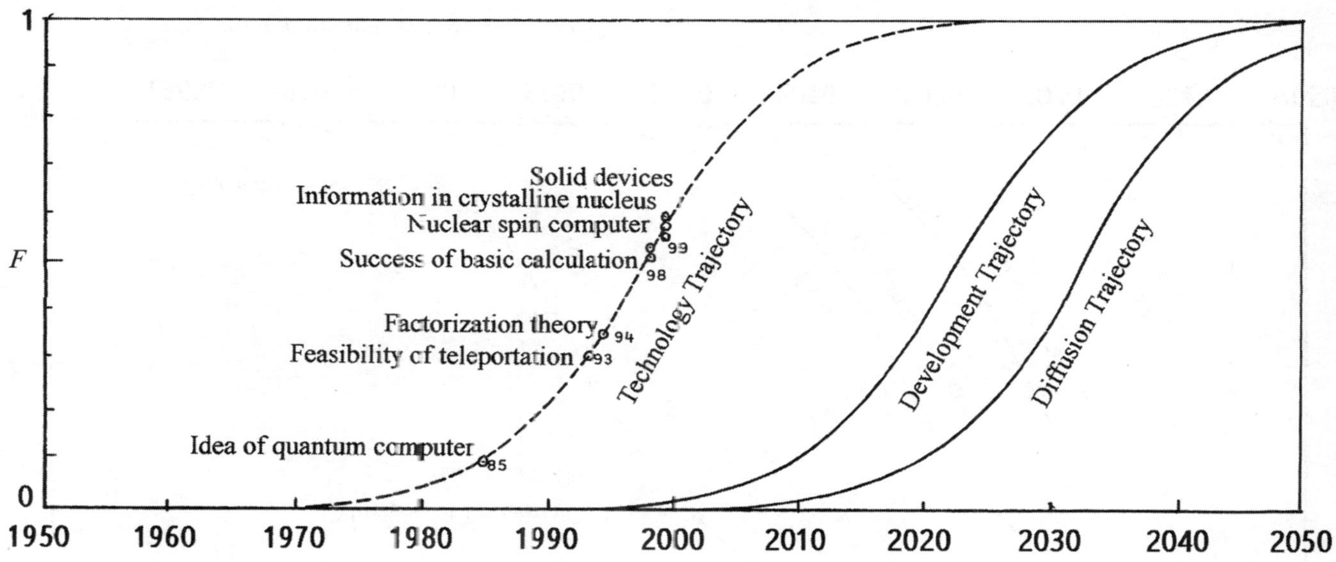

Figure 11.6 Innovation paradigm of quantum computers

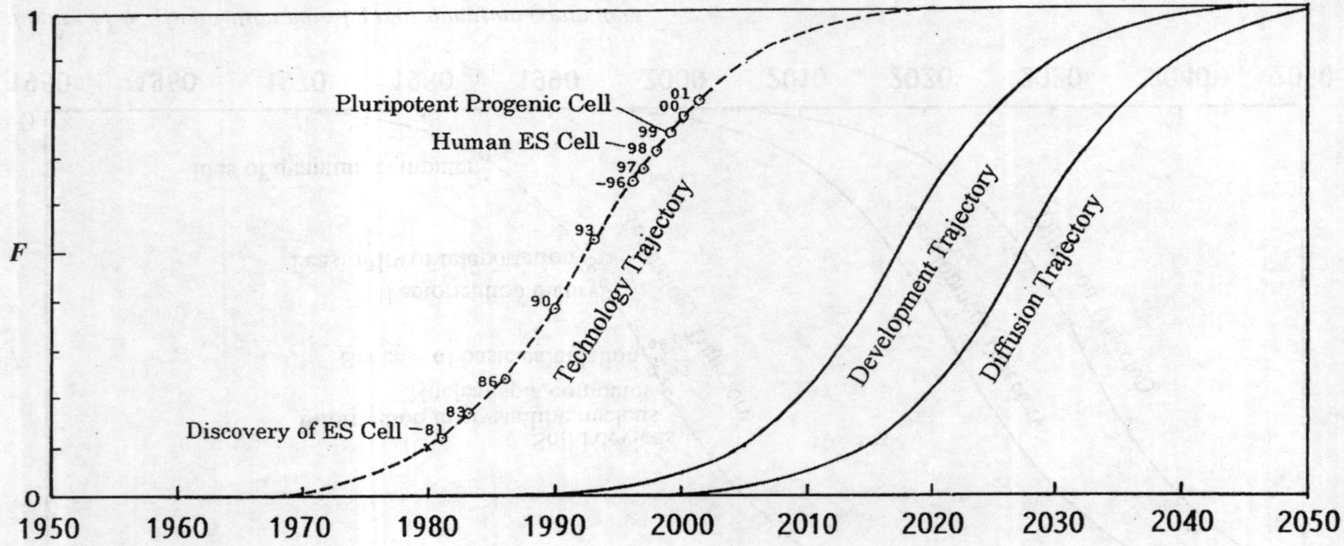

Figure 11.7 Innovation paradigm of regeneration technology

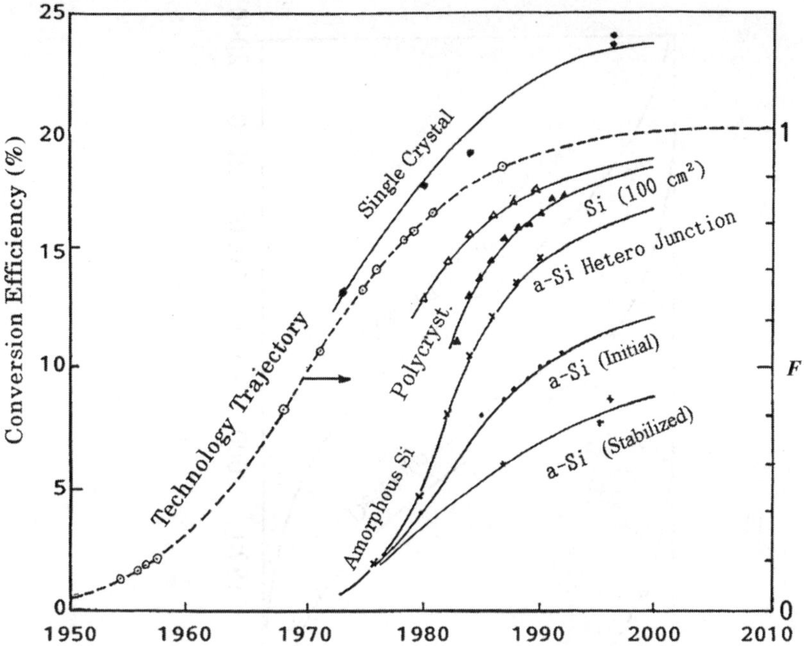

Source: Hirooka (1997b).

Figure 11.8 Conversion efficiencies of solar cells

Figure 11.8. The conversion efficiencies of single crystalline, polycrystalline and amorphous silicon change in parallel along the technology trajectory and reach a saturation level. This means that a further big jump cannot be expected. The actual paradigm for solar cells is shown in Figure 11.9. We are now in the diffusion trajectory; the world production of solar cells was equivalent to 740 000 KWh in 2003, a large jump from the 150 000 KWh in 2000. Extrapolation towards 2030 indicates the maturated market will be around 1 million KWh, which is negligible compared to the total expected consumption of energy worldwide.

Fuel Cells

The history given here of fuel cells comes mainly from Takahashi (1994), Hirose (2002) and Avadikyan et al. (2003). The fuel cell originated some time ago and, surprisingly, the first fuel cell, which was proposed by Davy in 1801, used carbon as the fuel. It followed very soon after the electric pile was invented by Volta in 1800. Grove proposed a hydrogen/oxygen

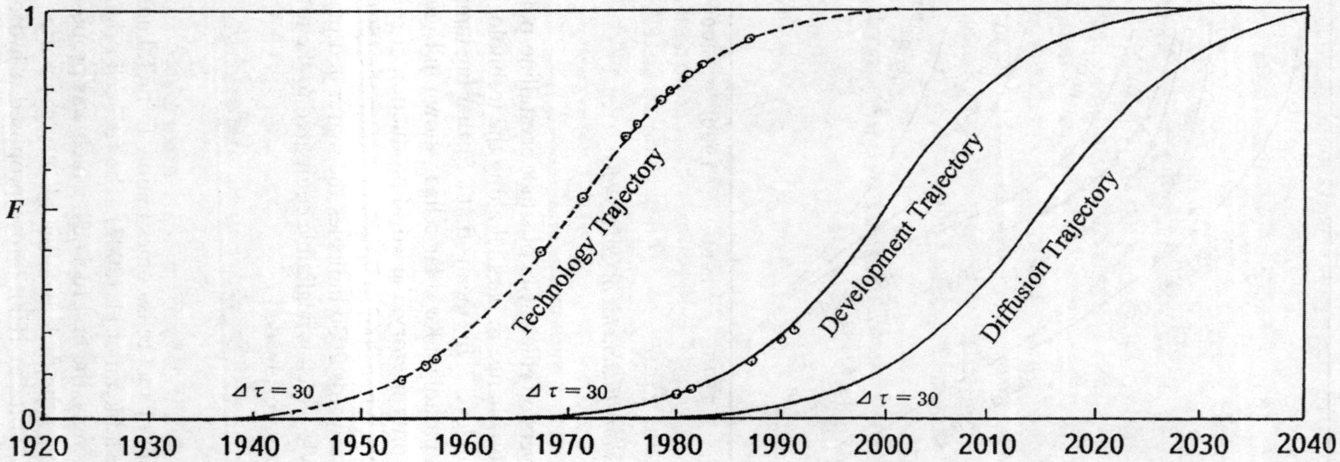

Source: Hirooka (1999d).

Figure 11.9 Innovation paradigm of solar cells

fuel cell that was used to successfully generate power in 1839. Mond and Langer promoted research on the Grove-type fuel cell and constructed a prototype in 1889. Jacques demonstrated an air battery made of molten caustic potash with a carbon electrode in an iron pot in 1896. Significantly, Nernst pointed out in 1899 the possibility of a solid fuel cell based on the ionic-conductivity of stabilized zirconia. Teitelbaum and Bauer proposed the principle for a molten carbonate fuel cell in 1921, which is the prototype of the modern carbonate fuel cell. The basis of today's alkali fuel cell is the Bacon cell, the concept of which was described by Bacon in 1932. Bauer and Preis prepared a sintered solid fuel cell in 1937. The Bacon fuel cell was patented in 1952 and was subsequently developed into an alkali aqueous solution type, which was used for the Apollo space project. A molten carbonate cell was developed by Ketelaar and Broers in 1958 and by Chambers and Tantram 1959. These fuel cell developments comprise the technology trajectory, which has a time span of 60 years.

The development trajectory started after the Second World War and was quickly accelerated by the stimulus of the space program in the United States. A solid polymer fuel cell using fluororesin Nafion of DuPont was used on Gemini 3 to Gemini 12 in 1965. Alkali fuel cells developed by General Electric and based on the Bacon patent were carried on Apollo 7 to Apollo 17. Stimulated by these space applications, private firms continued the development of fuel cells. Various projects for developing new ways to generate electric power have been carried out. The TARGET project for developing a phosphate fuel cell was undertaken by a US research consortium in 1967. Exxon started evaluating the use of alkali fuel cells as a part of the TARGET project in 1975.

Fuel cells for the engines of transportation vehicles also began to attract a great deal of attention following the development of a solid polymer fuel cell, PEFC, that uses methanol, by the US Department of Energy and Italy's Fiat Corporation in 1975. Ballard Power Systems of Canada developed a high-performance PEFC using Dow Chemical's fluororesin ionic exchange membrane, DEM, in 1983 and developed a high-power PEFC in 1987. They introduced an experimental bus with a PEFC in 1993. Daimler/Chrysler tied up with Ballard to develop an PEFC.

Japan started the Moonlight project in 1981 to develop a wide range of fuel cells, including alkali, phosphoric acid, molten salt and solid fuel cells. The Ministry of Economy, Trade and Industry in Japan launched a project in 2001 to develop low-cost fuel cells especially for automobiles and set up a three-year plan in 2003 to develop practical fuel cells as a next-generation energy source, aiming at the practical use of fuel cell vehicles, hydrogen supply facilities and in-house power generation facilities. The US terminated the Partnership for a New Generation of Vehicles (PNGV) project for enhancing the fuel efficiency of gasoline vehicles in 2002 and implemented a national plan for supporting the development of fuel cell vehicles.

The innovation paradigm for fuel cells has a rather long technology trajectory of 60 years, and the development trajectory is now almost completed. The diffusion trajectory has not yet begun, which differs from the trend for recent advanced technologies. The main reasons are those of cost competition with other energy resources and incomplete technological development. The practical use of fuel cells in general purpose utilities should begin around 2010, as shown in Figure 11.10.

Superconducting Materials

Superconductors are closely related to energy and are therefore discussed in this section. The superconductor was first discovered in mercury by Onnes in 1911. Since then various metallic superconductors have been discovered. Along with the series of superconducting metals discoveries, various theories for superconducting phenomenon, including BCS theory, have been developed, and the technology trajectory lasted until 1965, a time span of 55 years. The development trajectory started with the success of preparing lead wires, and a series of developments – superconducting magnets, magnetic levitation, electric power generation, electrical transmission and electric storage and so on – followed, covering a time span of 45 years. Nonmetallic superconductors discovered since the 1960s include graphite, polythiazyl, organic superconductors and perovskite-type ceramics. Most importantly, several high-critical-temperature ceramic materials have been discovered since 1986. A series of these nonmetallic superconductors form another technology trajectory, running from 1965 to 2010; it is located at the same timing as the development trajectory of the metallic superconductors.

Now we are in the diffusion period of superconducting magnets. The other applications, such as SQUIDs, Josephson junction devices, electric storage, power generation and electric transmittance, have not yet been commercialized. Thus, the diffusion phase is separated into two trajectories as it was for steam engines, which were used for motive force in the spinning and iron industries in the early nineteenth century and for locomotive force in the railway industry in the latter half of the nineteenth century. This second round of diffusion should begin around 2010 and last for about 40 years, until 2050. The innovation paradigm for superconductors is illustrated in Figure 11.11.

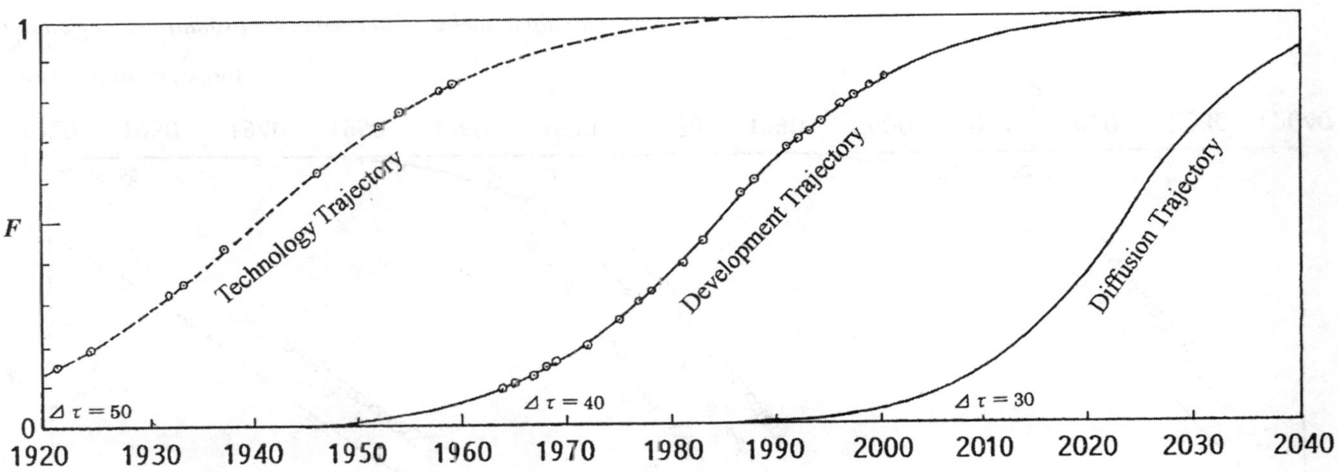

Source: Hirooka (1999d).

Figure 11.10 Innovation paradigm of fuel cells

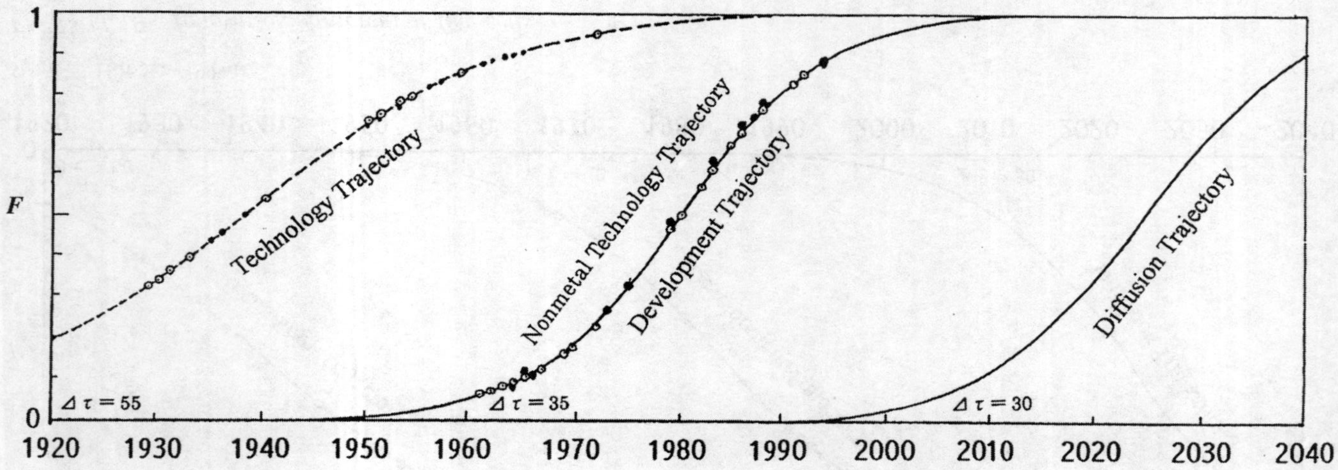

Source: Hirooka (1999d).

Figure 11.11 Innovation paradigm of superconductors

11.4 TECHNOLOGY FORESIGHT AND INNOVATION TRAJECTORIES

Technology foresight has been carried out regularly every five years using the Delphi process by the National Institute of Science and Technology Policy (NISTEP) of the Science and Technology Agency in Japan (which has been under the Ministry of Education and Science since 2000). The Delphi process is a technology foresight method using questionnaire-based enquiry related to future technological trends in advanced technology fields. This process has attracted much attention in many countries and has been carried out internationally in recent years. The questionnaires are carried out in three categories: (1) when will the technology be developed, (2) when will it be commercialized, and (3) when will it be commonly used? This process does not account for the scientific or technological background: it focuses on just the year of realization. The enquiry is repeated referring to the previous statistics. The second answer expects a more accurate convergence of estimated timing.

In order to convert the results of this technology foresight into our procedure of innovation trajectory dynamics, the following replacement of timing is carried out: (1) 'developed' means a timing of $F=0.1$ on the development trajectory, (2) 'commercialized' means a timing of $F=0.1$ on the diffusion trajectory, and (3) 'be commonly used' means a timing of $F=0.2$ on the diffusion trajectory. These interpretations make it possible to convert the Delphi data into logistic trajectories. Figure 11.12 shows the transformation of the technology foresight data on energy development collected by the seventh foresight in 2001 into innovation trajectories. Note that these data are just estimates of the realization timings; they are not estimates based on the scientific and technological background, so the timings are not always realized. These data indicate that almost all the energy technologies will be realized within the coming 50 years, with the peak in 2040–2050.

While the Delphi process highlights technologies that might be realized in the near future, there is a risk that some may never be realized, like chasing a rainbow. In this case, the Delphi process never converges to a realization time. Two areas, solar cells and superconductors, were examined to clarify this problem. There have been some questionnaires that were repeatedly used in every foresight. These are the same as or similar to ones used before. For solar cells, three questionnaires have been commonly used: (1) commercialization of large surface amorphous solar cells with convergence efficiency exceeding 20 per cent, (2) commercialization of layer-built cells with more than 50 per cent convergence efficiency, and (3) commercialization of space power generation by solar cells. For superconductors, seven questionnaires have been repeatedly used: (1) commercialization of JJ devices, (2) common use of superconductors having more than 77K critical temperature, (3) development of polymeric superconductors, (4) development of room temperature superconductors, (5) common use of superconductor generators, (6) commercialization of superconductor electric storage facilities, and (7) commercialization of power transmission. Comparison of the technology foresights for 1982, 1987, 1982, 1997 and 2001 were made for both areas and the results are shown in Figures 11.13 and 11.14. It is interesting to see that the difficulty of the technological level by subject is comparable in the same order for every foresight, showing that the respondents consistently judged the technological difficulty. Responding to this trend, the realization times for each area form an S-shaped curve, as shown in Figures 11.13 and 11.14. Each estimated S-curve is shifted forward more than five years, which means that the estimated realization time is never shortened, like chasing a rainbow. Does this mean that these areas will never be realized? If the foresight of subjects is more than a decade ahead of us the respondent every time puts it more than five to ten years later. When a development approaches realization, the foresight suddenly becomes more accurate and the timing is properly estimated. The above estimations indicate that the last foresight is suddenly shortened, indicating that realization is imminent. On the other hand, the logistic trajectories always indicate the timings. It is clear that the logistic paradigm with three trajectories provides appropriately more information about the near future. Of course, technology foresight using the Delphi process is very important in terms of identifying critical areas in each field, but the complementary use of logistic trajectory analysis is more promising.

11.5 ENERGY-RELATED RESOURCES AND GLOBAL ISSUES

We have discussed foresight of near future technologies using logistic trajectory analysis. Here we discuss global issues by applying the same procedure.

Overview of Global Issues

More than 30 years have passed since *The Limits to Growth – a Report for the Club of Rome's Project on the Predicament of Mankind* (Meadows et al., 1972) was published. The follow-up report, published 20 years later, was *Beyond the Limits – Confronting Global Collapse, Envisioning a Sustainable Future* (Meadows, 1992), and the global situation has grown worse since.

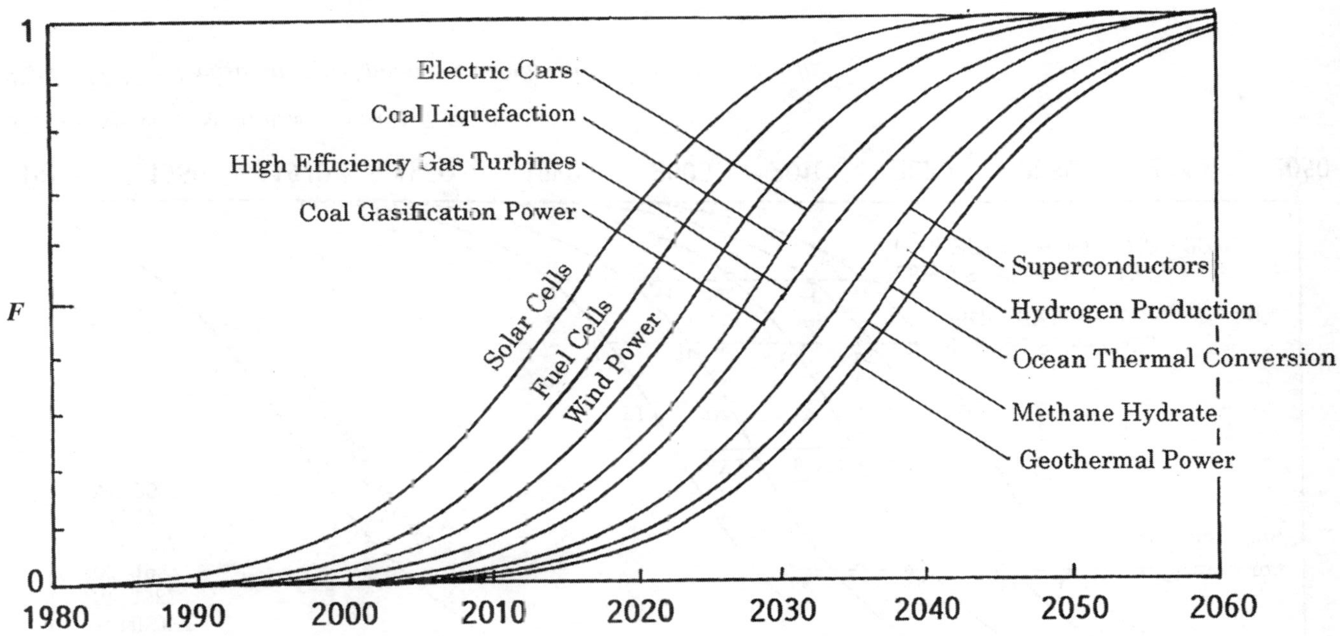

Sources: Hirooka (2003b); data from NISTEP (2001).

Figure 11.12 Diffusion trajectories derived from Delphi foresight

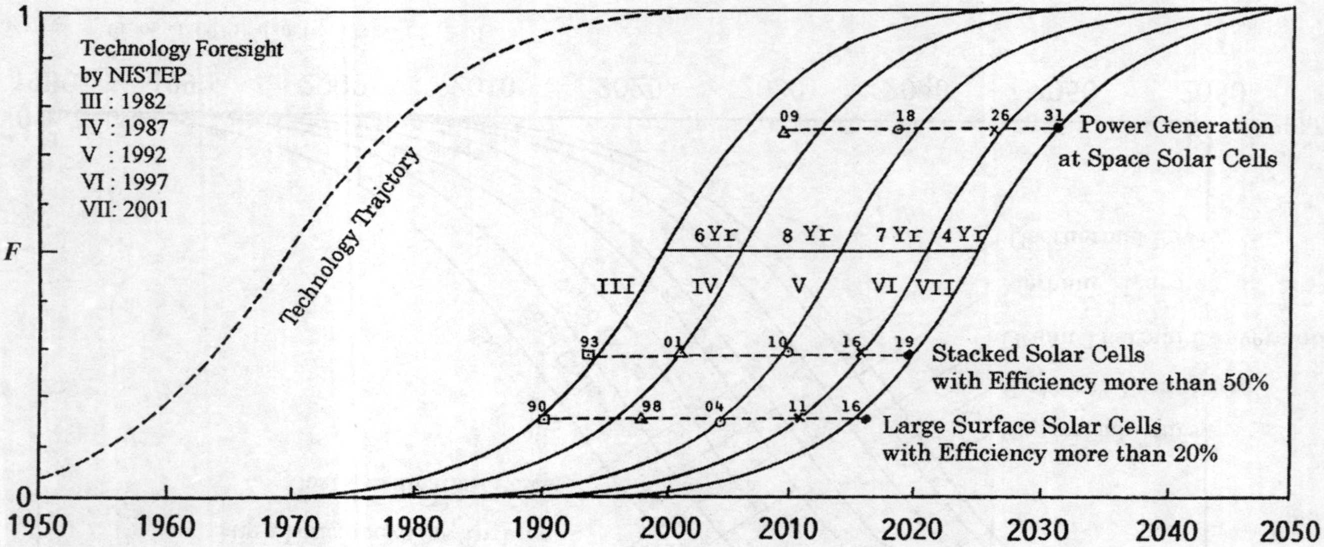

Sources: Hirooka (1999d); data from NISTEP (2001).

Figure 11.13 Transition of technology foresights in the case of solar cells

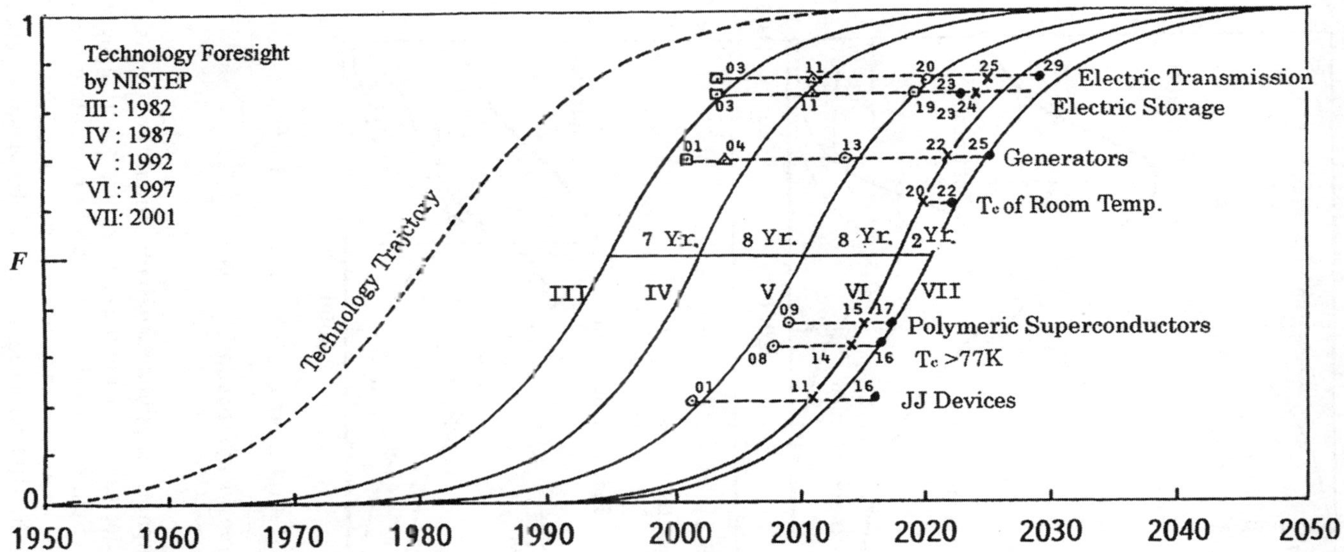

Sources: Hirooka (1999d); data from NISTEP (2001).

Figure 11.14 Transition of technology foresights in the case of superconductors

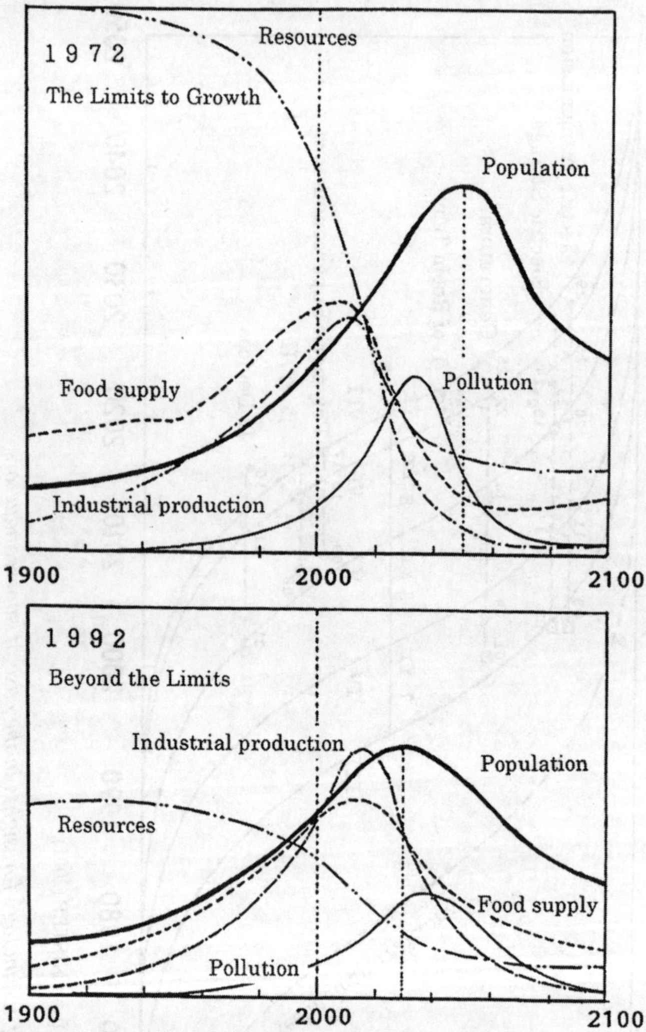

Sources: Meadows et al. (1972, 1992).

Figure 11.15 Transition of the world scenarios

Figure 11.15 compares the estimates presented in the two reports. The population seems to decrease sooner in the later report, reflecting the worsened global situation. Figure 11.16 shows the world population extrapolated through 2100. The simulation indicates that the soft landing at a population of 10 billion in 2100 will become impossible. Meadows et al. (1992)

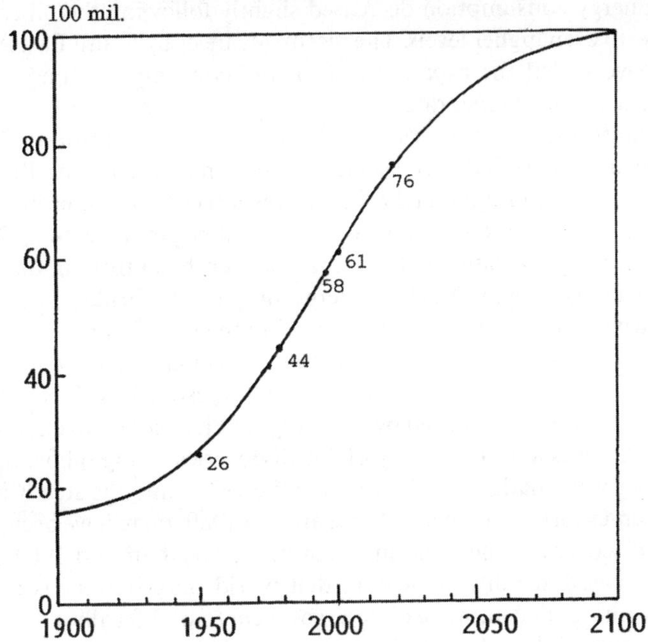

Source: UN Statistics (1990).

Figure 11.16 *Trend of the world population*

explained the reason for this by describing two traps from which we cannot escape. On the one hand, advanced countries never stop their economic growth in order to maintain sound finance. On the other hand, the least developing countries have to increase their population in order to gain more subsistence. After the oil crises, there was a period when the developed countries focused on expanding renewable energy resources; however, this movement has now come to a halt, and the situation will probably continue to deteriorate in the future.

Prospects for Energy Resources

The main player in the development of modern industrial society has been innovation, and a key supporter has been energy. As shown in Figure 3.6, the GNP in the United States increased by 30 times over the past 100 years, and energy consumption increased in parallel with the elasticity of unity. This data was presented in a report by the US government, *The Global 2000 Report to the President: Entering the Twenty-First Century* (US Government,

1980). Energy consumption decreased slightly following the oil crises but then rose to even higher levels. The recent advances by South East Asia and China have fueled the expansion of energy consumption, and it is now growing at an even higher rate.

Marchetti and Nakićenović (1979) analyzed the transition of energy resources in modern industrial society, as shown by the Fisher–Pry plot of the logistic curve in Figure 11.17. Energy resources were primarily charcoal before the Industrial Revolution, and then coal began to be used. Oil came into use during the latter half of the nineteenth century as the oil and natural gas age began. Nuclear energy may be the bridge to renewable energy, which hopefully will be developed in the near future.

Figure 11.18 shows the long-term trend in primary energy consumption worldwide. It indicates that oil and natural gas resources will become tight around 2020 and be exhausted by 2070. Figure 11.19 shows the trend in ultimate oil reserves and the history of the discoveries of big oil fields. Major discoveries were made from the 1920s to the 1970s, and the steady increase in oil reserves forms an S-curve. Since around 1980, there have been no substantial discoveries. The trend in discoveries peaked around 1960 and has since decreased steadily, indicating that world oil discovery is now in a maturity phase. The ultimate oil reserves seem to be 2.5 trillion barrels.

According to Hatfield (1997) and Koyama (2001), the ultimate amount (a total of 2.35 trillion barrels, as estimated on data for 1996) breaks down into cumulative production of 800 billion barrels, residual proved reserves of 1 trillion barrels, and estimated discoverable reserves of 550 billion. Thus, recoverable reserves could be 1.55 trillion barrels. If oil consumption worldwide increases by 2 per cent per annum, the available oil will be consumed within about 40 years.

A more important perspective is Hubbert's analysis of oil depletion timing (Hubbert, 1956, 1969). He examined the history of oil discoveries in the United States and found that the trend in oil production tends to follow a bell-shaped curve and that production peaked when almost exactly half the total oil reserves had been extracted and some oil was still left in the ground. In 1956, he estimated that oil production would peak between 1966 and 1972; it actually peaked in 1970. He also forecast the peak in global production and estimated that it would come between 1990 and 2000. This turned out to be much too pessimistic, partly because of inadequate data and partly because of minor flaws in his method. His method has since been refined and is now broadly accepted as a reliable prediction method.

Hubbert's method has been used by many petroleum geologists to predict the peak in world oil production. They have updated the data and their work figures prominently in the current discussion about petroleum depletion, as described by Heinberg (2003). Campbell has taken over

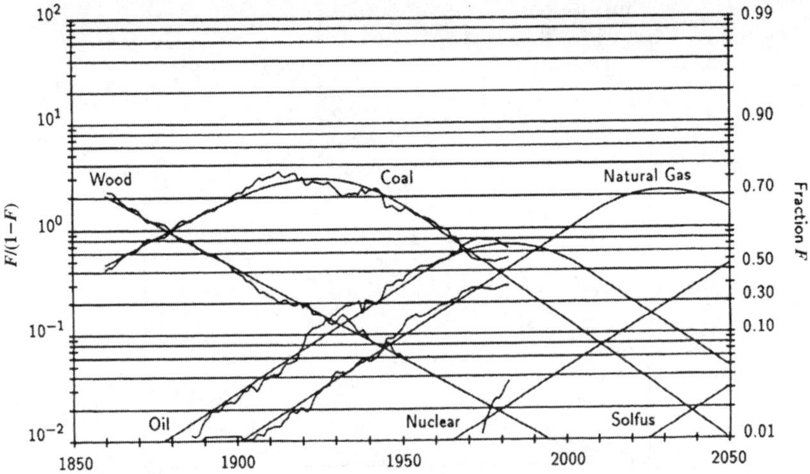

Source: Marchetti and Nakićenović (1979).

Figure 11.17 World primary energy substitution

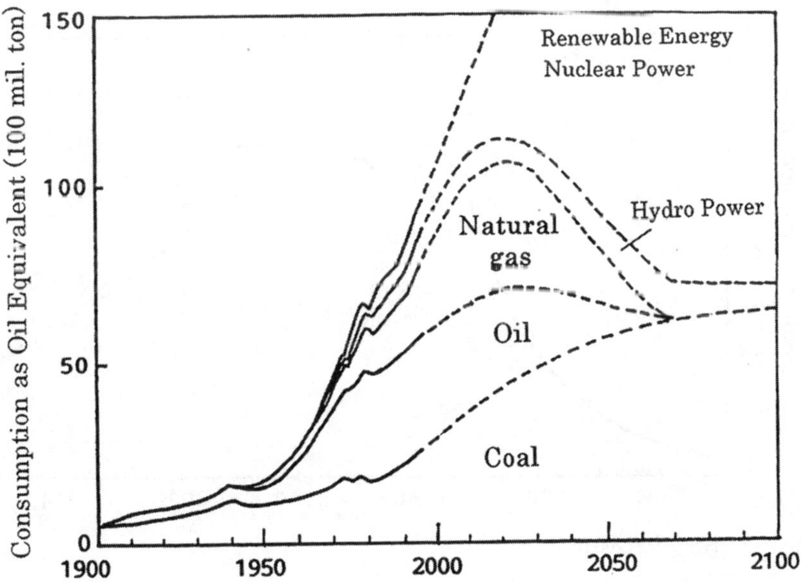

Source: Yoda (1996).

Figure 11.18 Consumption of the world primary energy

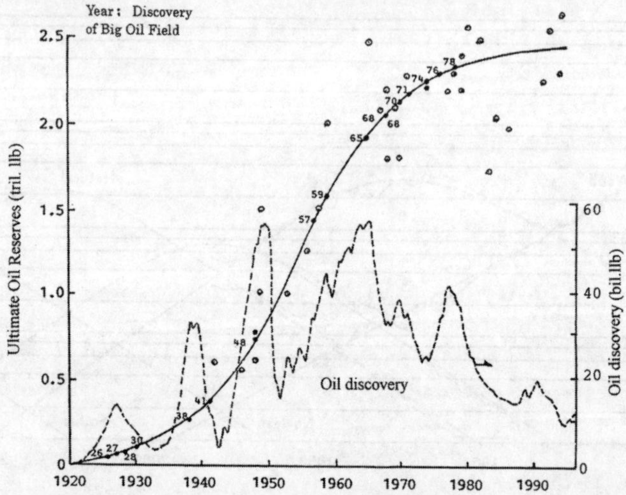

Sources: Hirooka (1997b); discovery data from Campbell (2005).

Figure 11.19 Past discovery and ultimate reserves of oil

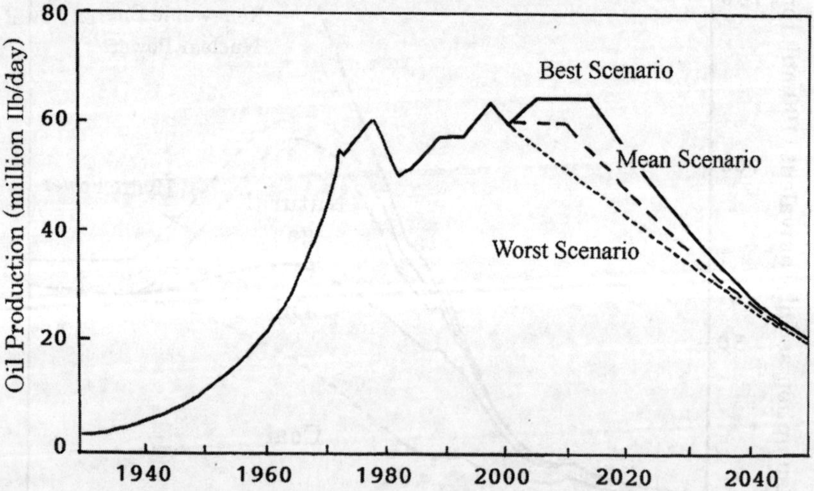

Source: Heinberg (2003).

Figure 11.20 Hubbert scheme of world oil depletion

Hubbert's method and refined the estimates. He set up the Association for the Study of Peak Oil (ASPO) and published extensively on the subject of petroleum depletion, for example, Campbell (2005), Campbell and Laherrère (1998). Based on their analysis, Campbell and Laherrère (1998) predicted that the decline in world oil production would begin before 2010 and at the latest in 2015, as shown in Figure 11.20. Using Hatfield's data of 1.55 trillion barrels, they also estimated that production would peak within 15 years, that is, by 2013; using the 2.036 trillion barrel estimate of Edwards at the University of Toronto, they estimated that the peak would occur in 2020. Deffeyes (2001) estimated that world production would peak in 2003. Duncan (2000) at the Institute of Energy and Man compiled data on projected oil production peaks for 44 nations in seven regions in the world. The peaks furthest in the future were for Saudi Arabia in 2017, Kuwait in 2010, and the United Arab Emirates in 2009. The average for the Middle East was 2009. In short, the estimated peaks furthest in the future lie between 2003 and 2020, meaning that, in the very near future, oil depletion will become obvious. Recent rises in the oil price could reflect this trend.

These estimates seem realistic, and there are no countermeasures available. How can we switch to other energy resources under such a tight schedule so that we can eliminate the present deep dependence on oil? Since the development of renewable energy resources is not far enough along to fulfill the prerequisite for replacing oil, will we have to depend on nuclear energy? While nuclear energy is primarily produced using light water reactors, other possibilities will have to be reconsidered. One previously discarded possibility is the thorium molten breeder reactor. Starting in 1965, the Oak Ridge National Research Laboratory of the US Department of Energy tested one successfully for 3.5 years as the Molten-Salt Reactor Experiment. This homogeneous molten system is attractive because a meltdown cannot occur, and weapons-grade material is not produced. Moreover, thorium is widely available worldwide. Another alternative is the use of hydrogen. Since the decomposition of water is an energy-consuming process, it is not easy to obtain hydrogen from water. It may become feasible, however. There is, for example, catalytic photodecomposition by sunlight using a photosensitive catalyst, such as the Honda–Fujishima effect with an improved efficiency.

11.6 GLOBAL PERIODICITY AND SOCIAL MEGA-TREND

We have discussed innovation paradigms and expanded subjects beyond simple innovations towards global issues. Although we ascribed the phenomenon of long business cycles to the clustering of innovations, we have not yet clarified why innovations form clusters. To get more information about cluster formation and periodicity, we need to examine the phenomenon from a wider perspective, one that is beyond the coverage of this book.

Global Periodicity and Evolution of Life

Earth formed about 4.6 billion years ago. The land areas have repeatedly joined together and split apart about every 400 million years. Super continents have periodically formed and broken apart: Nuna, 1.9 billion years ago, Rodinia, 1 billion years ago, Gondwana, 550 million years ago, and Pangea, 250 million years ago. Violent volcanic eruptions accompanied the breaking apart of the continents, causing the extinction of a great number of organisms. Various new organisms then emerged in a spurt of evolution. Interestingly, Ikeda (2004) claims that the evolution of living organisms has almost finished. He wrote that the first living organism emerged about 3.8 billion years ago as a procaryote and that two domains, Archaea and Eubacteria, formed about three billion years ago. A third domain, Eukarya, emerged through cell fusion between aerobes and anaerobes two billion years ago. Eukarya was the last domain to have evolved, and, within it, four kingdoms have developed Protists, Fungi, Plantae and Animalia. A total of six kingdoms, including Archaea and Eubacteria, appeared one billion years ago. These six kingdoms were the last kingdoms – no new kingdom has emerged since then. After a long glacial era starting about 800 million years ago and lasting for about 200 million years, the number of types of Ediacara organisms expanded explosively 600 million years ago. After Gondwana formed, mass extinction took place. Then, another explosive evolution took place during the Cambrian era, and the last Phylum in Animalia arose 550 million years ago. These periods around 500–600 million years ago could be the peak of evolution. Mass extinction again occurred about 251 million years ago after Pangea formed and about 65 million years ago following the impact of a huge meteorite. Mammals developed after this period. Every mass extinction led to a violent change in the evolutionary situation. These historical transitions indicate that while biological evolution is not simple, the evolutionary scenario is a one way street and is almost at the final stage.

Human Prehistory and Social Dynamism

Human beings began splitting from the ape about 6 million years ago in four evolutionary steps: apeman (homo habilis, australopithecus), homo erectus (pithecanthropus), hominid (homo neanderthalensis) and homo sapiens. Glacial and interglacial periods have repeatedly occurred from about two million years ago. There are Milankovitch cycles comprising three periodicities of eccentricity cycle of about 100 000 years, obliquity cycle of 41 000 years, and precession of the equinoxes of 19 000 and 23 000 years. Now we are in an interglacial period following the Würm glacial period, which lasted 60 000 years and ended a little over 10 000 years ago. It followed a short cold period called the 'younger Dryas', which ended 15 000 years ago. The present interglacial period has been unusually warm, with stable temperatures. This comfortable climate has greatly affected humankind's activities, promoting unbelievably robust development. The transition from a hunter-gatherer lifestyle to a farming lifestyle 10 000 years ago was certainly not accidental. Global warming must have been deeply implicated.

Human society began to develop, but it took more than 5000 years before the appearance of the four ancient civilizations: Mesopotamia, Egypt, Indus and China. Greece should be added to this list because the Trojan and Cretan city-states were founded around 4500 years ago, and the Mycenaean city-states converted to Greek city-states. The city-states of these five civilizations thrived for almost 2500 years and then began to decline. The synchronized emergence of these five civilizations is quite interesting. Following the decline of these city-states, much wider territories were conquered by Babylonia, the empire ruled by Alexander the Great, and the Roman Empire. Qin and Han conquered all of China. The Maurya dynasty unified northern India.

It is particularly interesting to note that when these civilizations had just about run their course and were in a state of decline, they gave birth to religions that have endured to this day and have greatly influenced many people. In China, the teachings of Confucius established Confucianism around 550 BC. In India, the teachings of Buddha led to the creation of Buddhism in 486 BC. The Hindu religion, on the other hand, is estimated to have been created during the early years of the Indus civilization, around 3000 BC. Zoroastrianism was predominant in the early years of the Mesopotamian civilization, around 600 BC. It later declined, and the teachings of Muhammad led to the creation of Islam around AD 600. In Palestine, Judaism emerged around 600 BC with the initial writing of the Old Testament; it branched off into Christianity following the death of Jesus. While the ancient city-states were governed on a magical-religious basis and were strongly influenced by spiritualistic mediums, modern states are governed under the strong influence of their own religion. Now that scientific thought has progressed, the religious influence has been diminishing.

Loci of Social Development and Knowledge Development Dynamics

Modern economic society has been in existence for the 2000 years after Christ. The collapse of the Roman Empire in 476 following the invasion by the Germanic tribes led to the Dark Ages, which lasted for about half a century. The Holy Roman Empire that subsequently arose launched a total of seven crusades against the Muslim caliphates of the Near East between 1096 and 1270. Although the failure of the crusades undermined the strength of the Christian world, the crusades did promote trade between the Orient and the Occident. Italian merchants in particular benefited from the increased trade and accumulated immense wealth. This wealth became an incubator for the Renaissance. It was used to fund the development of new shipbuilding and navigation technologies, resulting in the Caravel ships that were sturdy enough to cruise the oceans. Portugal and Spain put these ships to good use exploring for new routes to India, which brought about the Age of Great Voyages.

We have discussed the development of innovations after the Industrial Revolution using an analytical process of logistic dynamics in which the distribution of discrete innovation elements is found within a definite time period and their time spans of the elements are determined. This process makes it possible to quantitatively estimate the spread of a series of knowledge activities and to describe them using a logistic curve. We could expand this process to a more general analysis of sciences and economics beyond innovations. Such an expansion seems reasonable given that any knowledge development should reach a maturation determined by the growth function. In this wider sense, we have termed it 'knowledge development dynamics'. We will now expand the process to other knowledge development fields such as art, philosophy and societal affairs, which can be summarized as social knowledge activities including human behavior and actions based on a definite purpose and aim brought about by knowledge. Using this process, let's now consider the background to the Industrial Revolution.

Figure 11.21 gives a bird's-eye view of the European continent after the crusades. It illustrates not only the trends in science and technology but also the trends in scientific thought, arts and societal actions in parallel. The Age of Great Voyages is represented by a logistic curve with a time span of about 100 years, from the exploration of the Azores islands and African coastline by Prince Henrique Navegador (Henry the Navigator) of Portugal, starting in 1418, to the discovery and exploration of America by Columbus, Vasco da Gama, and Amerigo Vespucci, from 1492 to 1502, and to Magellan's circumnavigation of the globe in 1522. Prior to this Great Voyages trajectory, there was a corresponding technology trajectory that starts with the invention of the compass in 1180 and covers the inventions of rudders, terrestrial magnetic devices and navigation compasses and the development of technologies for building Caravel ships during the fourteenth century, a time span of about 170 years. This trajectory has the same timing as the crusades trajectory. It is interesting to learn that the logistic nature of the history of Columbus' voyages has been clearly recognized by Marchetti (1986).

The technology trajectory of military weaponry also has this timing. That is, various basic technologies of weapons form a technology trajectory composed of the development of gunpowder, firearms, handguns and artillery, that runs from the middle of the thirteenth century to the end of the fourteenth century. This technology trajectory ignited the weaponry diffusion trajectory, which starts with the mass production of handguns, harquebus and artillery for use in the battles of the Hundred Years' War. This trajectory has the same timing as the Great Voyages trajectory. In the middle of this weaponry diffusion trajectory, Leonard da Vinci actively worked on the invention of weapons. Paralleling these two diffusion trajectories is another important trajectory, that of Renaissance art, from the construction of the great cathedral in Florence with the work of great Florentine artists such as Fra Angelico and Donatello, through the creation of various great masterpieces by Botticelli, Leonardo da Vinci, Michelangelo and Raphael, among others. These works of art form a trajectory with a time span of about 150 years, centred around the fifteenth century. These three fifteenth-century trajectories compose a second cluster of social knowledge activities. Before this art trajectory, there is another trajectory of Gothic architecture which started in the middle of twelfth century and had a time span of about 200 years. This trajectory has been estimated by Marchetti (1990) and Modis (1992).

In the seventeenth century there was another cluster of knowledge activities composed of the Newtonian dynamics trajectory, which started with Copernicus' discovery of a heliocentric universe, the first scientific thoughts trajectory and the theatrical arts trajectories. The art trajectory includes theatrical dramas (represented by Shakespeare's plays), court ballets in France and baroque music operas in Italy. Musical instruments were also developed in this period, forming an orchestra trajectory. Interestingly, there are no technology trajectories within this cluster age. This is ascribed to the grave ecological and political circumstances of the 'General Crisis' in Europe in the seventeenth century, which thwarted the development of new technologies. The gravest problems were the prevalence of infectious disease (pests) and the enduring famine caused by unusual cold waves. Furthermore, the political situation in Europe was very unstable, with wars occurring frequently.

Although new technologies were not being developed under these difficult conditions, there was a robust tide of science. As new scientific concepts sometimes contradicted the tenets of Christianity, pioneers of the new sciences faced severe resistance by the Church, as symbolized by the inquisitions for Copernicus and Galileo Galilei. Newton escaped from disease-

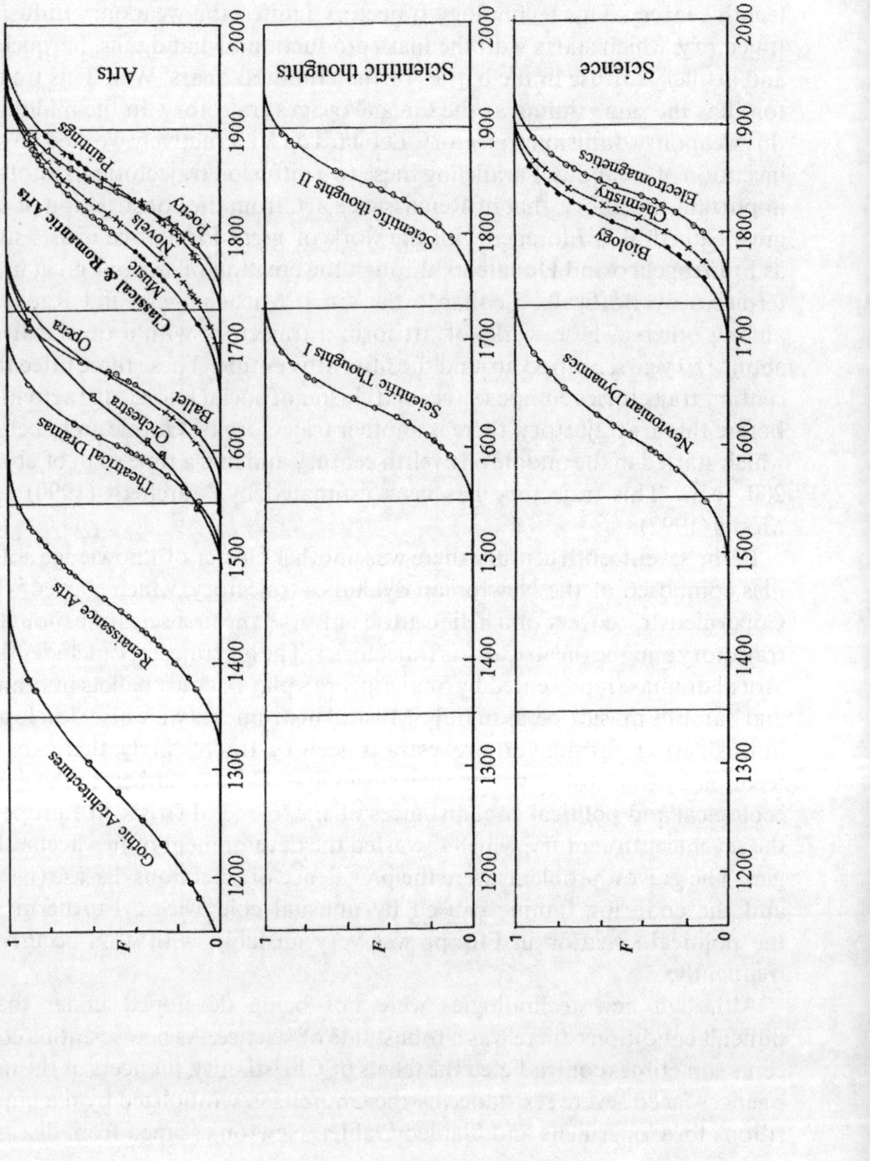
Arts
Gothic Architectures
Renaissance Arts
Theatrical Dramas
Orchestra
Ballet
Opera
Classical & Romantic Arts
Music
Novels
Poetry
Paintings
Scientific thoughts
Scientific Thoughts I
Scientific thoughts II
Science
Newtonian Dynamics
Biology
Chemistry
Electromagnetics
F
0
1
1200
1300
1400
1500
1600
1700
1800
1900
2000

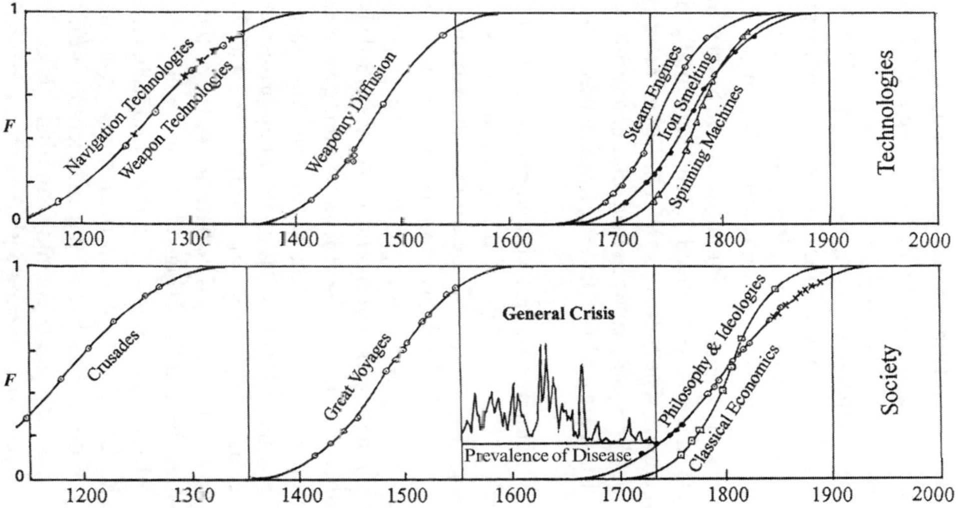

Figure 11.21 *Knowledge development trajectories in civilized society*

ridden London to his home town, where he was prompted towards his theory of gravity. The scientific thoughts of such great thinkers as Francis Bacon and René Descartes were combined with scientific developments, for example, Newtonian dynamics and vacuum theories. Many universities were established in Europe from the twelfth to the fifteenth centuries, including those of Oxford, Cambridge, Paris, Bologna, Rome and Lisbon. Urbanization progressed, and many European cities arose from the eleventh century. Commerce gradually took hold, led by the great Italian cities of Florence, Pisa, Rome, Siena and Venice.

Recovery from the General Crisis happened first in the Netherlands, then in England. The Industrial Revolution then took hold, and the technology trajectories of the steam engine, iron production and the spinning machines started in the early eighteenth century. The steam engine trajectory began with the inventions of Papin in 1688, Savery in 1698 and Newcomen in 1705 and was completed when Watt finally developed a practical, efficient engine. The S-curve shown in Figure 6.4 is the development trajectory for the steam engine. Iron manufacturing technologies started in 1709 with Darby's blast furnace, which used coke instead of charcoal.

As pointed out by Schumpeter, the technology trajectory for the spinning machine went back to the invention of the flying shuttle by Kay in 1733. Thus, the technology trajectories of these three innovations have a time span of about 90 years, the same as those of the science trajectories of chemistry, biology (embryology) and physics (electromagnetism), which have the same timing in the eighteenth century. As described above, the trajectory of classical economics also has the same timing as those of science and industrial technologies. Interestingly, the arts trajectories also have the same timing. The period from 1750 to 1900 comprises various highly exalted arts (music, paintings, novels and poetry), and was followed by a golden age in the nineteenth century. The baroque music of the seventeenth century was followed by the flowering of classical and romantic music by Bach, Haydn, Mozart, Beethoven, Schubert and Tchaikovsky among others. Starting with Goethe and Wordsworth, many great poets formed a definite trajectory. From the middle of the eighteenth century, a series of outstanding novelists appeared, forming another trajectory. Modern painters also form a trajectory, starting with the romantics through the realists to the impressionists. These classical and romantic arts form a cluster of bundle fractals.

The second trajectory of scientific thought starting with Laplace in 1799 reflects the explicit recognition of science and self-establishment apart from Christian belief. Laplace's demon was the universal recognition of science as a kind of dynamic understanding of nature. After their long struggle against Christian authority, such as Darwin's debates with religious authorities about evolution, the natural scientists reached a firm conviction about the truth of their discipline. Since natural science became an established discipline, scientists have been able to carry out their studies without having to consider the potential role of religious beliefs. In 1878 and 1888, Engels (after Hegel) summarized the role of natural science and put forth the idea of dialectical materialism. He asserted that nature does not work independently but mutually interacts resulting in historically duplicated and hierarchical structures.

Along with the development of scientific thought, modern philosophies and political ideologies developed, forming another trajectory, from enlightenment to idealism, materialism and socialism.

These trends clearly indicate that cluster formation is not limited to technological innovations; social sciences and arts are also involved as societal knowledge activities having a kind of synergistic effect. The clustering of knowledge trajectories is considered to be the origin of the dynamic pulsation of society as well as the origin of the dynamism of the economy. Such concerted societal behavior seems to be a kind of dissipative system governed by a self-organizing mechanism.

Great Surge of Innovations and Mega-structure of Society

The above examination indicates that trajectory formation is not limited to innovations; it includes common elements of the knowledge society. Trajectories have formed regularly, forming a cluster of various knowledge activities in each era. Many economists have recognized this cyclical formation, including Kondratiev, Juglar and Kitchin, as described in this book. Now, let's look at this cyclical formation in terms of natural phenomenon.

From his observations of economic activities in England, Clarke (1847) identified two cycles of famine and depression: a short one with a duration of 10–11 years and a longer one of 54–55 years. He ascribed these cycles to climatic conditions and called their study 'physical economics'. Jevons, a founder of the economic school of marginal utility and an expert in meteorology, continued the development of this idea (Jevons and Foxwell, 1909). He wrote about the relationship between commercial recession and sunspots in the journal *Nature* in 1875 and wrote various papers about the relationship between sunspots and economic activities between 1875 and 1882. He found that 10–11 sunspot cycles corresponded to one cycle of economic activity, resulting in his coining of the term 'solar economics'. Using statistical data, Juglar (1862), a French economist, showed that economic activity has a cycle of 10–11 years.

While the solar cycle of 10–11 years has been well accepted, it was recently discovered by Yoshimura of the University of Tokyo that five such cycles form a longer cycle of 52–55 years, the so-called 'Yoshimura' cycle. This cycle corresponds to the Kondratiev cycle. It was also found that five or six Yoshimura cycles form an even longer cycle of 272–335 years. Shimanaka (1987) summarized the solar economics and inferred from them that the fluctuation in solar activity affects the economic cycle. Shimanaka (1987) introduced John Eddy's analysis of the unusual climatic conditions that endured from the fourteenth to the early eighteenth century, for 400 years, in Europe. This period includes the Maunder minimum period of solar activity from 1645 to 1714 and is considered a short glacial period. The cold conditions were the result of relatively weak solar activity producing less solar energy. This well explains the occurrence of the 'General Crisis' in the seventeenth century. Sumida (1993, 2000) also discussed the correlation between solar cycles and business cycles.

Cycles are also observed in other natural phenomenon. The haul of sardines has a cycle similar to the Kondratiev business cycles. A study by Hiramoto (1996) revealed that the sardine cycles have a time span close to that of the Kondratiev cycles over the last 500 years. The present analysis indicates that the peaks in the sardine cycles are by chance near those of the Kondratiev cycles. Sumida (2000) discussed the correlation between fishery and solar spot cycles.

A long overview of the history of civilization shows that civilizations first emerged in four or five areas about 5000 years ago, and those first civilizations endured for more than 2000 years. We are now in the second civilization period, which started to flower at around the time of the Renaissance. This flowering was initiated by scientific and technological innovations, the pace of which has continued to accelerate.

Analogous to the fractal structure of sunspots having a unit cycle of 11 years, longer cycles of 50–60 years, and megacycles of 300 years, modern industrialized society must have a mega-structure of about 300 years. Various factors support this hypothesis. Modern industrialized society has grown along a long logistic curve, shown in Figure 3.6, that reflects the actual trends in the growth of GNP and energy consumption from 1800 to 2100. The growth curve of the global population shows the same trend as the logistic curve shown in Figure 11.16. As shown in Table 3.1 and Figure 3.5, innovation paradigms have evolved stepwise in each industrial sector, and almost all possibilities of trunk innovation have apparently already appeared, so that not many remain. Thus, innovation paradigms are considered to form a longer trajectory with a time span of about 300 years. As these trends correspond to the actual diffusion of innovations, the mega-structure of innovation trajectory must be located say in the 300 years from

1750 to 2050. Of course, we can expect many sophisticated innovations in the near future, but as far as trunk innovations are concerned, we cannot expect so many innovations to add much more value to the economy. Furthermore, the weight of the new innovations will generally be lower than the ones that have already appeared because the mass of the economy is already large enough.

In short, modern civilized society has already reached an impasse in terms of energy consumption, food supply and environmental problems to which solutions are not available.

11.7 SUMMARY

In the study described in this chapter, innovation dynamism was expanded to incorporate a future perspective and to fields beyond innovation.

I. Science paradigms

1. Science paradigms can be described by a logistic equation, using the same procedure as that for innovations.
2. Starting with Newtonian dynamics, many scientific disciplines can be identified as trajectories, and the time span of these trajectories is shortening year by year.
3. A science paradigm evolves from a conventional one to an advanced one.
4. The paradigms of all economics schools have a time span of 90 years, differing from those of the natural sciences in that there is no shortening in the time spans.

II. Technology forecasting and the fifth Kondratiev cycle

5. A series of science and technology trajectories forming a cluster is identified as the origin of the fifth Kondratiev cycle.
6. Ongoing science and technology trajectories can be extrapolated into the near future to estimate the trends of next-generation industries.
7. Technology foresights using innovation trajectories are more informative than those obtained using the Delphi process.

III. Energy-related resources

8. Energy-related resources were studied and paradigms for solar cells, fuel cells and superconductors were analyzed by the innovation dynamics.

9. Global problems were examined by a trajectory analysis. Analysis indicated that oil depletion is an urgent problem.

IV. Global periodicity and mega-structure of the economy

10. Analysis of civilizations revealed that various knowledge activities, not only technological innovations and scinces but also artistic and social advances, form trajectories with definite time spans which are termed 'knowledge development dynamics'.
11. The trajectories of various knowledge activities form a common surge that includes innovation clusters and are the origin of the dynamic pulsation of human society.
12. Innovation paradigms are considered to form a mega-structure covering about 300 years, from 1750 to 2050, comprising five to six Kondratiev cycles.

PART IV

New Insight into the Economics of Technological Change

The analysis of innovation dynamism presented in this book has revealed several new features of the innovation process, and we have shown that nonlinear dynamics can be used to examine the economics of technological change from a new angle. This part summarizes these features and clarifies what meaning the analysis of the innovation paradigm will have and what kind of knowledge will be added to the analysis of economic development.

This is the first attempt to describe the complete picture of innovation, from the first basic invention or discovery to the maturation of the market, as a consistent single paradigm comprising three trajectories. It is especially interesting to describe such an abstract idea as technological development by using visible representation, that is, specific trajectories. One of the important aspects of this analysis is its use of trajectories having a nonlinear nature to express the progress of innovation. Although this is an analysis of innovation itself, it can also be used to analyze the economics of technological change.

While the product diffusion of innovation is known to obey a logistic equation, it is an interesting finding that the trend of technological development can be described by a logistic equation as well. This trajectory of the technological development period is formed by gathering information about innovation elements scattered within a time span into a concrete string of knowledge developments. This formation results from the nonlinear nature of the locus of knowledge development. This formation of trajectories reveals various interesting features about innovation, and the concept of innovation can thus be further interpreted from different angles. Furthermore, various knowledge developments such as science, scientific thought, economics, art, philosophy, exploration and various social actions also can be described by a logistic trajectory as a knowledge development dynamics. Such wider concepts are also discussed.

12. Nonlinear dynamics of economic development

This is the final summary of this book, focusing on the new insights into the economics of technological change on the basis of the study of innovation dynamics using trajectories.

There are three features of particular interest raised by this unique study. One is the nonlinear nature of innovation, which means that each of these trajectories has an ultimate level, reaching saturation. This implies that timing during innovation is very crucial. The second feature is the exogenous nature of the technology trajectory that triggers innovation, which means that the origin of an innovation often comes from outside the economy, especially from academia. The third feature is the importance of knowledge transfer among intra- and inter-trajectories. The concept of '*field for knowledge transfer*' is particularly important because knowledge transfer is essential to the progress of innovation.

12.1 RELATION TO THE EXISTING ECONOMICS

Emergence of Innovation

The economics of technological change has progressed since Schumpeter postulated in 1912 that the driving force of economic development is technological innovation.

Looking back at the development of economics since Adam Smith, most economists have discussed causal relations between labor, money and goods – technological contributions have been regarded as exogenous factors. Schumpeter first asserted that technological innovation plays a decisive role as an engine in economic development. As already discussed, Schumpeter recognized the existence of various related inventions made long before an innovation emerged. He first treated these inventions as exogenous but later considered them endogenous because research and development within firms plays an important role in economic development. This was the changeover from the Mark I to Mark II models. Schumpeter precisely examined inventions that came before the innovation and discussed the nature of their background.

Schumpeter's concept was inherited by the neo-Schumpeterians, who extensively examined and analyzed innovations and economic development, leading to the present economics of technological change. Enos (1962) estimated the time lag between invention and innovation for the oil refinery industry and other industries. Kuznets (1966, quoted from Rosenberg, 1982, p. 141) pointed out that 'the distinctive feature of modern industrial societies is their success in applying systematized knowledge to the economic sphere, knowledge derived from scientific research'. Rosenberg (1974, 1976, 1982, 1994a, 2000) extensively studied the contribution of science and technology to economic development. Schmookler (1966, 1972) pointed out the importance of demand for the emergence of innovation. This is the counterpart to Schumpeter's description of the contribution of innovation to economic development. Schumpeter stressed the emergence of innovation from the supply side and barely described demand-pull. The empirical core of Schmookler's work is an attempt to demonstrate that the demand-side considerations are the major determinant of variations in the allocation of inventive effort to specific industries. Rosenberg (1974, p. 97) criticized this point: 'Many important categories of human wants have long gone either unsatisfied or very badly catered for in spite of a well-established demand'. For example, progress in medicine had to await the development of the science of bacteriology in the second half of the nineteenth century. Rosenberg contended that we need to understand which scientific and technological discoveries are needed before key breakthroughs in invention can be made.

Freeman pointed out the increasing scientific character of technology and described the technological development of innovations in an article entitled 'The Rise of Science-related Technology', in which he focused on process innovation, synthetic materials, and electronics (Freeman, 1982). He also developed the notion of a 'new technology system' (Freeman, 1984b) based on the concept of the techno-economic paradigm (Perez, 1983). Freeman and Perez argued that new technologies can form new paradigms that not only bring new technologies to the economy but also induce institutional changes that reform society itself.

Evolutionary Economics

The evolutionary approach to economics proposed by Nelson and Winter (1982) has received much attention. Nelson broke from neoclassical economics and sought a new concept that focused on technological change as the key to long-term economic development and took into account Schumpeterian dynamic change. Winter had been interested in the evolutionary development of firm behavior. These two approaches were combined, leading to an evolutionary theory of economic change. This theory addressed the performance and behavior of firms under a restricted condition and how they struggle to exist, using behavioral patterns called 'routines'. Nelson and Winter's approach is characterized by the notion that the behavior of firms is never the same, and each firm is in a different selection environment. This idea originated with questions raised about uncertainty and evolution by Alchian (1950) and is related to the bounded rationality of Simon (1959, 1982), which means that firms have to make selective decisions without fully understanding all aspects of their environment. This notion about firm behavior accords with the idea of how the environment affects human behavior and decision-making and is related to the recognition of the 'field' of developing innovation.

The important advance in the evolutionary theory of firms made by Nelson and Winter is the recognition of market competition as dynamic evolution; it goes beyond the conventional framework based on perfectly competitive equilibrium. The evolutionary theory, from the viewpoint of the behavioral science of firms, reveals an endogenous understanding of research and development in firms and thus endogenous recognition of the field of innovation. A key insight is the dynamic behavior of firms; that is, individual firms intentionally link technological advances with making decisions in a selective environment and evolve by commanding their behavioral routines. While this is surely a Schumpeterian approach, as Nelson and Winter described it, it seems to go further.

Present Study Based on Inductive Analysis

The present study departs from the scientific and technological background of innovation and instead relies on fact-finding by describing the trends in science and technology developments and on pursuing criteria by identifying common features among innovations. This is an inductive approach based on natural science and technology.

The most characteristic feature of this analysis is the recognition of the nonlinearity of innovation. This is based on the observation that an innovation proceeds based on a logistic equation not only in its diffusion period but also in the technological development period. We have observed various characteristic features based on the nonlinearity of innovations. This means that innovation reaches a maturity level within a definite time span. This leads to the observation that innovation is a specific function of time.

Another point to be stressed is the exogenous characteristic of the technology trajectory, which is a development within a field of scientific knowledge. If innovation development is regarded as a field of knowledge development including academic activities, innovation should be discussed within the more pervasive institutional sphere of society.

The present analysis assumes that innovation is a dynamic phenomenon brought about through knowledge transfer from person to person and aims to clarify the mechanism and the actual nature of innovation on the basis of such knowledge transfer. The recognition of a selection environment by Nelson and Winter is similar to ours in the sense of human relations. In particular, we accept the notion that an innovation evolves by trial and error, a Markov process, as described below. The present study, however, differ's in that it recognizes that the innovation stage ends with a deterministic process, forming a trajectory. Further, the present study demonstrates that the stage of technological development can also be described by an equation of physical diffusion. The recognition of the 'field' of knowledge transfer is discussed in detail later on.

We now focus on the above three features: nonlinearity, exogeneity and concept of field.

12.2 NONLINEAR NATURE OF INNOVATION

The present study has clarified the nonlinear nature of innovation. Two important aspects arise from this: the importance of time and the judgment of timing. When an innovation reaches maturity, the technology confronts a boundary it cannot cross – there is a limit on its contribution to and impact on the economy. The rate of development of innovation is also finite – it cannot accelerate infinitely. These limits result from the nonlinear nature of innovation.

Nonlinear Nature with Maturation in Innovation

The nonlinear nature of innovation means that each trajectory reaches a saturation level within a definite time. This makes it possible to identify each trajectory by simply determining the time span of the trajectory. This also indicates the timing for innovation, as shown by the timings of venture businesses, government policies and technology fusion. Thus, innovation should be recognized as a function of time.

Each trajectory has its own limitation. The technology trajectory, which consists of core technologies, reaches maturity at the time immediately before diffusion starts. The development trajectory has a limit on the introduction of new products, which cannot be expected after the maturity period. The diffusion trajectory has a limit on growth, which cannot be expected in a saturated market.

In short, the innovation paradigm has a nonlinear nature and has the following characteristics.

1. The innovation paradigm is composed of three nonlinear trajectories: technology, development and diffusion.
2. Each trajectory has its own dimension. If they are normalized to be dimensionless, they can be expressed in the same way irrespective of the dimension, and their interrelationships can be compared.
3. The rate of innovation development is determined by the time span of the logistic curves, which correspond to the rate coefficients of technology development and diffusion.
4. The fact that a constant-rate coefficient of diffusion is observed beyond the economic turbulence indicates that the diffusion of an innovation is a kind of physical phenomenon having an intrinsic rate coefficient.
5. It is possible to determine the appropriate timing of participation for various innovation aspects from the nonlinear nature of innovation.
6. The mature state of the innovation paradigm lasts until an alternative innovation emerges. If an alternative innovation takes hold and the performance of the new product exceeds that of the old one, the old innovation paradigm becomes obsolete and extinct. This is a kind of natural selection, similar to biological evolution, particularly Gouldian evolution. That is, if an alternative competitive innovation does not emerge, the innovation holds the market forever.

The determining procedure of the innovation paradigm is reviewed below. The logistic nature indicates that the curve converges to a limited value within a definite time span. If the curve is normalized to unity at the full scale of the vertical axis, the time span is a priori determined irrespective of the dimension of the scale. The time span is the value corresponding to the rate coefficient of the diffusion for the diffusion trajectory and to the rate coefficient of the technology development for the technology and development trajectories. As the actual trajectory is composed of discrete elements of innovation, it can be estimated simply by examining the cluster of a series of innovation elements lined up in the corresponding time sequence. That is, as individual elements are gathered within a certain time span, the actual trajectory can be quantitatively identified by determining the interval between the first and last elements. While this procedure seems to have some ambiguity, many trials have shown that a well-defined time span can be determined with reasonable accuracy if enough element points are collected. Although this is rather unexpected, actual use has shown that this procedure is sufficient for determining the time span of a trajectory. This is attributed to the fact that the trajectory has a nonlinear nature with a definite saturation and elements are intensively gathered around the mid-point of the logistic curve due to the logistic nature of innovation. Of further interest is the observation that the diffusion trajectory starts only when the technology trajectory has reached the mature stage, as shown in Chapter 6. In short, we can describe quantitatively the whole figure of an innovation paradigm by using the three trajectories.

Recognition of Time Dependence in the Innovation Paradigm

Innovation as a function of time

Innovation proceeds, changing the situation every minute as a function of time. Although conventional economics neglected the time factor, Schumpeter considered innovation to be a 'gale of creative destruction', moving dynamically through time, altering the economy. This is certainly recognition of the time factor. A major innovation appears suddenly, like a typhoon, and differs greatly from the ordinary activities of the economy. An innovation in progress is always growing and penetrating into the economy. Firms must thus recognize the timing of innovations. An appropriate response should be taken based on the timing along either the development or diffusion trajectory, and strategies should be adapted to the timing. We thus cannot discuss innovation without considering the time factor.

Value-added and timing of innovation

An innovation affects not only the relevant industry but also the whole economy. At maturity, it is fused into the economy, becoming a part of it. Economies grow, and the bigger the economy, the smaller the relative effect of the innovation.

The value added by an innovation is relatively small during its initial stage, that is, at the beginning of the diffusion period. There are three reasons for this. As the innovation product is initially rough and not particularly useful, the value it adds to the economy during the first-stage of products is low. Second, production costs are high and profitability is low due to the small scale of production. As the innovation progresses, economies of scale and scope kick in, and the profitability of the innovation increases. Third, and the most important factor, is the size of the market. It is small during the first stage, but over time demand increases and the market grows. The value added by an innovation thus increases by a multiplicative factor over time as a function of these elements. For example, although multimedia will eventually add much value to the economy, the amount is still low. The reason lies not in the performance of multimedia itself, but in the fact that it is still in the early stage of diffusion, too soon for profitability.

Inherent Potential of Innovation

An innovation product is improved and refined and eventually reaches its ultimate performance during the course of the innovation. This means that the performance of an innovation product has an intrinsic limit that can be attained – performance never exceeds this limit. An innovation product becomes refined as its possible performance become utilized, and the performance of the innovation becomes fully realized by which time the markets for the products mature. Thus, the market size is limited, and the impact of an innovation on the economy is limited. That is, each innovation has its own potential, which is never exceeded.

Innovations have contributed to the economy through their own unique performances. Innovation products each have their own level of performance, which is thoroughly utilized by the economy in the course of the innovation. The steam engine, for example, has been utilized in various ways. During the Industrial Revolution, steam engines were used as motive power for the textile industry, by iron manufacturers, and by iron works and then for steamships and steam locomotives. They made a significant contribution to economic development. They were, however, never made practical enough for automobiles and airplanes. That is, the steam engine has its own potential and cannot exceed it. Every innovation product has its own potential and economic niche. Consequently, an innovation can last 'forever' unless a competitive one emerges, or the innovation products are made obsolete by an alternative innovation that comes later.

Concept of Economic Impact Index

The existence of intrinsic potential in an innovation means that the impact of the innovation on the economy is ultimately limited, and each innovation has its own particular effect. Many innovations are not trunk innovations as defined in Chapter 5, and their impact is limited. In particular, when the population to be served is limited, the impact of the innovation is smaller. Various innovations that have received much attention have not always had a large impact on the economy. We have to estimate the size of the population to determine the potential impact. We can define an 'economic impact index' as the product of the valued added by an innovation and the potential population to be served.

In the discussion in Chapter 5, it was argued that many trunk innovations contributed to the economy through the construction of infrastructures. There have been many trunk innovations since the Industrial Revolution, as shown in Table 3.1. These innovations added much value to the economy by producing the upswings in the Kondratiev cycles. While many advanced technologies are expected to result in innovations within the next few years, their impact on the economy will not always be huge. Rather, the impacts seem to be getting smaller and smaller. Many advanced technologies have emerged over the years since the 1980s, among them biotechnology, fine ceramics, engineering plastics and advanced composites. The resulting products were surely high-technology products, and they were clustered in the same time frame. But their combined impact was not enough to generate an upswing in the Kondratiev cycle. One possible reason for this is that these innovations are derivatives of trunk innovations that appeared during the upswing of the fourth Kondratiev cycle after the Second World War. While engineering plastics are surely advanced technology products, they are simply a branch of general-purpose plastics. Therefore, while the value added by engineering plastics is high, the demand is limited due to the higher costs, so their total sales value is far less than that of general-purpose plastics. Biotechnology has been widely developed, but, at the moment, its impact is limited because the population of potential users of biotechnology products is limited. While anti-cancer drugs are highly appreciated by cancer victims, they are relatively few in number. We thus have to take into account the population of potential users when analyzing the impact of an innovation. Unfortunately, it is often difficult, if not impossible, to estimate the number of people who might benefit from an innovation. When, for example, Imperial Chemical Industries invented polyethylene in 1932, they were looking for an excellent electrical insulator for radar. They never expected it to find a general application because its production required extremely high pressure and high temperature, making it apparently uneconomical for general use.

Growth Rates of Innovation and the Rate-determining Step

Another interesting factor in the nonlinear nature of innovation is the spread of the logistic curve. The time span of the logistic curve is an index of the maturation time – how long does it take for the innovation to mature?

The growth rates of innovation and science paradigms have increased step by step since the Renaissance, around 500 years ago. The time span dropped from 150 years for Newtonian dynamics, to 90 years for electromagnetics, spinning machines and steam engines in the Industrial Revolution 230 years ago, to 40 years for synthetic dyestuffs, and now to 25 years for biotechnology and electronics. How much further will they fall?

Thanks to the recent development of information and communication technologies, the rate of information transmission is no longer the rate-determining step. Therefore, the rate-determining step of innovation must be determined by brain processing capacities because innovation comes about through knowledge transfer from person to person. Information passed from person to person is digested mentally by each person, eventually resulting in the creation of an innovation concept. Thus, as long as overall mental performance is not improved, the growth rate of innovation is restricted. What is the threshold?

The atomic bomb is the best example of something that has been developed without regard to expense or human resources and within the shortest possible time. Chadwick discovered the neutron in 1932 and in 1938 estimated the nuclear fission of the uranium atom by neutrons. In 1939, Meitner and Frisch predicted that a considerable amount of energy would be emitted by the nuclear fission of uranium. In 1939, two days before the outbreak of the Second World War, Bohr and Wheeler discovered that this process is a consecutive chain reaction induced by neutrons. The Manhattan Project was immediately instigated. Fermi and his team designed and constructed the first nuclear reactor, which produced plutonium, at Chicago University. It reached critical state in 1942. Two atomic bombs were dropped on Japan in 1945, the first on Hiroshima and the second on Nagasaki. The United Kingdom succeeded in producing plutonium in 1946 and began developing a nuclear bomb and a reactor for generating electric power. The nuclear reactor at Calder Hall, England commenced generation of electric power for commercial use in 1956. The United States tested a hydrogen bomb in 1952. This is the development history of nuclear fission.

The technology trajectory of nuclear fission apparently started with the discovery of radiation from uranium by the Curies in 1898 and ended with the discovery of the neutron in 1932, a span of 35 years. From the discovery of nuclear fission to the dropping of the first atomic bomb took only seven years. The development trajectory lasted for 24 years, from the discovery of the neutron in 1932 to the generation of nuclear power in 1956. This indicates that the time span of innovation was not shortened as had been expected for commercial development.

The term 'dog year' is now often used in reference to the short development period of an innovation. (A dog year is the equivalent of seven human years.) The actual development of the innovation trajectory, however, takes a rather longer time from the first basic invention or discovery. Innovation paradigms have various fractals. Each fractal has a comparatively short time span, like a dog year, as shown in the case of silicon cycles. This is the fractal in the electronics paradigm as discussed in Chapter 10. Looking at such fractals gives meaning to the expression 'dog year'. The important point is how to estimate the trajectory and paradigm. We have discussed the innovation paradigm, which has three trajectories that should be governed by a clear-cut discipline. This discipline should be made clear. This will be treated in the next section in more detail.

Infra-trajectories and Institutional Change

Chapter 5 described how the innovation paradigm has a more dynamic feature in the case of trunk innovations. After maturation of a trunk innovation paradigm during the upswing of a Kondratiev cycle, institutional change proceeds gradually across the downswing, and a greatly enhanced market develops during the upswing of the next Kondratiev cycle. The trunk innovation fully blooms during the next cycle following the completion of a robust infrastructure accompanying institutional change in the economy and even supports the development of new innovations. That is, the trunk innovation not only develops further during the next cycle but also induces various subsequent innovations. This longer trajectory across two Kondratiev cycles is called an 'infra-trajectory'; it forms a bundle fractal of infra-trajectories with induced innovations. This pervasive influence of trunk innovations brings about further institutional change in society. This can be expressed as 'exaltation of the economy'.

12.3 EXOGENEITY IN INNOVATION – A WIDER CONCEPT

The second feature of the present study on innovation is the wider concept leading up to the technological development period, starting from the original basic invention or discovery. This means that the study goes beyond the existing economics of technological change, which mainly focuses on the behavior of firms. In other words, this study is related to the recognition of exogenous factors of innovation. The innovation paradigm consists of three trajectories, of which the technology trajectory is the topic of discussion here. Simply speaking, conventional discussion regarding the economics of technological change is mostly restricted to the development and diffusion trajectories. The present study especially considers the technology trajectory.

Emergence of Innovation and Non-routine Nature

Schumpeter (1934a) described economic life as a 'circular flow', running in channels in an economic equilibrium system, in the way that blood flows through an animal. The equilibrium is altered by innovations, and changes in the framework of the economy appear discontinuously and revolutionarily. As Schumpeter put it, 'add successively as many mail coaches as you like, you will never get a railway thereby' (1934a, pp. 61–4). Innovation is not a routine part of the economy; it comes as an intruder. The economic system is in equilibrium, resulting from the accumulation of previous innovations, most of which have been absorbed into the economic system and become routine. When an innovation emerges, it perturbs the orbit of the economic systems, modifying them. Of course, the economic system is dynamic, always growing and changing. However, the above description indicates a relative situation between the economy and innovation.

Exogeneity of Technology Trajectories

As discussed in Chapters 6 and 7, the technology trajectory is mostly a science trajectory and has an exogenous nature. This is because in the early stage of the trajectory, recognizing the value of the innovation is difficult, as shown by the polyethylene example above. In this sense, the technology trajectory has an exogenous nature even within an industry.

In the case of steel-making innovations, as pointed out by Rosenberg (2000), the innovations of Bessemer, Siemens, and Thomas form a development trajectory of steel innovation, and the technology trajectory is the science trajectory of chemistry through which chemical analysis had first been made possible, which triggered the innovations. In this sense, the technology trajectory was exogenous to the iron manufacturers. Immediately after the recognition of chemistry as a tool for chemical analysis, the development trajectory commenced. In the course of the development trajectory of steel-making innovations, the iron manufacturers endogenized the chemical analysis and developed steel-making innovations.

In the case of the synthetic polymer industry, research people at DuPont and ICI carried out basic research in their laboratories concurrently with the technology trajectory of polymer science, which started with the postulation of the macromolecular theory by Staudinger in 1920. They succeeded in synthesizing nylon and polyethylene in 1928 and 1932, respectively. They worked during the course of the technology trajectory. As there had been no substantial synthetic polymer industry, there was little chance of knowledge transfer within the industry, but nevertheless they gleaned ideas from polymer science.

In the case of the electronics innovation paradigm, the vacuum tube industry was the forerunner, and Bell Laboratories intended to invent a device to replace the vacuum tube. The research, however, progressed in a direction different from vacuum tube technology. The researchers obtained knowledge of semiconductors, which had been almost completed as a science trajectory of semiconductors, and this led to the invention of the transistor using exogenous knowledge of semiconductor science, not as elongation from the knowledge of the vacuum tube. This is an example of bifurcation – the transition from vacuum tube technology to the semiconductor technology by a trigger from another side.

In the case of the steam engine, James Watt developed a steam engine practical enough for application, but he never intended to construct railways and steamships. The technology trajectory started with the work of von Guericke, Huygens, Papin, Savery and Newcomen. Watt's steam engine triggered the development trajectory of steam engines, which encompassed various applications of steam engines reflecting the demands of various industries.

In short, as far as the technology trajectory is concerned, inventions emerge from an initial trigger invention or discovery independently of the development and diffusion trajectories or industry. This means that inventions have an exogenous nature – they emerge independently of the later innovation realm. Epoch-making discoveries or basic inventions that trigger a technology trajectory do not arise from the demands of industry; they arise serendipitously. In this sense, the initiation of the technology trajectory must be exogenous.

Linear Model of Innovation Revisited

The development process of innovation has been discussed for a long time. A linear model was first proposed by Price and Bass (1969); it was then criticized by Kline (1985), who proposed a 'linked-chain model'.

Our analysis has been carried out within an innovation scheme consisting of three trajectories. The trajectories in the innovation paradigm are always arranged in the order of technology, development and diffusion, so far as we have examined. This means that basic research, new product development and market formation occur in that order. This is the concept of the linear model of innovation, which seems to differ from the accepted feedback model, especially the 'linked-chain model'.

There is, however, no contradiction between our three trajectories model and the feedback model. The difference lies in the essential role of the technology trajectory in our model, while the feedback model is limited to the development and diffusion trajectories. In our model, knowledge transfer between the three trajectories is indispensable, and the feedback effect is understood to be an essential factor. Knowledge transfer occurs among intra- and inter-trajectories, as it does in the feedback model. However, the technology trajectory is considered to trigger innovation, and knowledge transfer is mostly one way, from the technology trajectory to the development trajectory. This relationship requires revisiting the linear model of innovation.

Knowledge transfer is discussed in more detail in the next section. The knowledge about an innovation diffuses freely in related fields with feedback and the innovation process proceeds irreversibly, step by step, developing with the accumulation of knowledge. This is the origin of the linear model. Time's arrow never points back. In short, innovation has two faces: knowledge transfer in the innovation paradigm is essential with a link-chain model, and the process of innovation proceeds irreversibly with a linear model.

As described by Rosenberg (1976), innovation is not always initiated by science; many innovations have been triggered by something else. Nevertheless, without a trigger, an innovation cannot arise. The steam locomotive and subsequent railways would never have emerged without the invention of the steam engine. Cures for bacterial diseases depended on discoveries in bacteriology. Insulin could be synthesized due to the outstanding achievement of recombinant DNA technology. These inventions and discoveries triggered the technology trajectory and preceded the development of innovation products. This is simply a linear model of innovation.

Mowery and Rosenberg (1989), referring to Kuznets (1966), pointed out that there is general agreement with the view that the dependence of technological progress on science has increased substantially in the course of industrialization. We have observed that the number of scientific contributions by industrial researchers has greatly increased, even on the technology trajectory, as seen in the contributions by DuPont and ICI in the case of the polymer industry and by Bell Laboratories, Texas Instruments and Fairchild Corp. on the technology trajectory of the electronics paradigm, as described in Chapter 6. These contributions clearly indicate that basic scientific research has been intensively carried out in industry.

Later, Rosenberg (2000) recognized that the inventions of steel-making technologies by Bessemer, Siemens and Thomas were achieved based on the science of analytical chemistry, which was the technology trajectory according to our scheme. It is certainly true that scientific research has been deeply endogenized within industries, as described by Rosenberg (2000).

Rosenberg (1974) also criticized Schmookler's (1966) assertion of 'demand side innovation'. Rosenberg pointed out that:

> Progress in medicine had to await the development of the science of bacteriology in the second half of the nineteenth century. Although the field of medicine was one which attracted great interest, considerable sums of money, and large numbers of scientifically-trained people, medical progress was very small until the great breakthroughs of Pasteur and Lister. (1974, p. 97).

The most important thing is that an innovation paradigm is never initiated without a great achievement, a creative invention or a scientific discovery. This is an irreversible aspect, and the process is absolutely linear. The endogenizing tendency of science is different from the exogeneity of the initiation of the technology trajectory.

Vannevar Bush (1945) clearly recognized the importance of science in terms of national welfare and international competitiveness for enhancing the likelihood for innovation and recommended that the President set up an agency to promote scientific knowledge. The National Science Foundation was subsequently set up in 1950. Most scientific research is carried out in academia and is mostly exogenous to innovation. President Carter sent a message to Congress in 1979 regarding innovation that emphasized the importance of knowledge transfer from academia to industry. The Stevenson–Wydler Technology Innovation Act of 1980 has promoted such transfer. President Reagan authorized the establishment of science centers in universities that would facilitate the transfer of knowledge from academia to industry, thereby fostering innovation based on basic research. This is a typical recognition of the exogeneity of science. These actions from Vannevar Bush through Carter to Reagan are a typical linear model approach. The United States recognized the importance of basic research in academia as the origin of new industries and took actions to trigger the emergence of new industries and foster new ventures. There is no doubt that this accounts for the international competitiveness of the United States.

Two Concepts of Evolutionary Theory

The development of biological evolutionary theory was substantially started by the work of Lamarck in 1809; it continued with Wallace's natural selection theory and Darwin's *The Origin of Species* (1859). Both Darwin and Wallace had been affected by Malthus's 'An Essay on the Principle of Population' (1803), which predicted a catastrophe resulting from the geometric increase in population against the arithmetical increase in subsistence. While Darwin adopted the idea of natural selection by which species variation is continuous, that is, the process of evolution, de Fries adopted the idea of mutation in 1901 by which species variation is discontinuous. The unified theory is called Neo-Darwinism.

Many economists have been affected by biological evolutionary theory, especially by Darwinism. Hodgson (1993), Andersen (1994) and Witt (2003) summarized the evolutionary approach in economics. Hodgson classified the behaviors of economists into developmental evolution and genetic evolution. The developmental evolution group is represented by Marx and Hobson, and the genetic evolution group includes two subgroups: the ontogenetic group of Smith, Menger, Walras, Marshall and Schumpeter and the phylogenetic group of Spencer, Hayek, Malthus, Veblen and Nelson and Winter (Hodgson, 1993, p. 39).

Veblen (1898) applied Darwinian thought to his evolutionary economics, which became the manifesto of American Institutionalism. Schumpeter's 'gale of creative destruction', can be interpreted as an evolutionary theory of economics. Alchian (1950) drew an analogy between the evolutionary development of the economy and biological evolution. Winter (1964) discussed economic natural selection and the theory of firms. Nelson and Winter (1982) combined Schumpeter's ideas with the natural selection metaphor to skillfully describe the innovative behavior of industrial firms as the same process as the evolution of life. In this evolutionary theory of economics, they described the behavior of firms as routines, which play the same roles as genes. Routines evolve through regeneration, species variation and natural selection. Firms having supreme routines survive in the struggle for existence, as in the process of natural selection. This scenario is a much wider description of the evolutionary change of the economy than the mere development of the innovation process itself. Metcalfe (1994a, 1998) formulated natural selection as a competitive process by using Fisher's principle of natural selection and increasing return in the selection process under the notion of Schumpeter's creative destruction. Fisher (1930) proposed a principle of population genetics based on Darwinian theory that relates genetic variance in terms of behavior within a given population to the speed with which selection eliminates inferior traits from the population. Foster and Metcalfe (2001) summarized the recent developments in evolutionary economics.

An organism evolves, and its development consists of two processes: ontogeny and phylogeny. Each organism has its own life, as termed by ontogeny, in which the phylogenic history is revived within the embryonic stage. Phylogenic evolution is achieved through repeated stages of ontogeny. As described by Hall (1998), an organism has two kinds of evolution: micro-evolution, which is the evolution within a species (more properly, within a family in terms of biological taxonomy), and macro-evolution, which is the evolution of species. If we apply this taxonomy to the innovation process, micro-evolution corresponds to the progress of innovation proceeding along the innovation trajectories. This is the development of an innovation paradigm corresponding to the evolution of human beings from apemen to homo erectus and homo sapiens. Macro-evolution corresponds to the jump of an innovation paradigm to another paradigm through evolution. This can be described as 'paradigm evolution' or a 'paradigm shift'. It corresponds to the bifurcation of mankind from apes. Hall expressed a more striding evolution in his concept of 'mega-evolution', which corresponds to the evolution accompanying a big bifurcation, for example from reptiles to mammals. This might correspond to the evolution of modern civilization that started with the Industrial Revolution in comparison with the ancient civilization.

The evolutionary theory of Nelson and Winter (1982) is deemed 'micro-evolution' because it focuses endogenously on the development of firms within the same domain of routines; however, this is a much wider concept than mere technological change. Nevertheless, the process of economic change could be better discussed in terms of our innovation dynamics, which has a nonlinear nature, with time dependence.

Schumpeter (1939) coined the term 'railroadization' to represent the railway innovation paradigm, in which the steam locomotive was invented, railways were developed, and stagecoaches were replaced. He referred to this innovation process as an evolutionary process and recognized maturation as the final stage. This concept was described as 'punctuated evolution' by Awan, as described in Andersen (1994). As mentioned above, every innovation paradigm we examined had a nonlinear nature and reached maturity within a definite time span. An innovation paradigm evolves intermittently, jumping to a newer paradigm if there is a trigger. The iron-making paradigm emerged during the upswing of the first Kondratiev cycle and evolved into the steel-making paradigm during the third Kondratiev cycle. The electromagnetic science paradigm was followed by the vacuum tube paradigm and the semiconductor paradigm, which will be followed by the nanodevice paradigm, all of which form a series of evolved surges. Science paradigms also have definite time spans, reach maturity, and evolve from time to time. In theoretical physics, we have Newtonian dynamics in Euclidian space and relativity in non-Euclidian space; the ultimate unified theory is still pending. These paradigms have intermittently emerged through a macro-evolutionary process.

Evolutionary theory in biology has evolved through the introduction of knowledge from molecular biology and genetics. Knowledge of genes inspired the formation of a more sophisticated mechanism of evolution and led to 'the neutral theory of molecular evolution' by Kimura (1968). Another interesting theory, 'punctuated equilibrium theory', was proposed by Gould (1977) and Gould and Eldredge (1977). They asserted that evolution takes place within a fixed time period, and, after it is completed, the species remains unchanged. According to this theory, the timing of evolution coincides with the timing of species bifurcation. The coelacanth evolved 350 million years ago and has remained as it was. Stromatolite (cyanobacteria) emerged 3.5 billion years ago and still live along seashores in Australia. Innovations also emerge during a fixed period of time, creating a paradigm, and then mature. After maturation, the innovation products survive until a stronger competitor comes along. The steam locomotive still has the same structure and functions as when it was invented 200 years ago. Of course, innovation products are incrementally improved through micro-evolution along the trajectories of the paradigm during the course of the innovation.

12.4 THE FIELD OF KNOWLEDGE TRANSFER AND INNOVATION

Division of Labor in Innovation and Knowledge Transfer Fields

A key feature of the innovation paradigm is the transfer of knowledge about the innovation within each trajectory and across trajectories. Such transfer takes place within the field of knowledge transfer, which has been formulated by various 'actors' in innovation trajectories. Nelson and Winter (1982) defined 'routines' that occur in the course of economic change, and various divisions of labor can be identified that carry out these routines in the course of innovation. Each trajectory has its own 'actors'. In the case of the technology trajectory, the actors are often research people in academia who transfer knowledge to research and marketing people in industry, which results in the formation of the development trajectory. In the diffusion trajectory, knowledge is transferred, for example, from the research people to the production and sales people. That is, an innovation develops through the efforts of many actors in various areas.

An innovation takes place through the emergence of new knowledge within the mass of people in the field. New knowledge is created in the course of information transfer from person to person. Similarly, the species information of genes is transferred to descendants through DNA. Evolution takes place through gene mutation, while innovation takes place through information transfer, person-to-person. Each person combines the information with previous experience, which results in the creation of tacit knowledge. Innovations emerge through this process. Innovation never takes place based on the ability of a single individual; great inventions and novel discoveries emerge through the varied experiences of many people. The accumulation and incubation of knowledge are thus crucially important.

The paradigm of innovation is described as follows. The technology trajectory tends to emerge in academia, and core technologies develop along the science trajectory, substantially apart from the demands of the economy. During the maturity period of the technology trajectory, substantially the science trajectory, the core technologies become almost completed. They then begin migrating to related industries, or entrepreneurs set up venture businesses to exploit them. The firms involved compete to succeed, and micro-evolution proceeds. Various new products appear and are gradually squeezed towards a single product or dominant design due to lock-in, as described in Chapter 7. During the course of an innovation, knowledge about it spills over, making it hard for the inventing firm to maintain its proprietary rights. The innovation paradigm develops into a robust entity, which cannot be easily controlled by any one entity.

Uncertainty and the Markov Chain Process

Nelson and Winter (1982) described an evolutionary model in which firm behavior is a kind of Markov process (pp. 19, 147, 151, 175). That is, the evolutionary system of innovation progresses through trial and error – a kind of stochastic system with high uncertainty. Tracing of the preceding phenomena produces a Markov chain. This reflects the causal relationships between the series of inventions in the innovation paradigm. The resulting paradigm in our study, in contrast, is produced by well-defined trajectories in a deterministic process. There is no contradiction, however, because although there is much uncertainty during the course of an innovation due to its stochastic nature, it ultimately reaches a deterministic logistic curve. This is ascribed to the self-organizing nature of innovation – a deterministic trajectory is finally achieved following a stochastic phase of trial and error. This is similar to the two phases of light, a stochastic phase when it is a wave and a deterministic phase when it is a particle.

Complexity in Innovation Systems and Empirical Evidence

Recent discussions of complexity in innovation systems have not cited enough empirical evidence. We have tackled this issue and have examined complexity in innovation systems to confirm its existence.

Existence of fractals

We have observed various kinds of fractals in the innovation paradigm, as shown in Figures 10.3, 10.4 and 10.5, which illustrate chain fractals, bundle fractals and system fractals, respectively, in the electronics paradigm. Our observations clearly indicate the existence of fractals along the development and diffusion trajectories. The development trajectory itself is composed of fractals of new products; the application of these products forms fusion fractals, and the combination of these products with software forms system fractals. These fractal formations are related to the conventional economy and are the origin of institutional change.

Existence of self-organization

We also observed self-organization phenomena, evidenced by cluster formation during the upswings of Kondratiev cycles, fusion trajectory formation along the development trajectory, and trajectory formation itself in the innovation paradigm. These attractors of self-organization and fractal formation are ascribed to the value added by innovation.

Acceptable Difference Equation and its Unique Feature

The field of knowledge transfer in the innovation paradigm is actually a discrete system because of the person-to-person transfer of knowledge. This phenomenon in a discrete system cannot be expressed by a logistic equation; a difference equation must be used. It is, however, given the mathematical convention of continuity, permissible to apply a logistic equation to a system consisting of a large number of people as a social community of a homogeneous system. The innovation paradigm is usually applied to this kind of system. In the case of a discrete system, May's mapping (May, 1976), for example, corresponds to a logistic equation. When we apply this difference equation to the actual diffusion of innovation products, as shown in Table 2.1, the magnitude of the diffusion coefficients indicates an ordinary condition tending to $1 - 1/\beta$, which is out of the ergodic and chaotic regions, as described in Section 10.4.

May's mapping, however, is not a monotone increasing function but widely changeable depending on *a* value. This is not the case of the actual innovation systems. A practically applicable logistic difference equation is derived through bilinear analysis in a condition of gauge invariance. This difference equation has a unique property in which any solution of the mapping fulfills the solution of the original differential equation. This unique feature suggests a specific nature of the logistic equation that any fractals can be mathematically included within the original logistic trajectory irrespective of any time span of the fractal s-curve.

Knowledge Development Dynamics

Innovation is a nonlinear and discrete system created by knowledge transfer from person to person and such discrete knowledge dynamics can be generalized to cover a much wider range of human behavior, including various human knowledge activities such as science, art, ideology, exploration and various social behaviors. Thus, human knowledge behavior can generally be termed 'knowledge development dynamics' by applying logistic dynamics to all the fields of knowledge transfer. That is, this dynamics is applicable to all human knowledge activities. As discussed in Chapters 8 and 11, the governmental actions taken by countries to foster electronics innovation, the history of the Crusades, and the history of the Great Voyages can all be described by a logistic dynamics with a finite development locus because human activities reach a limit within a finite time. These analyses were made feasible by the estimation of the time span of the distributed cluster of the relevant elements of the logistic wave.

The fact that human knowledge activities reach a limit or mature within a finite time means that these behaviors have a life cycle and can be expressed by a sigmoid curve. This is very important because it means that human knowledge activities can be placed on a historical time scale and historical footprints can be identified. As pointed out by Winston (1974), Marchetti (1985, 2002) and Modis (1992) various human activities can be described by logistic sigmoid curves; acquisition of children's vocabulary and the life works of composers, novelists and great scholars. While these descriptions are of the life cycles of individuals, the trajectories of arts and sciences shown in Figure 11.21 are cumulative trajectories composed of individual fractals. Marchetti (1986, 1990) found that the history of exploration of the Western hemisphere by Columbus and the history of the building of Gothic cathedrals in the twelfth to the fourteenth centuries can be described by a logistic equation. These studies indicate that human activities can be described using a logistics curve.

Why should this sigmoid curve be a logistic one? One piece of evidence comes from the Fisher–Pry plot of a series of relevant products and patents in an innovation paradigm, as shown in Figure 6.1. And various trends in innovation have been described using a logistic equation, as done, for example, by Marchetti (1979, 1980, 1996) and Andersen (2001). That is, knowledge of an innovation is transferred from person to person and eventually reaches a finite level at maturation. This phenomenon can be most simply described by a logistic equation, in the same way as an epidemic model can describe the prevalence of infectious disease and a growth model can describe the growth of organisms. That is, a logistic equation is the only equation that can represent simple interaction between possessors and non-possessors without any modification, as discussed in Chapter 10. This means that a logistic equation should be the most appropriate equation for expressing knowledge transfer between persons, and 'knowledge development dynamics' can be applied to a wide variety of knowledge transfer behaviors in human society. The important thing, however, is the sigmoid nature of knowledge activities – they converge at maturity within a definite time span.

In knowledge development dynamics, the slowdown of the growth rate in the later period is ascribed to the decrease among the population of knowledge non-possessors within a finite population of capacity, not to any external depression factor. In the case of innovation, the technology trajectory descends during the later period due to technological maturation, that is, new knowledge exhaustion for core technologies corresponding to the potential of the innovation itself. The Age of Great Voyages ended when Magellan circumnavigated the globe, and thus the history of that age can be described using logistic dynamics as a knowledge exploring process.

12.5 INDUCTIVE THINKING AND ENTITY OF INNOVATION

The study described in this book was a comprehensive analysis of innovation dynamism and a description of the innovation paradigm with three trajectories. This groundbreaking analysis revealed several new insights. This final part summarizes the specific features of the study and compares them with those of other economic approaches.

Characteristic Nature of Innovation Dynamism

The innovation paradigm has been interpreted in various ways. Andersen (1994) examined the methodologies of innovation studies and compared various ways of thinking and approaches – mostly focusing on evolutionary economics. We have discussed most of them in this study. The key words in studies of innovation behavior are summarized as follows:

1. Empirical inductive approach or deductive thinking
2. Discrete system – heterogeneous principle and population thinking
3. Evolutionary thinking – routines, variance of species, and natural selection
4. Field of knowledge transfer and path dependence
5. Markov chain and self-organization in dissipative systems
6. Nonlinearity and complex systems
7. Punctuated evolution versus gradualism.

Way of thinking

Schumpeter (1939) took the standpoint of 'empirical analysis' throughout his studies and discussed the innovation paradigm on the basis of fact-finding. He precisely analyzed the transition of innovations, and, based on his analysis, he developed a theory of business cycle formation. Many economists followed his inductive approach and developed economic theories to explain technological change. We followed the same inductive approach using empirical data and uncovered the nonlinear dynamics of the innovation paradigm.

Most economists, however, have developed theories using a deductive process. Nelson and Winter, for example, used deductive reasoning and an analogy to biological evolution to develop their theory of evolutionary economics. This was an important step towards describing the behavior of firms. This theory has been widely accepted and used to explain the role of firms in the innovation process. Evolutionary economics follows the Darwinian ideas of gradualism. On the other hand, the theories of complexity developed by

Goodwin, Peters and Lorenz are punctuated consideration under a nonlinear concept, but they have not been verified using concrete data. That is, they used a deductive approach. Our approach is inductive based on empirical data, and the nonlinear logistic approach corresponds to punctuated consideration.

Complexity and cumulative nature of innovations

Another point is that innovations independently emerge through serendipity. Economic development has been arbitrarily brought about by such innovations. Goodwin developed an important approach to complex systems. He interpreted the long business cycle by analyzing complex systems based on the assumption that the nonlinear logistic dynamics of innovation could have a cyclical nature. He also recognized the importance of the self-organizational nature of nonlinear dynamics in producing a 'swarm' of innovations. These analyses by Goodwin, Peters and Lorenz indicate that there is a universal criterion for innovation dynamics, but they have not been verified empirically. In contrast, our study presents strong support for their propositions. That is, innovation dynamics exhibits a logistic nature, the innovation clusters formed during the upswings of the Kondratiev cycles indicate a self-organizational characteristic, and substantial fractal formations have been observed. The present study revealed a complex system in a discrete entity based on concrete evidence.

The field of innovation is not homogeneous; rather, innovations intermittently and spontaneously emerge in a discrete system. Economic society is fulfilled through various innovations that have been developed in the past. Each innovation has its own pace and timing of development, so that each time frame in the economy has various stages of innovations. Each innovation paradigm can be recognized by its position on the trajectories. There is no universal phase of innovation; rather, each innovation has its own unique timing. Another characteristic of innovation systems is their cumulative nature, which is the origin of economic development. The clusters of innovations that emerge during the upswing are cumulatively incorporated into the economic entities, becoming new components of the economy. This is also the origin of the hierarchical complex structure of the economy. Various causes of business cycle formation have been proposed, but only innovations generate the accumulation of value-added to the economy. This is a unique feature of innovations.

Innovation Paradigms and Hierarchy

Our innovation paradigm consisting of three trajectories has certain phases. The technology trajectory is initially vague and not foreseeable, but after completion it is easy to identify and has a fairly rigid structure, from which the time span is easily determined. The transition of the time span is well defined, as shown in Figure 6.4. The development trajectory has almost the same time span as the technology trajectory and is composed of several stages of fractals. Innovation products are developed through the formation of chain fractals, as shown for IC chips in Figure 10.3. The application of innovation products forms bundle fractals, as shown for electronics fusion in Figure 10.4. Elementary technological systems form chain fractals, as shown in Figure 10.5. These fractals emerge on the development trajectory and result in a series of corresponding small diffusion trajectories that lead into the diffusion trajectory of the original paradigm. The infra-trajectories extending to the next Kondratiev cycle form bundle fractals with a longer time span. The occurrence of Kitchin and Juglar cycles are sometimes brought about by the formation of various types of fractals. These structures are discrete and quantumized: they have a definite time span and are incorporated into a single parent logistic curve, which is an intrinsic feature of the logistic equation, as discussed in Chapter 10.

Thus, an innovation paradigm has a complex hierarchical structure. The most important point in innovation dynamism is that each innovation paradigm has its own clear-cut discipline on which it depends. This is the basis of the establishment of the innovation paradigm with a definite time span. Within a paradigm, there is the quasi-continuous field of micro-evolution, which is composed of development and diffusion trajectories, as described by Nelson and Winter. The jump from one paradigm to another occurs intermittently as part of a macro-evolution.

Mega-Structure of the Modern Industrialized Society

Since the Industrial Revolution, various innovations have emerged, from which our sophisticated industrialized society has developed. The innovations that cluster during each upswing of the Kondratiev cycle mature, leading towards a bubble economy caused by the excitement generated by the effect of the higher added value brought by the innovations. This in turn leads to depression, which dampens the take-off of new innovations. Innovations, especially trunk innovations, have benefited the economy by constructing new infrastructures. Trunk innovations have constructed a highly sophisticated economic society, which could be now entering a maturity phase. That is, the innovation paradigms have evolved several times, and almost all possible paradigms have already been played out. Actually, several generations of evolved paradigms could form a mega-structure, which is a larger S-curve having a time span of about 300 years, corresponding to five or six Kondratiev cycles. This might be the S-curve of the modern industrialized society.

While the growth rate of modern civilized society will soon decrease, there will still be enthusiasm for growing the economy. Modern society was limited to Europe in the nineteenth century; it subsequently expanded to North America, and then to Japan and Asia. This trend indicates that the growth rate of the human economy has risen year by year and that we have entered an accelerating phase as China and India modernize. This might be a kind of bubble phenomenon, which we have experienced several times in the past. As seen in Chapter 4, just before the maturity phase in an innovation paradigm, a bubble economy develops, followed by a crash. This must be the overshoot phenomenon pointed out by Meadows et al. (1992). We may suffer from a sudden catastrophe following the escalation of industrialization. We will likely encounter various difficulties, such as shortages of energy and resources, and face various environmental problems. I believe that the only way to solve them is the timely development of advanced innovations.

Bibliography

Abramovitz, M. (1989), *Thinking about Growth*, Cambridge: Cambridge University Press.

Achilladelis, B. (1990), 'The Dynamics of Technological Innovation: The Case of the Chemical Industry', *Research Policy*, **19**, 1–34.

Achilladelis, B., A. Schwarzkoph and M. Cines (1987), 'A Study of Innovation in the Pesticide Industry: Analysis of Innovation Record of an Industrial Sector', *Research Policy*, **16**, 175–212.

Adachi, T. (1981), 'Taking off Japanese Aerospace Industry', Long-Term Credit Bank of Japan, Research Report No. 85 [10], 3–52 (in Japanese). Original source: *Aeronautical Journal*, March 1976.

Aftalion, F. (1991), *A History of the International Chemical Industry*, Philadelphia: University of Pennsylvania Press.

Aida, H. (1997), *Development of Japanese Electronics Industry, 2nd Version, No. 5 Huge Electronics Systems*, Tokyo: NHK Publications: (in Japanese).

Akimoto, E. (1999), *A History of American Economy 1492–1993*, Tokyo: University of Tokyo Publishing Ltd (in Japanese).

Alchian, A.A. (1950), 'Uncertainty, Evolution and Economic Theory', *Journal of Political Economy*, **58**, 219–22.

Alchian, A.A. (1953), 'Biological Analogies in the Theory of the Firm: Comment', *American Economic Review*, **3**, 600–603.

Andersen, B. (1999), 'The Hint for S-Shaped Growth Paths in Technological Innovation. A Patent Study', *Journal of Evolutionary Economics*, **9**, 487–526.

Andersen, B. (2001), *Technological Change and the Evolution of Corporate Innovation*, Cheltenham, UK and Northampton, MA: Edward Elgar.

Andersen, E.S. (1991), 'Techno-Economic Paradigms as Typical Interfaces between Producers and Users', *Journal of Evolutionary Economics*, **1**, 119–44.

Andersen, E.S. (1994), *Evolutionary Economics – Post Schumpeterian Contributions*, New York: Pinter.

Aoyama, S. (1999), *Division of Labor on High-tech Networks – How Strong is Taiwanese Electronics Industry?*, Tokyo: Hakuto Publishers (in Japanese).

Arora, A., R. Landau and N. Rosenberg (1992), *Chemicals and Long-Term Economic Growth – Insights from the Chemical Industry*, New York: John Wiley & Sons, Inc.

Arrow, K.J. (1962a), 'The Economic Implications of Learning by Doing', *Review of Economic Studies*, **29**, June, 155–73.

Arrow, K.J. (1962b), 'Economic Welfare and the Allocation of Resources for Invention', in *University–National Bureau Committee for Economic Research, The Rate and Direction of Inventive Activity*, Princeton, NJ: Princeton University Press.

Arrow, K.J. (1969), 'Classificatory Notes on the Production and Transmission of Technical Knowledge', *American Economic Review*, **59** (2), 29–35.

Arthur, W.B. (1985), 'Competing Technologies and Lock-in by Historical Small Events: The Dynamics of Choice under Increasing Returns' C.E.P. Research Paper 4, Stanford. (Revised Version of IIASA, Paper WP-83-90, in 1983.)

Arthur, W.B. (1987), 'Urban Systems and Path-Dependence', in R. Herman and J. Ausubel (eds), *Urban Systems and Infrastructure*, Washington, DC: National Academy of Science/National Academy of Engineering.

Arthur, W.B. (1988), 'Self-Reinforcing Mechanisms in Economics', in P.W. Anderson K.J. Arrow and D. Pines (eds), *The Economy as an Evolving Complex System*, Redwood City: Addison-Wesley, pp. 9–31.

Arthur, W.B. (1989), 'Competing Technologies, Increasing Returns, and Lock-in by Historical Events', *Economic Journal*, **99**, 116–31.

Arthur, W.B. (1994), *Increasing Returns and Path Dependence in the Economy*, Ann Arbor, MI: The University of Michigan Press.

ATSS (1987), *Recent Traffic Signal Technologies in National Railways*, Tokyo: Association for Traffic Signals and Safety (in Japanese).

Auto, Japan (2001), *The Motor Industry of Japan*, Japan Automobile Manufacturers Association (in Japanese).

Avadikyan A., P. Cohendet and J.-A. Heraud (eds) (2003), *The Economic Dynamics of Fuel Cell Technologies*, Berlin: Springer-Verlag.

Baarspul, M. (1990), 'A Review of Flight Simulation Techniques', *Progr. Aerospace Sci*, **27**, 1–120.

Barro, R.J. (1991), 'Economic Growth in a Cross-Section of Countries', *Quarterly Journal of Economics*, May.

Barro, R. and X. Sala- I- Martin (1995), *Economic Growth*, New York: McGraw-Hill, Inc.

Bass, F.M. (1969), 'A New Product Growth Model for Consumer Durables', *Management Science*, 15, 215–27.

Beckman, R. (1988), *Crashes*, London: Sidgwick & Jackson Ltd.

Berhulst, P.F. (1838), 'Notice sur la loi que la population suit dans son accroissement', *Correspondances Mathematiques et Physiques*, **10**, 113–21.

Bernhardt, I. and K.M. MacKenzie (1972), 'Some Problems in Using Diffusion Models for New Products', *Management Science*, **19**, 187–200.

Betz, F. (1987), *Managing Technology*, Engle wood Cliffs, NJ: Prentice Hall.

Betz, F. (1998), *Managing Technological Innovation – Competitive Advantage from Change*, New York: John Wiley & Sons, Inc.

Blackman, A.W. (1974), 'The Market Dynamics of Technological Substitutions', *Technological Forecasting and Social Change*, **6**, 41–63.

Bresnahan, T.F. and M. Trajtenberg (1995), 'General Purpose Technologies: Engines of Growth?', *Journal of Econometrics*, **65** (1), 83–108.

Brown, L.A. (1981), *Innovation Diffusion: A New Perspective*, New York: Methuen.

Brown, W.A. (1940), *The International Gold Standard Reinterpreted, 1914–34*, New York: National Bureau of Economic Research.

Bush, V. (1945), *Science – the Endless Frontier: A Report for the President*, Washington DC: United States Government Printing Office.

Cameron, H.M. and J.S. Metcalfe (1987), 'On the Economics of Technological Substitution', *Technological Forecasting and Social Change*, **32** (2), 147–62.

Campbell, C.J. (1988), *The Coming Oil Crisis*, Brentwood, Essex: Multi-Science Publishing Co. Ltd.

Campbell, C.J. (2005), *Hubbert Peak of Oil Production*, http://www.hubbertpeak.com/campbell/.

Campbell, C.J. and J. Laherrère (1998), 'The End of Cheap Oil', *Scientific American*, March 1998, 78–83.

Carter, C.F. and B.R. Williams (1957), *Industry and Technical Progress*, Oxford: Oxford University Press.

Casetti, E. and R.K. Semple (1969), 'Concerning and Testing of Spatial Diffusion Hypotheses', *Geographical Analysis*, **1**, 245–59.

Census, USA (2002), *US, Statistical Abstract of the United States, Motor Vehicle Registrations*, Bureau of the Census, Washington DC, G.P.O.

Chancellor, E. (1999), *Devil Take the Hindmost: A History of Financial Speculation*, London: Gillon Aitken Associates Ltd.

Ching, H. and T. Hayashi (1995), *Technological Development and Technogogy Transfer in Asia*, Tokyo: Bunshindo (in Japanese).

Clarke, H. (1847), 'Physical Economy – a Preliminary Inquiry into the Physical Laws Governing the Periods of Famine and Panics', *Railway Register*, reprinted in Ichikawa (1984), *Long Waves in World Business Activities*, Tokyo: Aki Shobo.

Coleman, J.S., E. Katz and H. Menzel (1966), *Medical Innovation: a Diffusion Study*, Indianapolis: Bobbs-Merrill.

Converse, P. (1959), *Fifty Years of Marketing in Retrospect*, University of Texas: Bureau of Business Research.

Coombs, R., P. Saviotti and V. Walsh (1987), *Economics and Technological Change*, London: Macmillan Publishers.

Coombs, R., P. Saviotti and V. Walsh (1992), *Technological Change and Company Strategies*, London: Harcourt Brace Javanovich.

Cowan, R. (1990), 'Nuclear Power Reactors: A Study in Technological Lock-in', *Journal of Economic History*, **50** (3), 541–67.

Cusumano, M., Y. Mylonadis and R. Rosenbloom (1992), 'Strategic Maneuvering and Mass-Market Dynamics: The Triumph of VHS over Beta', *Business History Review*, **66**, Spring, 51–94.

David, P.A. (1975), *Technical Choice, Innovation and Economic Growth*, New York: Cambridge University Press.

David, P.A. (1985), 'Clio and the Economics of QWERTY', *American Economic Review*, **75**, May, 332–7.

David, P.A. (1986), 'Understanding the Economics of QWERTY: The Necessity of History', in William N. Parker (ed.), *Economics, History and the Modern Economist*, Oxford: Basil Blackwell.

David, P.A. (1988), 'Path Dependence: Putting the Past into the Future of Economics', Discussion Paper, Stanford University.

David, P.A. (1991), 'Computer and Dynamo: The Modern Productivity Paradox in a Not-too-distant Mirror in Technology and Productivity', in *The Challenge for Economic Policy*, Paris: OECD.

David, P.A. (1992a), 'Path-dependence in Economic Processes: Implications for Policy Analysis in Dynamical Systems Context', Discussion Paper, Center for Economic Policy Research, Stanford University.

David, P.A. (1992b), 'Heros, Herds, and Hysteresis in Technological History', *Industrial and Corporate Change*, **1** (1), 139.

David, P.A. (1993), 'Path-dependence and Predictability in Dynamical Systems with Local Network Externalities: A Paradigm for Historical Economics', in D. Foray and C. Freeman (eds), *Technology and Wealth of Nations*, London: Pinter, pp. 208–31.

Davies, S. (1979), *The Diffusion of Process Innovations*, New York: Cambridge University Press.

Davis, L.E. et al. (1972), *American Economic Growth: An Economist's History of the United States*, New York: Harper and Row.

De la Monte, J. and G. Paquet (eds) (1996), *Evolutionary Economics and the New International Political Economy*, London: Pinter.

Deffeyes, K.S. (2001), *Hubberts Peak: The Impending World Oil Shortage*, Princeton, NJ: Princeton University Press.

Denison, E. (1962), 'United States Economic Growth', *Journal of Business*, **35**, 109–21.

Dertouzos, M.L., R.K. Lester and R.M. Solow (1989), *Made in America*, Cambridge, MA: MIT Press.

Dickinson, F.G. (1940), 'An Aftercost of the World War to the United States', *American Economic Review*, **XXX** (1), Supplement (Part 2), 326–39.

Dixon, R. (1980), Hybrid Corn Revisited, *Econometrica*, **6** (9), 1451–61.

DOC, USA (1940), *Historical Statistics of the United States*, US Department of Commerce.

DOC, USA (1986), *Historical Statistics of the United States*, Vol. 2, US Department of Commerce.

Domar, D. (1946), 'Capital Expansion, Rate of Growth and Employment', *Econometrica*, **14**, 137–47.

Domar, D. (1957), *Essay in the Theory of Economic Growth*, New York: Oxford University Press.

Dosi, G. (1982), 'Technological Paradigms and Technological Trajectories: A Suggested Interpretation of the Determinants and Directions of Technical Change', *Research Policy*, **11**, 147–62.

Dosi, G. (1984), *Technological Change and Industrial Transformation*, London: Macmillan Press Ltd.

Dosi, G., C. Freeman, R. Nelson, G. Silverberg and L. Soete (eds) (1988), *Technological Change and Economic Theory*, London: Pinter Publishers.

Duncan, R. (2000), 'World Oil Forecast #5', Geographical Society of America, Summit 2000, Pardee Keynote Symposia, Reno, 13 November 2000, reprinted in Heinberg (2003), pp. 103–4.

DuPont (1952), *DuPont – The Autobiography of an American Enterprise*, Wilmington, Delaware: E.I. DuPont de Nemours & Company.

Eckermann, E. (1981), *vom Dampfwagen zum Auto*, translated into Japanese by R. Matsumoto (1996), *World History of Automobiles*, Tokyo: Grand Prix Publishers.

Endo, M. et al. (1986), 'Electronic Control and Reliability of Engines', *Journal of Aerospace Association Japan*, **34** (388), 256–61 (in Japanese).

Energy Shigen Gakkai (1991), *Energy and Future Society*, Tokyo: Energy and Resources Saving Center (in Japanese).

Enos, J.L. (1958), 'A Measure of the Rate of Technological Progress in the Petroleum Refining Industry', *Journal of Industrial Economics*, **6**, June.

Enos, J.L. (1962), *Petroleum Progress and Profits: A History of Process Innovation*, Cambridge, MA: MIT Press.

Feigenbaum, M. (1978), 'Quantitative Universality for a Class of Non-linear Transformations', *Journal of Statistical Physics*, **19**, 25–52, and **21**, 669–706.

Fisher, J.C. and R.H. Pry (1971), 'A Simple Substitution Model of Technological Change', *Technological Forecasting and Social Change*, **3**, 75–88.

Fisher, R.A. (1930), *The Genetical Theory of Natural Selection*, Oxford: Oxford University Press.

Floyd, A. (1968), 'A Methodology for Trend Forecasting of Figures of Merit', in L. Bright (ed.), *Technological Forecasting for Industry and Government: Methods and Applications*, Inglewood Cliffs, NJ: Prentice Hall, pp. 95–105.

Forrester, J.W. (1977), 'Growth Cycle', *De Economist* (The Netherlands), **125** (4), 525–43.

Forrester, J.W. (1979), 'Innovation and Economic Long Wave', *Management Review*, June, 16–24.

Foster, J. and J.S. Metcalfe (ed.) (2001), *Frontiers of Evolutionary Economics – Competition, Self-Organization and Innovation Policy*, Cheltenham, UK, and Northampton, MA: Edward Elgar.

Freeman, C. (1977), 'The Kondratiev Long Waves, Technical Change and Unemployment', in *Structural Determinants of Employment and Unemployment*, Vol. II, Paris: OECD, p. 183.

Freeman, C. (1982), *The Economics of Industrial Innovation*, London: Frances Pinter.

Freeman, C. (1984a), *Long Waves in the World Economy*, London: Frances Pinter Publishers.

Freeman, C. (1984b), 'Keynes or Kondratiev? How Can We Get Back to Full Employment?', in P. Marstrand (ed.), *New Technology and the Future of Work and Skills*, London: Frances Pinter.

Freeman, C. (ed.) (1986), *Design, Innovation and Long Cycles in Economic Development*, London: Frances Pinter.

Freeman, C. (1987), *Technology Policy and Economic Performance*, London: Pinter Publishers.

Freeman, C. (1994), 'Rise of East Asian Economies and the Computerisation of the World Economy', *Journal of Economics & Business Administration, Kobe University*, **170** (6), 1–21.

Freeman, C. and F. Louçã (2001), *As Time Goes By, From the International Revolutions to the Information Revolution*, Oxford: Oxford University Press.

Freeman, C. and L. Soete (eds) (1990), *New Explorations in the Economics of Technological Change*, London: Pinter Publishers.

Freeman, C. and L. Soete (1997), *The Economics of Industrial Innovation*, Third Edition, London: Continuum.

Freeman, C., J. Clark and L. Soete (1982), *Unemployment and Technological Innovation: A Study of Long Waves in Economic Development*, London: Frances Pinter.

Fujimoto, H. (1994), *Changing Manufacturing Systems*, Tokyo: Ohm-sha (in Japanese).

Fukagawa, Y. (1989), *Korea – A Development History of an Industry*, Tokyo: JETRO, (in Japanese).

Furukawa, K. (2001), *Revolution of Nuclear Reactors*, Tokyo: Bunshun-shinsho, Bungei Shunjyu (in Japanese).

Galbraith, J.K. (1956), *The Great Crash 1929*, Boston: Houghton Mifflin.

Galbraith, J.K. (1990), *A Short History of Financial Euphoria*, Tennessee: Whittle Direct Books.

Gill, R.T. (1967), *Evolution of Modern Economics*, Englewood Cliffs, NJ: Prentice-Hall.

Gold, T. (1989), *Power from The Earth*, London: George Weidenfeld and Nicolson Ltd.

Goldstein, J.S. (1985), 'Kondratieff Waves as War Cycles', *International Studies Quarterly*, **29**, December, 411–44.

Goodwin, R.M. (1950), 'A Non-linear Theory of the Cycle', *Review of Economics and Statistics*, **32**, 316–20.

Goodwin, R.M. (1982), *Essay in Economic Dynamics*, London: Macmillan.

Goodwin, R.M. (1988), 'Walras and Schumpeter: The Vision Reaffirmed', Conference of the International Schumpeter Society, Siena, Italy, May 24–27. Reprinted in A. Heertje and M. Perlman (eds) (1990), *Evolving Technology and Market Structure – Studies in Schumperterian Economics*, Ann Arbor: The University of Michigan Press.

Goodwin, R.M. (1990), *Chaotic Economic Dynamics*, Oxford: Oxford University Press.

Gordon, D.M., (1978), 'Up and Down the Long Roller Coaster', in *US Capitalism in Crisis*, edited by the Crisis Reader Editorial Collective, New York: Union for Radical Political Economists.

Gould, S.J. (1977), *Ontogeny and Phylogeny*, Cambridge, MA: Harvard University Press.

Gould, S.J. and N. Eldredge (1977), 'Punctuated Equilibria: The Tempo and Mode of Evolution Reconsidered', *Paleobiology*, **3**, 115–51.

Gregory, G. (1986), *Japanese Electronics Technology: Enterprise and Innovation*, New York: John Wiley.

Griliches, Z. (1957), 'Hybrid Corn: An Explanation in the Economics of Technological Change', *Econometrica*, **25** (4), 501–22.

Grübler, A. (1997), 'Time for a Change: On the Patterns of Diffusion of Innovation', International Institute for Applied Systems Analysis, Austria.

Grübler, A. (1999), 'The Rise and Fall of Infrastructures – Dynamics of Evolution and Technological Change in Transport', International

Institute for Applied Systems Analysis, Austria, 1st edn, Heidelberg: Physica-Verlag, 1990.

Hagerstrand, T. (1967), *Innovation Diffusion as a Simple Process*, Chicago: University of Chicago Press.

Haiteku (1990), 'Technological Competitiveness of Japan and the United States', Research Association for High-technology Strategy, Nikkei Science Ltd. (in Japanese).

Hall, B.K. (1998), *Evolutionary Developmental Biology*, London: Chapman & Hall.

Hall, N. (1991), *Guide to CHAOS*, IPC Magazines New Scientist, London: Penguin Books Ltd.

Hamblin, R.L., R.B. Jacobson and J.L.L. Miller (1973), *A Mathematical Theory of Social Change*, New York: John Wiley.

Hansen, A.H. (1932), *Economic Stabilization in an Unbalanced World*, New York: Harcourt, Brace and Co.

Hansen, A.H. (1941), *Fiscal Policy and Business Cycles*, translated into Japanese by S. Tsuru (1950), Tokyo: Nippon Hyoron Sha.

Harrod, R.F. (1949), 'An Essay in Dynamic Theory', *Economic Journal*, **49** (1), 14–33.

Hatfield, C.B. (1997), 'Oil Back on the Global Agenda', *Nature*, **387**, 8 May, 121.

Hatsumei Kyokai (1997), *Survey Report on the Development History of Industrial Technologies – Camera*, Survey Group for Japan Patent Office, Tokyo: Japan Institute of Invention and Innovation (in Japanese).

Hawthorne, N. (2001), *Twice-Told Tales*, New York: Modern Library, Random House.

Hertje, A. and M. Perlman (eds) (1990), *Evolving Technology and Market Structure – Studies in Schumpeterian Economics*, Ann Arbor: University of Michigan Press.

Heinberg, R. (2003), *The Party's Over*, Forest Row, East Sussex, UK: Clairview Books.

Helpman, E. and M. Trajtenberg (1994), 'A Time to Sow and a Time to Reap: Growth Based General Purpose Technologies', National Bureau of Economic Research, Working Paper Series, No. 4854.

Helpman, E. and M. Trajtenberg (1996), 'Diffusion of General Purpose Technologies', National Bureau of Economic Research, Working Paper Series, No. 5773.

Hicks, J.R. (1950), *A Contribution to the Theory of the Trade Cycle*, Oxford: Oxford University Press.

Hiramoto, K. (1996), *Natural Cycles of Sardine Hauls*, Tokyo: Chuko Shinsho (in Japanese).

Hirooka, M. (1992a), 'Evaluation of Technological Trajectory and Impact of Technological Innovation on the Development of Economy – Examination of Schumpeter's Concept by Product Life Cycle', the fourth Conference of the International J.A. Schumpeter Society, Kyoto, 20 pages.

Hirooka, M. (1992b), 'An Analysis of Innovation Process in Terms of Product Cycle and Technoeconomic Paradigm Change', in *Proceedings of The Third International Conference on Science and Technology Policy, NISTEP, Japan at Oiso, March, 1992* (1993), 207–14.

Hirooka, M. (1992c), 'Frontiers of Generic Technology and Chemical Industry', in *Plastics Age Encyclopedia 1993*, Tokyo: Plastics Age, pp. 15–30 (in Japanese). The English version in *Japan Plastics Industrial Annual* (1993), **36**, 13–25.

Hirooka, M. (1993a), 'Paradigms of Technological Innovation and Industrial Development', *Chemical Economics*, **40** (1), 14–24 (in Japanese).

Hirooka, M. (1993b), 'Life Cycles of Polymer Industry and Market Perspectives', in *Plastics Age Encyclopedia 1994*, Tokyo: Plastics Age, pp. 15–30 (in Japanese).

Hirooka, M. (1994a), 'Innovation Dynamism – Collapse of Bubble Economy and Tide of Technoglobalism', *Chemical Economics*, **41** (1), 25–33 (in Japanese).

Hirooka, M. (1994b), 'Innovation Paradigm and Business Cycles – Impact Analysis of Product Cycles', *Journal of Economics and Business Administration, Kobe University*, **169** (2), 57–77 (in Japanese).

Hirooka, M. (1994c), 'Dynamism of Technological Innovation and Economic Development – A Trial of Schumpeter's Paradigm Reconsidered', the Fifth Conference of the International J.A. Schumpeter Society, Münster, Germany, August, 23 pages.

Hirooka, M. (1994d), 'Trend of the World Economy and Chemical Industry – Future Perspectives of Plastic Industry', in *Plastics Age Encyclopedia 1995*, Tokyo: Plastics Age, pp. 15–28 (in Japanese).

Hirooka, M. (1994e), 'Technoglobalization and Regional Issues in the Dynamism of Technological Change', in *Proceedings of the Third Euro-Japan Conference on New Frontiers for Regional Economies*,(1996), Sunderland, UK, pp. 13–16.

Hirooka, M. (1994f), 'Global Production towards Problem-Solving Growth', Paper at Session 3, Mastering Technology, 94 Seminar, FINE Project, EC government, 21 December, Brussels, 11 pages.

Hirooka, M. (1995a), 'Innovation Dynamism and Economic Change – Perspectives of Plastics Industry and Problems', *Plastics Age*, **41** (4), 119–131 (in Japanese).

Hirooka, M. (1995b), 'Innovation Dynamism and R&D Strategies', *Science and Industry*, **69** (11), 451–67 (in Japanese).

Hirooka, M. (1995c), 'Dynamism of Technological Innovation and the Chemical Industry in the Pacific Basin', in *Proceeding of the Seminar on the Changing Chemical Scene in the Pacific Basin, the Symposium of PACIFICHEM'95, Hawaii, Business Development and Management Division* (1997), American Chemical Society, pp. 92–102.

Hirooka, M. (1995d), 'Limits of Technological Development and Growth of East Asia', Workshop on Problem-solving Growth at the Wuppertal Institute, Wuppertal, Germany, 2–4 November, 6 pages.

Hirooka, M. (1996), 'Dynamism of Technoglobalization and Regional Economies in Japan and East Asia', in *Workshop on Regional Strategies for an Innovative Economy, Estersund, Sweden, on 12–15, June (1996), Proceedings* (1998), pp. 66–70.

Hirooka, M. (1997a), 'Industrial Development and Global Issues in the 21st Century in Terms of Innovation Dynamism', *Journal of Ryutsu Kagku University – Information, Economics & Management Science*, **5** (2), 53–68 (in Japanese).

Hirooka, M. (1997b), 'Indusrial Development towards 21st Century – Innovation Dynamism and Global Issues', *Energy and Resources*, **18** (5), 421–6 (in Japanese).

Hirooka, M. (1998a), 'Positioning the Evolutionary Economics – Analysis by a Logistic Equation', in *Collected Papers of the Japan Association for Evolutionary Economics No. 2*, pp. 112–23 (in Japanese).

Hirooka, M. (1998b), 'Innovation Dynamism and Technological Opportunities for Venture Businesses', *Journal of Ryutsu Kagaku University – Information, Economics & Management Science*, **6** (2), 55–67 (in Japanese).

Hirooka, M. (1998c), 'Innovation Dynamism – Technology Fusion by Existing Industries and Complex Production Systems', *Journal of Ryutsu Kagaku University – Information, Economics & Management Science*, **7** (1), 49–70 (in Japanese).

Hirooka, M. (1998d), 'Innovation Dynamism and Evolutionary Economics', in Japan Association for Evolutionary Economics (ed.), *What's Evolutionary Economics*, Tokyo: Yuhikaku Publishers, pp. 151–62 (in Japanese).

Hirooka, M. (1998e), 'Innovation Dynamism and Chemical Industries', *Chemical Economics*, **45** (12), 2–11 (in Japanese).

Hirooka, M. (1999a), 'Dynamism of Retail Business and Innovation', *Journal of Ryutsu Kagaku University – Information, Economics & Management Science*, **7** (2), 73–86 (in Japanese).

Hirooka, M. (1999b), 'Logistic Innovations and Logistic Equation', *Journal of Ryutsu Kagaku University, Special Issue in Honor of Mr Nakauchi for his 77th Birthday*, 187–97 (in Japanese).

Hirooka, M. (1999c), 'Business Opportunity and Technology Fusion in Terms of Innovation Dynamism', in *Proceedings of 1998 Conference on Pioneering New Technologies: Management Issues and Challenges in the Third Millennium, (IEMC'98)*, IEEE EMS Engineering Management Society, pp. 1–6.

Hirooka, M. (1999d), 'Innovation Dynamism and Kondratiev Business Cycles', in *Plastics Age Encyclopedia 2000*, Tokyo: Plastics Age, pp. 31–44 (in Japanese).

Hirooka, M. (1999e), 'Innovation Dynamism of Electronics (1) – Behavior of Firms and National Systems of Innovation', *Journal of Ryutsu Kagaku University – Information, Economics & Management Science*, **8** (2), 11–24 (in Japanese).

Hirooka, M. (2000a), 'Cycles of Bubble Economy and Depression – Interpretation by Innovation Dynamism', *Journal of Ryutsu Kagaku University – Information, Economics & Management Science*, **9** (2), 1–19 (in Japanese).

Hirooka, M. (2000b), 'Innovation Dynamism and Market Paradigm in Chemical Economics', in *Proceedings of Pachifichem 2000, Area 4, Symposium 51, The Changing Chemical Scene in the Pacific Basin* (2001), pp. 163–72.

Hirooka, M. (2001a), 'Logistic Dynamism of Technological Innovation and Role of Government in the Case of Electronics', *IEEE EMS Conference, Albany, NY, October 7–9*, 6 pages.

Hirooka, M. (2001b), Logistic Dynamism of Science and Technologies – Mathematical Description of Knowledge Transfer', *Journal of Ryutsu Kagaku University – Humanities, Arts & Sciences*, **14** (1), 13–29 (in Japanese).

Hirooka, M. (2002a), 'Nonlinear Dynamism of Innovation and Business Cycles', *Ninth Conference of the International J.A. Schumpeter Society, Gainesville, FL, March*, 24 pages.

Hirooka, M. (2002b), 'Exchange Rate and International Competitiveness', *Journal of Ryutsu Kagaku University – Information, Economics & Management Science*, **10** (3), 71–86 (in Japanese).

Hirooka, M. (2002c), 'Dynamism of Japan's Economy and Frontier of New Industries', in *Plastics Age Encyclopedia 2003*, Tokyo: Plastics Age, pp. 15–26 (in Japanese).

Hirooka, M. (2003a), 'Nonlinear Dynamism of Innovation and Knowledge Transfer', The Conference of ASEAT and the Institute of Innovation Research on Knowledge and Economic and Social Change: New Challenges to Innovation Studies, April 7–9, Manchester, 21 pages.

Hirooka, M. (2003b), *Technological Innovation and Economic Growth –*

Elucidation of Nonlinear Dynamism, Tokyo: Nihon Keizai Shimbun Inc. (in Japanese).

Hirooka, M. (2003c), 'Nonlinear Dynamism of Innovation and Business Cycles', *Journal of Evolutionary Economics*, **13** (5), 549–76.

Hirooka, M. (2004a), 'Institutional Impact of Trunk Innovation and Long Business Cycles', *Tenth Annual Conference of the International J.A. Schumpeter Society, Bocconi University, Milan on 9–12 June 2004*, 25 pages.

Hirooka, M. (2004b), 'Complexity in Discrete Innovation Systems', *Conference on Organisations, Innovation and Complexity: New Perspectives on the Knowledge Economy, University of Manchester, September 9–10, 2004*, ESRC Nexus Priority Network, and Complexity Society, 19 pages.

Hirooka, M. (2005a), 'Nonlinear Dynamism of Innovation and Business Cycles', in U. Cantner, E. Dinopoulos and R.F. Lanzillotti (eds), *Entrepreneurship, the New Economy and Public Policy, Schumpeter Perspectives*, Berlin: Springer, pp. 289–316.

Hirooka, M. (2005b), 'Nonlinear Dynamism of Innovation and Knowledge Transfer', in K. Green, M. Miozzo and P. Dewick (eds), *Technology, Knowledge and the Firm: Implication for Strategy and Industrial Change*, Chltenham, UK, and Northampton, MA: Edward Elgar, pp. 427–86.

Hirooka, M. and T. Hagiwara (1990), 'Growth of Technological Innovations and Market Development of New Technology – An Interpretation of Technoeconomic Paradigm Shift', in *Proceedings of the First International Conference on Science and Technology Policy, NISTEP, Japan, at Shimoda, February, 1990* (1991), pp. 425–32.

Hirooka, M. and T. Hagiwara (1991), 'Present Status of Japanese Chemical Industry – Competitiveness and Diversification', in *Proceedings of the Second International Conference on Science and Technology Policy, NISTEP, Japan at Oiso, January, 1991* (1992), pp. 149–57.

Hirooka, M. and T. Hagiwara (1992), 'Characterization of Diffusion Trajectory of New Products in the Course of Technological Innovation', *Kobe University Economic Review*, **38**, 47–62.

Hirooka, M. and M. Nitta (2000), 'Innovation Dynamism of Electronoics (2) – Development of Electronics Industry and Actions by Government', *Journal of Ryutsu Kagaku University – Information, Economics & Management Science*, **8** (3), 47–62 (in Japanese).

Hirooka, M. and M. Nitta (2002), 'Innovation Dynamism of Electronics (3) – Behavior of Firms in the Mature Period and Evolution of Paradigms', *Journal of Ryutsu Kagaku University – Information, Economics & Management Science*, **10** (3), 87–100 (in Japanese).

Hirose, K. (2002), *Story of Solar Cells*, Tokyo: Japan Standard Association (in Japanese).

Hirota, R. (2000), *Lecture on Difference Equations, Mathematical Science, SGC Library 8*, Tokyo: Saiensu-Sha (in Japanese).

Hodgson, G.M. (1993), *Economics and Evolution: Bringing Life Back into Economics*, Ann Arbor: University of Michigan Press.

HTSRG (High Technology Strategy Research Group) (1990), *Technological Competitiveness of Japan and the United States*, Tokyo: Nikkei Science Inc. (in Japanese).

Hubbert, M.K. (1956), 'Nuclear Energy and Fossil Energy', American. Petroleum Institute, Drilling and Production Practice, Proceedings, Spring Meeting, San Antonio, Texas.

Hubbert, M.K. (1969), *Resources and Man*, Washington DC: US National Academy of Sciences and National Research Council.

Hutchinson, G.E. (1978), *An Introduction to Population Ecology*, New Haven: Yale University Press.

Ichikawa, T. (1984), *Long Waves in World Business Activities*, Tokyo: Aki Publishers (in Japanese).

Ijiri, K. (1994), 'Aging and Death', in *Life, Information and Mathematics, Mathematical Science*, Tokyo: Science Publishers.

Ikeda, K., (2004), 'Tapering-off of Versatility of Animals – A Consideration in Biological Evolution', *Nihon Keizai Shimbun*, 9 May, 31 (in Japanese).

Inoue, R. (1986), *Retail & Wholesale*, Tokyo: Nihon Keizai Shimbun Inc. (in Japanese).

Ishihara T., K. Ikeo and Y. Sato (1995), *Commerce*, Tokyo: Yuhikaku Publishers (in Japanese).

Ishiyama, J. (1993), *Canon Creating Multiphase Businesses*, Tokyo: Tokuma Shoten

Itoh, A. (1999), 'Development of Global Out Souring and Subjects to be Solved', *Shizuoka University Economic Studies*, **3** (4), 256–61.

Iwamoto, K. (2000), 'Development of New Solid Catalyst Reaction Systems Using Regular Meso-porus Nature', *Chemistry*, **55** (5), 64–5 (in Japanese).

JAMA (1988), *Chronicles of Japan Automobile Industry*, Tokyo: Japan Automobile Manufacturers Association, Inc. (in Japanese).

Japan Productivity Center (1998), *International Comparison of Labor Productivity*, Tokyo: Japan Productivity Center (in Japanese).

JBA (1999), Private Communications at the Japan Bio-industry Association, Tokyo.

Jevons, W.S. (1878), 'The periodicity of Commercial Crises and Sun-Spots – Part One', *Nature*.

Jevons, W.S. and H.S. Foxwell (eds) (1909), *Investigations in Currency & Finance*, London: Macmillan.

JISF (The Japan Iron and Steel Federation) (2003), 'The Steel Industry of Japan', http://www.jisf.or.jp.

JPCA (2001), *Present Situation of Petrochemical Industry in Japan*, Tokyo: Japan Petrochemical Industry Association (in Japanese).

Juglar, J.C. (1862), *Des Crises commerciales et leur retour periodique en France, Angleterre et aux Etats-Unis*, Paris: Guillaumin.

Kaku, S. (1986), *Introduction to the History of German Chemical Industry*, Tokyo: Minerva Publishers (in Japanese).

Kanamori, H. (1987), *Innovation and Industrial Structures*, Tokyo: Nihon Keizai Shimbun Inc. (in Japanese).

Kanazashi, M. (ed.) (1991), *Historical Approach of Analysis on Business Cycles by J.A. Schumpeter*, Collection of Articles by Schumpeter, including his lecture at the University of Tokyo in 1931, Tokyo: Hassaku Publishers (in Japanese).

Kaneko I., S. Momose and Y. Okamoto (1999), *Korean Economy – Industrial Development and Present Situation*, Tokyo: Keiso Publishers (in Japanese).

Kato, T. (1997), *Silicon Valley Waves*, Tokyo: NTT Publishers (in Japanese).

Kato, T. (1998), *Asian Network – Japan's Strategy for Information Society*, Tokyo: Nippon Economic Review Inc. (in Japanese).

Keirstead, B.S. (1948), *The Theory of Economic Change*, Toronto: Macmillan.

Kelly, T. and D. Fenton (2003), 'Iron and Steel Statistics', in T. Kelly, D. Buckingham, C. DiFrancesco, K. Porter, T. Goonan, J. Sznopek, C. Berry and M. Crane, *Historical Statistics for Mineral and Material Commodities in the United States*, US Geographical Survey Open-File Report 01–006, Histrorical Statistics for Commodities in the United States, USGS OFR 01–06, http://minerals.usgs.gov/minerals/pubs/of 01–006/.

Keynes, J.M. (1936), *General Theory of Employment, Interest and Money*, New York: Harcourt Brace.

Kihara, N. (1997), *Essence of Sony Technologies*, Sony Magazines, Tokyo (in Japanese).

Kimura, M. (1968), *The Neutral Theory of Molecular Evolution*, New York: Cambridge University Press.

Kindleberger, C.P. (1973), *The World in Depression 1929–1939*, Berkeley, CA: University of California Press.

Kisaka, S. (1984), *Science and Technology Chronicles Focusing on Electronics* (2nd edn), Tokyo: Daily Industrial News Ltd (in Japanese).

Krisaka, S. (2001), *Science and Technology Chronicles Focusing on Electronics* (5th edn), Tokyo: Daily Industrial News Ltd (in Japanese).

Kisimoto, M. (1987), 'The Present Situation and Future Trend of FADEC Development', *GTSJ*, 14–56, 28–35 (in Japenese).

Kitchin, J. (1923), 'Cycles and Trends in Economic Factors', *Review of Economic Statistics*, **5**, 10–16.

Kleinknecht, A. (1984), 'Innovation Patterns in Crisis and Prosperity: Schumpeter's Long Cycle Reconsidered', doctoral thesis, Free University of Amsterdam.

Kleinknecht, A., E. Mandel and I. Wallerstein (1992), *New Findings in Long-Wave Research*, New York: St Martin's Press.

Kline, S.J. (1985), 'Innovation is not a Linear Process', *Research Management*, **28** (4), 36–44.

Kline, S.J. and N. Rosenberg (1986), 'An Overview of Innovation', in R. Landau and N. Rosenberg (eds), *The Positive Sum Strategy – Harnessing for Economic Growth*, Washington DC: National Academy Press.

Kobayashi, N. and S. Sugihara (1989), *New Edition Economics History*, Tokyo: Yuhikaku (in Japanese).

Kodama, F. (1985), 'Mechatronics Technology as Japanese Innovation: A Study of Technological Fusions', Graduate School of Science Policy, Saitama University, Saitama, Japan.

Kodama, F. (1986), 'Japanese Innovation in Mechatronics Technology', *Science and Public Policy*, **13** (1), 44–51.

Kodama, F. (1991), *Analysing High Technologies: The Techno-paradigm Shift*, London: Pinter Publishers.

Kojima, S. (1995), *Indian Economy Changes Asia*, Kyoto: PHP Research Institute (in Japanese).

Kondratiev, N.D. (1926), 'Die Langen Wellen der Konjuktur', *Archiv fur Socialwissenschaft und Socialpolitik*, **56**, 573–606.

Koyama, S. (2001), *When the Oil is Exhausted?*, Tokyo: Jiji Press Inc. (in Japanese).

Krugman, P. (1996), *The Self-Organizing Economy*, Oxford: Blackwell.

Kuhn, T. (1962), *The Structure of Scientific Revolutions*, Chicago: The University of Chicago Press.

Kuo, K. and M. Tabata (1996), *Miracle in Taiwan, Great Country of Electronics*, Tokyo: Ascii Corp.

Kuznets, S. (1930), *Secular Movements in Production and Prices*, Boston: Houghton Mifflin.

Kuznets, S. (1940), 'Schumpeter's Buisiness Cycles', *American Economic Review*, **30** (2), 257–71.

Kuznets, S. (1953), *Economic Change*, New York: W.W. Norton & Co.

Kuznets, S. (1958), 'Long Swings in the Growth of Population and in

Related Economic Variables', *Proceedings of the American Philosophical Society*, **102**, 22–52.

Kuznets, S. (1959), *Six Lectures on Economic Growth*, Glencoe, NY: The Free Press.

Kuznets, S. (1966), *Modern Economic Growth*, New Haven: Yale University Press.

Landau, L. and N. Rosenberg (eds) (1986), *The Positive Sum Strategy – Harnessing Technology for Economic Growth*, Washington, DC: National Academy Press.

Lauwerier, H. (1987), *Fractals: Endlessly Repeating Geometric Figures*, Bloemendaal, the Netherlands: Utigevers.

Lebhar, G.M. (1963), *Chain Stores in America 1859–1962*, New York: Chain Store Publishing Corp.

Leontiev, W. (1953), 'Domestic Production and Foreign Trade: The American Position Re-examined', *Proceedings of the American Philosophical Society*, vol. 97.

Levin, R.C., W.Cohen and D. Mowery (1985), 'R & D Appropriability, Opportunity, and Market Structure: New Evidence on the Schumpeterian Hypothesis', *American Economic Review,* **75** (2), 20–24.

Levin, R.C., A.K. Klevorick, R.R.Nelson and S.G.Winter (1987), 'Appropriating the Returns from Industrial Research and Development', *Brookings Papers on Economic Activity*, 3, 783–831.

Li, T.Y. and J.A. Yorke (1975), 'Period Three Implies Chaos', *American Mathematical Monthly*, **82**, 271–3.

Lorenz, H.W. (1989), *Nonlinear Dynamical Economics and Chaotic Motion*, Berlin: Springer-Verlag.

Lorenz, H.W. (1993), *Nonlinear Dynamic Economics and Chaotic Motion,* Second revised and enlarged edition, Berlin: Springer-Verlag.

Lotka, A. (1924), *Elements of Mathematical Biology*, New York: Dover Publications (reprinted in 1956).

Lucas, R. (1988), 'On the Mechanics of Economic Development', *Journal of Monetary Economics*, **22** (1), 3–42.

Lundvall, B.-A. (1985), *Product Innovation and User–Producer Interaction*, Aalborg: Aalborg University Press.

Lundvall, B.-A. (ed.) (1992), *National Systems of Innovation*, London: Pinter Publishers.

Maddison, A. (1982), *Phases of Capitalist Development*, Oxford: Oxford University Press.

Mahajan, V. and E. Muller (1979), 'Innovation Diffusion and New Product Growth Models in Marketing', *Journal of Marketing*, **43**, 55–68.

Mahajan, V. and R.A. Peterson (1985), *Models for Innovation Diffusion*, California: Sage.

Mahajan, V. and R.A. Peterson (1979), 'Integrating Time and Space in Technological Substitution Models', *Technological Forecasting and Social Change*, **14**, 127–46.

Makino, N. and Y. Shimura (1984), *Technology War between Japan and the United States*, Tokyo: Nihon Keizai Shimbun Inc.(in Japanese).

Malthus, T.R. (1798), *An Essay on the Principle of Population*, London: Royal Economic Society.

Mandel, E. (1972), *Late Capitalism Spätkapitalismus* [1975], London: New Left Books.

Mansfield, E. (1961), 'Technical Change and the Rate of Imitation', *Econometrica*, **29** (4), 741–66.

Mansfield, E. (1963), 'Intra-firm Rates of Diffusion of an Innovation', *The Review of Economics and Statistics*, **45** (4), 348–59.

Mansfield, E. (1968), *Industrial Research and Technological Innovation: An Econometric Analysis*, New York: Norton & Co. Inc.

Marchetti, C. (1975), 'On Strategies and Fate', in W. Haefele et al., *Second States Report of the IIASA Project on Energy Systems 1975*, RR-76-1, International Institute for Applied Systems Analysis, Laxenberg, Austria.

Marchetti, C. (1979), 'Energy Systems – The Broader Context', *Technological Forecasting and Social Change*, **14**, 191–203.

Marchetti, C. (1980), 'Society as a Learning System: Discovery, Invention, and Innovation Cycles Revised', *Technological Forecasting & Social Change*, **18**, 267–82.

Marchetti, C. (1985), 'Action Curves and Clockwork Geniuses', WP-85-74, International Institute for Applied Systems Analysis, Laxenburg, Austria.

Marchetti, C. (1986), 'Cristoforo Colombo e il suo contesto: ovvero come il valore di ogni grande scoperta discende da una decisione sociale', *pubbl. Trimestrale Gruppo Italgas*, **16** (4), 40–45.

Marchetti, C. (1988), 'On Society and Nuclear Energy', Mimeo, International Institute for Applied Systems Analysis, Laxenberg, Austria.

Marchetti, C. (1990), 'Personal Memoir: From Terawatts to Witches. My Life with Logistics at IIASA', *Technological Forecasting and Social Change*, **37**, 409–14.

Marchetti, C. (1996), 'Looking Forward – Looking Backward: A Very Simple Mathematical Model for Very Complex Social Systems', the Conference 'Previsione Sociale e Previsione politica' in Urbino, 13–15 June.

Marchetti, C. (2002), 'Productivity versus Age', IIASA Report, International Institute for Applied Systems Analysis, Laxenburg, Austria.

Marchetti, C. and N. Nakićenović (1979), 'The Dynamics of Energy Systems and the Logistic Substitution Model', RR-79-13, International Institute for Applied Systems Analysis, Laxenburg, Austria.

Marchetti, C., P.S. Meyer and J.H. Ausubel (1995), 'Population Dynamics: How Much Can be Modeled and Predicted', Program for Human Environment, Rockefeller University.

Marchetti, C., P.S. Meyer and J.H. Ausubel (1996), 'Human Population Dynamics Revised with the Logistic Model', *Technological Forecasting and Social Change*, **52**, 1–30.

May, R. (1974), 'Populations with Non-overlapping Generation Stable Points, Stable Cycles, and Chaos', *Science*, **186**, 645–47.

May, R. (1976), 'Simple Mathematical Models with Very Complicated Dynamics', *Nature*, **261**, 459–67.

McKelvey, M. (1996), *Evolutionary Innovations – The Business of Biotechnology*, Oxford: Oxford University Press.

McKelvey, W. (1982), *Organizational Systematics: Taxonomy, Evolution, Classification*, Berkeley, CA: University of California Press.

McNeil, I. (1990), *Encyclopaedia of the History of Technology*, London: Routledge.

Meadows, D.H., D.L. Meadows, J. Randers and W.W. Behrens III (1972), *The Limits to Growth – A Report for the Club of Rome's Project on the Predicament of Mankind*, New York: Universe Books.

Meadows, D.H., D.L. Meadows and J. Randers (1992), *Beyond the Limits – Confronting Global Collapse, Envisioning a Sustainable Future*, Post Mills, VT: Chelsea Green Publishing Company.

Mensch, G. (1975), *Das Technologische Patt*, Frankfurt: Umschau Verlag.

Mensch, G. (1979), *Stalemate in Technology – Innovations Overcome the Depression*, New York: Ballinger Publishing Company.

Metcalfe, J.S. (1970), 'The Diffusion of Innovations in Lancashire Textile Industry', *Manchester School of Economics and Social Studies*, **2**, 145–62.

Metcalfe, J.S. (1981), 'Impulse and Diffusion in the Study of Technical Change', *Futures*, **13**, 347–59.

Metcalfe, J.S. (1982),'On the Diffusion of Innovation and the Evolution of Technology', Conference on Technological Change, January, London.

Metcalfe, J.S. (1988a), 'Evolution and Economic Change', in A. Silberston (ed.), *Technology and Economic Progress*, London: Macmillan, pp. 54–85.

Metcalfe, J.S. (1988b), 'The Diffusion of Innovation: An Interpretative Survey', in G. Dosi, C. Freeman, R. Nelson, G. Silverberg and L. Soete, *Technical Change and Economic Theory*, London: Pinter Publishers.

Metcalfe, J.S. (1994a), 'Industry Life Cycles and the Theory of Retardation: Towards an Ecology of Industrial Development', *Journal of Economics & Business Administration, Kobe University*, **170** (6), 41–61.

Metcalfe, J.S. (1994b), 'Competition, Evolution and Capital Market', *Metroeconomica*, **45**,127–54.

Metcalfe, J.S. (1994c), 'Evolutionary Economics and Technology Policy', *Economic Journal*, **104**, 931–41.

Metcalfe, J.S. (1994d), 'Competition, Fisher's Principle and Increasing Returns in the Selection Process', *Journal of Evolutionary Economics*, **4**, 327–46.

Metcalfe, J.S. (1995), 'The Economic Foundations of Technology Policy, Equilibrium and Evolutionary Perspectives', in P. Stoneman (ed.), *Handbook of the Economics of Innovation and Technological Change*, Oxford: Blackwell.

Metcalfe, J.S. (1998), *Evolutionary Economics and Creative Destruction: The Graz Schumpeter Lectures*, Graz Schumpeter Society, London: Routledge.

Metcalfe, J.S. (2001), 'Industrial Growth and the Theory of Retardation: Precursors of an Adaptive Evolutionary Theory of Economic Change', Conference in Celebration of the Work of Dick Nelson and Sid Winter, Danish Research Unit for Industrial Dynamics, University of Aalborg, 12–15 June.

Metcalfe, J.S. (2003), 'Industrial Growth and the Theory of Retardation: Precursors of an Adaptive Evolutionary Theory of Economic Change', *Revue Economique*, **54** (2), 407–31.

Metcalfe, J.S., J. Foster and R. Ramlogan (2002), Adaptive Economic Growth, Discussion Paper, Centre for Research on Innovation and Competition, December.

Mitchell, B.R. (1998), *International Historical Statistics: The Americas 1750–1993*, London: Macmillan.

MITI (Ministry of International Trade and Industry) (1998), *Yearbook of Industrial Statistics*, Research and Statistics Development, Minister's Secretariat, Japan.

MITI (Ministry of International Trade and Industry) (2001), *Yearbook of Industrial Statistics*, Research and Statistics Department, Minister's Secretariat, Japan.

Mitsubishi Chemical Corp. (1981), *Company Chronicle of Mitsubishi Chemical Corp* (in Japanese).

Mitsubishi Petrochemical Co. (1988), *The Thirty Years History of Mitsubishi Petrochemical Co.*, Chemical Economy Research Institute, Tokyo (in Japanese).

Mitsubishi Research Institute (1997), *Introduction to Asian Economy*, Tokyo: Tokuma Publishers (in Japanese).

Mitsui Petrochemical Industries Ltd (1988), *The Thirty Years Chronicle of Mitsui Petrochemical Industries Inc.*, Diamond Inc, Tokyo (in Japanese).

Mitsui Toatsu Chemicals Inc. (1994), *Chronicle of Mitsui Toatsu Chemical Co.*, Chemical Economy Research Institute, Tokyo (in Japanese).

Miyamoto, J. (1992), *Risk Assessment of Pesticides*, Sumitomo Chemical Co. Tokyo (in Japanese).

Miyazaki, K. (1995), *Building Competence in the Firm: Lessons from Japanese and European Optoelectronics*, London: Macmillan.

Modelski, G. (1987), *Long Cycles in World Politics*, Seattle: University of Washington Press.

Modis, T. (1992), *Predictions*, New York: Simon & Schuster.

Mowery, D.C. and N. Rosenberg (1989), *Technology and the Pursuit of Economic Growth*, Cambridge: Cambridge University Press.

Mowery, D.C. and N. Rosenberg (1998), *Paths of Innovation*, Cambridge: Cambridge University Press.

Murmann, J.P. (2003), *Knowledge and Competitive Advantage – The Coevolution of Firms, Technology, and National Institutes*, Cambridge: Cambridge University Press.

Murmann, J.P. and E. Homburg (2001), 'Comparing Evolutionary Dynamics across Different National Settings: The Case of the Synthetic Dye Industry, 1857–1914', *Journal of Evolutionary Economics*, **11**, 177–205.

Nabseth, L. and G. Ray (1974), *The Diffusion of New Industrial Processes*, Cambridge: Cambridge University Press.

Nakagiri, T. (1998), private communications.

Nakamura, T. (1987), *Kondratiev Business Cycles*, Aki Publishers, Tokyo (in Japanese).

Nakićenović, N. and A. Grübler (1991), *Diffusion of Technologies and Social Behavior*, Berlin: Springer-Verlag.

Nelson, R.R. (1959), 'The Simple Economics of Basic Scientific Research', *Journal of Political Economics*, 297–306.

Nelson, R.R. (1981), 'Research on Productivity Growth and Productivity Differences: Dead Ends and New Departures', *Journal of Economic Literature*, **19**, 1029–64.

Nelson, R.R. (1987), *Understanding Technical Change as an Evolutionary Process*, Amsterdam: North-Holland.

Nelson, R.R. (1988), 'Institutions Supporting Technical Change in the United States', in Dosi et al. (eds), *Technical Change and Economic Theory*, London: Pinter Publishers.

Nelson, R.R. (ed.) (1993), *National Innovation Systems*, New York: Oxford University Press.

Nelson, R.R. (1996), *The Sources of Economic Growth*, Cambridge MA: Harvard University Press.

Nelson, R.R. (2000), *The Sources of Economic Growth*, Cambridge, MA: Harvard University Press.

Nelson, R.R. (2004), 'The Market Economy, and the Scientific Commons', *Research Policy*, **33**, 455–71.

Nelson, R. and S. Winter (1974), 'Neoclassical and Evolutionary Theories of Growth: Critique and Prospectus', *Economic Journal*, **84**, 886–905.

Nelson, R. and S. Winter (1977), 'In Search of a Useful Theory of Innovation', *Research Policy*, **5**, 36–76.

Nelson, R. and S. Winter (1982), *An Evolutionary Theory of Economic Change*, Cambridge, MA: Harvard University Press.

Nicolis, G. and I. Prigogine (1977), *Self-organization in Non-equilibrium Systems*, Hoboken, NJ: John Wiley & Sons, NJ.

Nihon Motor (1988), *Industrial History of Japanese Automobiles*, Japan Automobile Manufacturers Association (in Japanese).

Nihonkeizaisinbun (1996), *The Silicon Valley Revolution*, Nihon Keizai Shinbun Inc. (in Japanese).

Niosi, J. and B. Bellon (1996), 'The Globalization of National Innovation Systems', in J. de la Monthe and G. Paquet (eds), *Evolutionary Economics and the New International Political Economy*, London: Pinter, pp. 139–59.

Nippon Steel (1981), *History of Yawata Steel Co.*, and *History of Nippon Steel Co. for the First 10 years* (in Japanese).

NISTEP (2001), *The 7th Technology Foresight*, Tokyo: National Institute of Science & Technology Policy.

Noguchi, Y. (1992), *Economics of Bubble Economy*, Tokyo: Nihon Keizai Shimbun, Inc. (in Japanese).

Nomura Research Institute (1996), *Industry Development Strategy in Asian Countries*, Nomura Research Institute (in Japanese).

Norris, K. and J. Vaizey (1973), *The Economics of Research and Technology*, London: George Allen & Unwin.

OECD (1980), *Technological Change and Economic Policy: Science and Technology in the New Economic and Social Context*, Paris: OECD.

OECD (2001), *National Accounts of OECD Countries*, Paris: OECD.

Okamoto, Y. (1998), *Technology Transfer by Japanese Industries*, Nippon Economic Review Inc. (in Japanese).

Okuno, T. (1991), *Transportation and Communities*, Tokyo: Taimei-do (in Japanese).

Onoda Cement (1981), *The Hundred Years History of Onoda Cement*, Japan Management History Research Institute (in Japanese).

Pasinetti, L.L. (1981), *Structural Change and Economic Growth*, Cambridge: Cambridge University Press.

Pavitt, K. (1980), *Technical Innovation and British Economic Performance*, London: Macmillan.

Perez, C. (1983), 'Structural Change and Assimilation of New Technologies in the Economic and Social System', *Futures*, **15** (4), 357–75.

Perez, C. (2002), *Technological Revolutions and Financial Capital – The*

Dynamics of Bubbles and Golden Ages, Cheltenham, UK and Northampton, MA: Edward Elgar.

Peters, E.E. (1991), *Chaos and Order in the Capital Markets: A New View of Cycles, Prices and Market Volatility*, New York: John Wiley & Sons.

Polanyi, M. (1966), *The Tacit Dimension*, London: Routledge & Kegan Paul.

Price, W.J. and L.W. Bass (1969), 'Scientific Research and the Innovation Process', *Science*, **164**, 802–6.

Prigogine, I. (1976), 'Order through Fluctuation: Self-organization and Social System', in E. Jantsch and C.H. Waddington (eds), *Evolution and Consciousness: Human Systems in Transition*, Reading, MA: Addison-Wesley.

Prigogine, I. and I. Stengers (1984), *Order out of Chaos – Man's New Dialogue with Nature*, New York: Bantam Books.

Ramlogan, R. and J.S. Metcalfe (2001), 'Limits to the Economy of Knowledge and Knowledge of the Economy', Workshop on 'Living with Limits to Knowledge', Central European University, Budapest, 22–23 June.

Ricardo, D. (1817), *On the Principles of Political Economy and Taxation*, London.

Robinson, B.C. and C. Lakhani (1975), 'Dynamic Price Models for New Product Planning', *Management Science*, **21**, 1113–22.

Rogers, E.M. (1983), *Diffusion of Innovations*, New York: Free Press.

Romer, P. (1986), 'Increasing Returns and Long Run Growth', *Journal of Political Economy*, **94**, October, S1002–37.

Rosenberg, N. (1969), 'The Direction of Technological Change: Inducement Mechanisms and Focusing Devices', *Economic Development and Cultural Change*, **18**, 26–42.

Rosenberg, N. (1974), 'Science, Invention and Economic Growth', *Economic Journal*, **84** (333), 90–108.

Rosenberg, N. (1976), *Perspectives on Technology*, Cambridge: Cambridge University Press.

Rosenberg, N. (1982), *Inside the Black Box: Technology and Economics*, New York: Cambridge University Press.

Rosenberg, N. (1994a), *Exploring the Black Box – Technology, Economics and History*, Cambridge: Cambridge University Press.

Rosenberg, N. (1994b), 'How the Developed Countries Became Rich', *Journal of Economics & Business Administration, Kobe University*, **170** (6), 23–39.

Rosenberg, N. (2000), *Schumpeter and the Endogeneity of Technology*, The Graz Schumpeter Lectures, London: Routledge.

Rosenberg, N. and C. Frischtak (1984), 'Technological Innovation and Long Waves', *Cambridge Journal of Economics*, **8** (1), 7–24.

Rosenberg, N. and R.R. Nelson (1994), 'American Universities and Technical Advance', *Research Policy*, **24**, 323–48.

Rosenberg, N. and M. Trajtenberg (2004), 'A General-purpose Technology at Work: The Corliss Steam Engine in the Late Nineteenth-Century United States', *Journal of Economic History*, **64** (3), 61–99.

Rosenberg, N., R. Landau and D.C. Mowery (eds) (1992), *Technology and the Wealth of Nations*, Stanford: Stanford University Press.

Rostow, W.W. (1952), *The Process of Economic Growth*, New York: W.W. Norton.

Rostow, W.W. (1960), *The Stages of Economic Growth*, Cambridge: Cambridge University Press.

Rostow, W.W. (1975), 'Kondratiev, Schumpeter, and Kuznets: Trend Periods Revisited', *Journal of Economic History*, **4**, 719–53.

Rostow, W.W. (1978), *The World Economy: History and Prospect*, London: Macmillan.

Rothwell, R. (1976), Innovation of Textile Machinery: Some Significant Actors in Success and Failure, SPRU Occasional Papers No. 2, University of Sussex.

Rothwell, R., C. Freeman, A. Horsley, V.T.P. Jervis, A.R. Robertson and J. Townsend (eds) (1974), 'SAPPHO Updated – Project SAPPHO Phase 2', *Research Policy*, **13** (3), 258–91.

Roughgarden, J. (1979), *Theory of Population Genetics and Evolutionary Ecology: An Introduction*, New York: Macmillan.

Sadler, M.T. (1830), *The Law of Population: A Treatise, in Six Books; in Disproof of the Super-Fecundity of Human Beings, and Developing the Real Principle of their Increase*, London: J. Murray.

Sahal, D. (1981), *Patterns of Technical Innovation*, Reading, MA: Addison-Wesley.

Sakuma, A. (1998), *Innovation and Market Structures*, Tokyo: Yuhikaku Publishers (in Japanese).

Scherer, F.M. (1984), *Innovation and Growth*, Cambridge: The MIT Press.

Schmookler, J. (1966), *Invention and Economic Growth*, Cambridge, MA: Harvard University Press.

Schmookler, J. (1972), *Patents, Invention and Economic Change*, Cambridge, MA: Harvard University Press.

Schumpeter, J.A. (1912), *Theorie der Wirtschaftlichen Entwicklung*, Leipzig: Duncker & Humbodt.

Schumpeter, J.A. (1926), *Theorie der Wirtschaftliche Entwicklung, 2 Aufl [The Theory of Economic Growth]*, Cambridge, MA: Harvard University Press (1st English-language edn 1934).

Schumpeter, J.A. (1931), 'The Theory of the Business Cycle', *Journal of Economics, the University of Tokyo*, **1** (1), 1–18.

Schumpeter, J.A. (1934a), *The Theory of Economic Development*, Cambridge, MA: Harvard University Press.

Schumpeter, J.A. (1934b), 'Depressions', in *The Theory of the Recovery Program*, New York and London: Whittlesey House, McGraw-Hill Book Co., pp. 3–21.

Schumpeter, J.A. (1935), 'The Analysis of Economic Change', *Review of Economic Statistics*, **17**, May, 2–10.

Schumpeter, J.A. (1939), *Business Cycles*, New York: McGraw-Hill.

Schumpeter, J.A. (1942), *Capitalism, Socialism and Democracy*, New York: Harper & Row.

Schumpeter, J.A. (1946), 'The Decade of the Twenties', *American Economic Review*, **36**, May, 1–10.

Schumpeter, J.A. (1947), 'Theoretical Problems of Economic Growth', *Journal of Economic History*, **7**, 1–9.

Schumpeter, J.A. (1948), 'There is Still Time to Stop Inflation', *Nation's Business*, **36,** June, 88–91.

Schumpeter, J.A. (1949), 'The Historical Approach to the Analysis of Business Cycles', Universities – National Bureau Conference on Business Cycle Research, New York, 25–27 November.

Sekioka, M. (1989), *Economics of the Great Depression*, Tokyo: Diamond Inc. (in Japanese).

Sharif, M.N. and C. Kabir (1976), 'A Generalized Model for Forecasting Tecnological Substitution', *Technological Forecasting and Social Change*, **8**, 353–64.

Shimanaka, Y. (1987), *Solar Activities and Business Movement*, Tokyo: Nihon Keizai Shimbun Inc. (in Japanese).

Shimojo, T. (1994), *Simulation Physics 6 – Introduction to Chaos*, Tokyo: Modern Science Inc. (in Japanese).

Shimura, Y. (1992), *Semiconductor Industry in the Year 2000*, Japan Management Association, Tokyo (in Japanese).

Shinohara, M. (1982), *Rise and Fall of Major Economic Power*, Tokyo: Toyo Keizai Inc. (in Japanese).

Shinozaki, Y. (1994), 'Growth Theory', in *Life, Information, and Mathematics, Special Edition: Mathematical Science*, Science Inc., Tokyo, pp. 19–27 (in Japanese).

Simon, H.A. (1957), *Models of Man: Social and Rational*, New York: Wiley.

Simon, H.A. (1959), 'Theories of Decision-making in Economic and Behavioral Sciences', *American Economic Review*, **49**, 253–83.

Simon, H.A. (1982), *Models of Bounded Rationality*, Cambridge, MA: MIT Press.

Simon, H.A. (1983), *Reason in Human Affairs*, Oxford: Blackwell.

Slobodkin, L.B. (1980), *Growth of Regulation in Animal Populations*, New York: Dover Books.
Smith, Adam (1776), *The Wealth of Nations*, Harmondsworth: Penguin.
Solow, R. (1957), 'Technical Change and the Aggregate Production Function', *Review of Economics and Statistics*, **39**, 312–20.
Spiethoff, A.A.C. (1923), *Krisen in Handwörterbuch der Staatswissenschaften*, 4th edn, vol. 6, Jena.
Stoneman, P. (1983), *The Economic Analysis of Technological Change*, Oxford: Oxford University Press.
Stoneman, P. (1984), 'Theoretical Approaches to the Analysis of the Diffusion of New Technologies', in S. Mansfield et al. (eds), *The Trouble with Technology*, London: Francis Pinter.
Stoneman, P. (2002), *The Economics of Technological Diffusion*, Oxford: Blackwell Publishers.
Sugiyama, N. (1996), 'Engine Control', *Journal of Gas Turbine Society, Japan*, **24** (93), 54–60; (94), 31–5; (95), 60–5.
Sumida, H. (1993), *Economic Change and Solar Spots*, Kyoto, Japan: Nakanishiya Publishers (in Japanese).
Sumida, H. (2000), *Global Environmental Change and Long Term Economic Fluctuations*, Tokyo: Doubunkan (in Japanese).
Sumitomo Chemical Co. (1981), *Chronicle of Sumitomo Chemical Company* (1997); *Recent Twenty Years of Sumitomo Chemical Company* (in Japanese).
Sumitomo Chemical Co. (2002), Private Communication.
Takahashi, T. (1994), *Fuel Cells*, 2nd Edition, Tokyo: Kyoritsu Publishers.
Takano, H. (1997), *Mechatronics*, Tokyo: Rikou Gakusha (in Japanese).
Tani, K. (1999), *Progress of Semiconductor Industry – Pioneers Developed the Huge Industry*, Tokyo: Nikkan Kogyo Shinbun Ltd (in Japanese).
Tanimitsu, T. (1999), *Genealogy of Electronics Industry – Pioneers Developed Huge Industry*, Tokyo: Nikkan Kogyo Shimbun Inc. (in Japanese).
Taniura, T. (1990), *Industrialization of Asia and Technology Transfer*, Tokyo: Research Institute of Asian Economy (in Japanese).
Tedlow, R.S. (1990), *The Story of Mass Marketing in America*, New York: Basic Books Inc.
Thirlwall, A.P. (2002), *The Nature of Economic Growth*, Cheltenham UK and Northampton, MA: Edward Elgar.
Toyota Motor (1991), *Toyota Car Electronics and an Outline of the Automobile Industry*, Toyota Motor Corp. (in Japanese).
Tsuji, K. (2002), private communication.
Tsunoda, M. (1992), *All of American Retail Businesses*, Tokyo: Keirin Publishers (in Japanese).

Tylecote, A. (1992), *The Long Wave in the World Economy – The Present Crisis in Historical Perspective*, London: Routledge.
UN Statistics (1990), *Industrial Statistics Year Book*, New York: United Nations.
US Census Bureau (1924), *Statistical Abstract of the United States.*
US Census Bureau (2002), *Statistical Abstract of the United States: 2000*, US Census Bureau, Department of Commerce.
US Government (1980), *The Global 2000 Report to the President Entering the Twenty-First Century.*
US Historical Statistics (1989), *Historical Statistics of the United States: Colonial Times to 1970*, Washington DC: Government Printing Office.
Utterback, J.M. (1994), *Measuring Dynamics of Innovation*, Boston: Harvard Business School Press.
Utterback, J.M. and W.J. Abernathy (1975), 'A Dynamic Model of Product and Process Innovation', *Omega*, **3** (6), 639–56.
Utterback, J.M. and F.F. Suárez (1993), 'Innovation, Competition, and Industry Structure', *Research Policy*, **22**, 1–21.
van Duijn, J.J. (1979), *De Lange Golf in de Economie*, Assen: Van Gorcum.
van Duijn, J.J. (1983), *The Long Wave in Economic Life*, London: Allen & Unwin.
van Duijn, J.J. (1984), 'Fluctuations in Innovations over Time', in C. Freeman (ed.), *Long Waves in the World Economy*, London: Frances Pinter.
Veblen, T. (1898), 'Why is Economics not an Evolutionary Science?', *Quarterly Journal of Economics*, **12**, 373–97.
Verhulst, P.F. (1838), 'Notice sur la loi que la population suit dans son accroissement', *Correspondences Mathématiques et Physiques*, **10**, 113–21.
Vernon, R. (1966), 'International Investment and International Trade in the Product Cycle', *Quarterly Journal of Economics*, **80**, 190–207.
Volterra, V. (1926), 'Variazionie fluttuazioni del numero d'individui in specie animali conviventi', *Mum. R. Acad. Naz., dei Lincei* (ser. 6) **2**, 31–113.
Volterra, V. (1927), 'Una teoria matematica sulla lotta per l'esistenza', *Scientia*, Vol. XLI.
von Bertalanffy, L. (1957), 'Quantitative Laws in Metabolism and Growth', *The Quarterly Review of Biology*, **32**, 217–31.
Walker, J.M. and E.B. Gingold (eds) (1985), *Molecular Biology and Biotechnology*, London: The Royal Society of Chemistry.
Wallerstein, I.M. (1979), *The Capitalist World-Economy: Essays*, Cambridge, MA: Cambridge University Press.
Watanabe, C., K. Miyazaki and M. Katsumoto (1998), *Economics of*

Technological Change, Tokyo: Union of Japanese Scientists and Engineers.
Webber, M.J. (1972), *Impact of Uncertainty on Location*, Cambridge, MA: MIT Press.
Werner, R.A. (2001), *Princes of the Yen*, Tokyo: Profit Research Center.
Winston, T.G. (1974), 'Life is Logarithmic', in J. Rose (ed.), *Advances in Cybernetics and Systems*, London: Gordon and Breach.
Winter, S.G. (1964), 'Economic "Natural Selection" and the Theory of the Firm', *Yale Economic Essays*, **4**, 225–72.
Winter, S.G. (1975), 'Optimization and Evolution in the Theory of the Firm', in H. Day and T. Grove, *Adaptive Economic Models*, New York: Academic Press, pp. 73–118.
Witt, U. (ed.) (1993), *Evolutionary Economics*, Ardershot, UK and Brookfield, US, Edward Elgar.
Witt, U. (2003), *The Evolving Economy*, Cheltenham, UK and Brookfield, US: Edward Elgar.
World Bank (2000), *World Development Indicators 2000*, Washington, DC: World Bank.
Yamaguchi, M. (1997), *Chaos and Fractals*, Tokyo: Kodansha (in Japanese).
Yamazaki, Y. (1997), *Introduction to Indian Economy*, Tokyo: Nihon Hyoron-Sha (in Japanese).
Yamazaki, T. and M. Hiromatsu (1990), 'Electronic Control of Jet Engines', *GTSJ*, **17–68**, 9–13 (in Japanese).
Yoda, S. (1996), 'Challenge the Human Crisis, the Trilemma', Lecture at IIASA Days in Japan, 22 April.
Yokomizo, M. (1991), *Japanese Economy and Business Cycles*, Tokyo: Nihon Keizai Shimbun Inc. (in Japanese).
Yoshimura, M. (1998), *Economic Development and Labour Structures in Malaysia*, Tokyo: Hosei University Press (in Japanese).
Yoshinaga, R. (1996), *What are Complex Systems?*, Tokyo: Kodansha Publishers (in Japanese).
Yoshinaka, T. (1994), *Development of Jet Engine*, Tokyo: Ohm-sha.
Yoshitomi, M. (1965), *The Great Depression in America,* Tokyo: Nippon Hyoron-Sha (in Japanese).
Yuasa, M. (ed.) (1989), *Chronological Table of Science*, Tokyo: Sannseido.
Zimmerman, M.M. (1955), *The Super Market – A Revolution in Distribution*, New York: McGraw-Hill.

Index